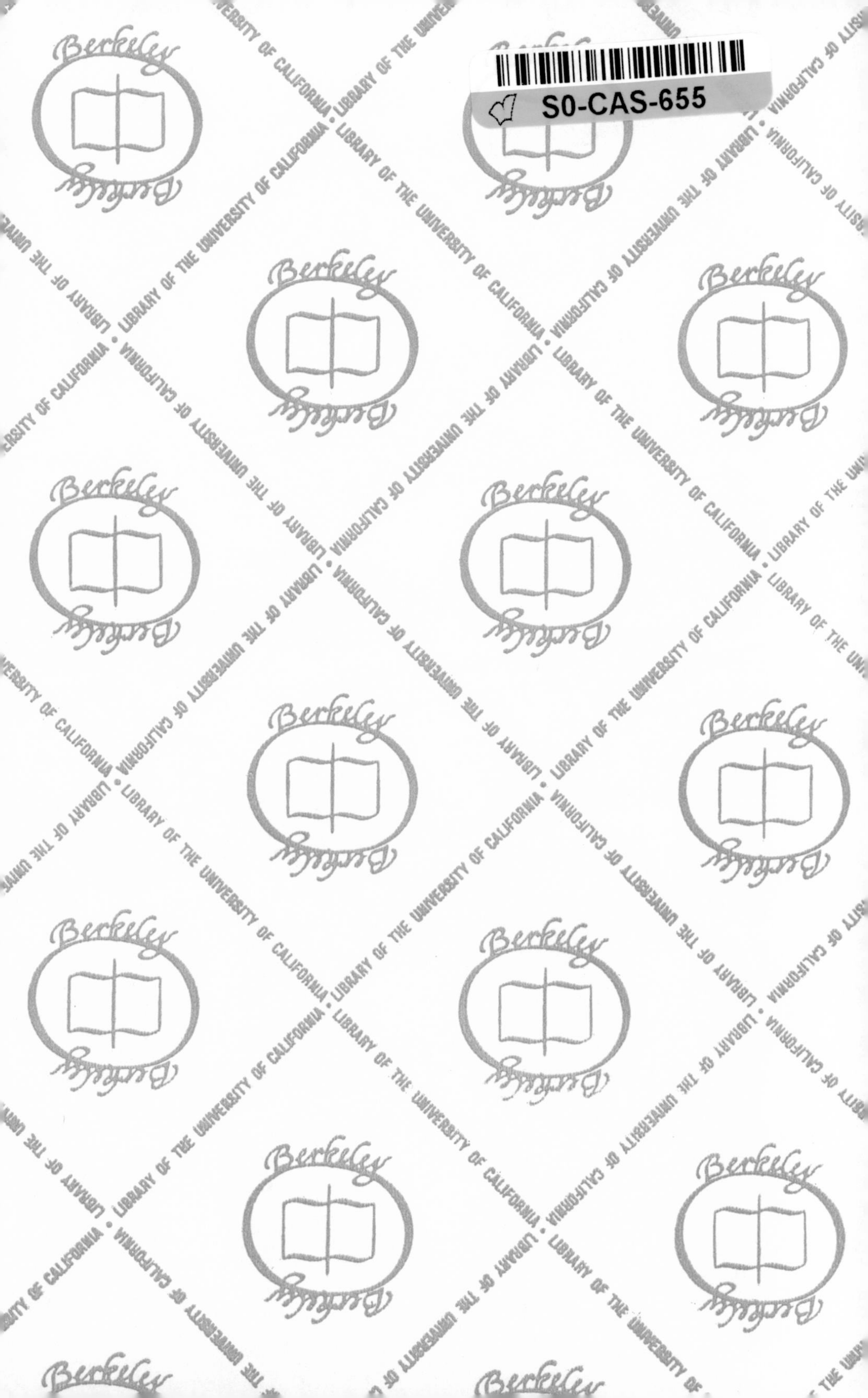
S0-CAS-655
Berkeley
LIBRARY OF THE UNIVERSITY OF CALIFORNIA

Scripts in Inorganic
and Organometallic Chemistry **1**

Gmelin-Institut (Ed.)

Wolfgang Petz

Iron-Carbene Complexes

With 23 Figures and 14 Tables

Springer-Verlag
Berlin Heidelberg New York
London Paris Tokyo
Hong Kong Barcelona Budapest

Editor:
Gmelin-Institut
Varrentrappstr. 40–42
6000 Frankfurt am Main 90, FRG

Authors
Prof. Dr. Wolfgang Petz (managing)
Dr. Jürgen Faust
Johannes Füssel
Varrentrappstr. 40–42
6000 Frankfurt am Main 90, FRG

ISBN 3-540-56258-3 Springer-Verlag Berlin Heidelberg New York
ISBN 0-387-56258-3 Springer-Verlag New York Berlin Heidelberg

Library of Congress Cataloging-in-Publication Data
Petz, Wolfgang, 1940 – Iron-carbene complexes/Wolfgang Petz. (Scripts in inorganic and organometallic chemistry; 1)
Includes index.
ISBN 3-540-56258-3 (Berlin: acid-free). – ISBN 0-387-56258-3
(New York: acid-free) 1. Organoiron compounds. I. Title. II. Series.
QD412.F4P48 1993 547'.05621 – dc20 92-46132 CIP

Printed in Germany

Typesetting: Macmillan India Ltd, Bangalore-25;
Printing: Saladruck, Berlin; Bookbinding: Lüderitz & Bauer, Berlin
51/3020-5 4 3 2 1 0 – Printed on acid-free paper

Preface

With this new edition the Gmelin Institute for Inorganic Chemistry of the Max-Planck Society starts a series of textbooks based on the Gmelin Handbook of Inorganic and Organometallic Chemistry. While the Gmelin Handbook aims at complete coverage of all material published on a certain subject and links each statement with its source, the new Gmelin series will review selected areas of inorganic and organometallic chemistry in textbook style. It will provide the lecturer, the advanced student, and the research chemist with a digest of the main features of each topic. More detailed information and numerical data on the subject may be found in the corresponding volume of the Gmelin Handbook and/or the Gmelin Databank.

The first book in this series covers compounds with a formal iron-to-carbon double bond. The scope is limited to species in which the iron atom of the $\mathrm{Fe{=}C}<$ skeleton is additionally bonded to a ^5L ligand.

For example, the ease of access to the base group, $C_5H_5(CO)_2Fe$, and the possibility of tailoring its electronic properties to specific uses by varying the two types of ligands at the iron atom has made these fragments the most used functional groups in organic and organometallic chemistry. This has encouraged us to select those cases where the appropriate fragments form Fischer-type carbenes and related species. Much of the chemistry at the iron moiety, e.g., metal chirality, cyclopropanation reactions, catalysis, etc., can be transferred to other metals such as Ru, Os, Mn, and Re, when they occur in a similar environment or in isoelectronic species. Thus, the chemistry of the iron compounds can serve as a model for many other organometallic compounds.

The original Gmelin Handbook was written by Wolfgang Petz and edited by Jürgen Faust and Johannes Füssel.

Frankfurt am Main, 1992

Ekkehard Fluck
Wolfgang Petz

Table of Contents

Introduction

This book deals with the chemistry and the structural aspects of all compounds in which a formally double-bonded carbene (chapter 1) or vinylidene (chapter 2) ligand is coordinated to a ^5LFe moiety.

A ^5L ligand represents an organic molecule which is coordinated to the metal by five carbon atoms in an η^5-manner. In this special type of compound the ^5L ligands are generally C_5H_5 (abbreviated as Cp) and subtituted or annelated C_5H_5 (abbreviated as Cp*); compounds with "open" ^5L ligands or more than a five-membered ^5L ring are not known in this series. The frequently appearing $C_5H_5(CO)_2Fe$ fragment in organometallic chemistry is usually abbreviated as Fp.

Electron-deficient carbene ligands of the general type CR_2 or similarly vinylidene ligands, $C{=}CR_2$, are derivatives of the formally divalent carbon atom and are considered as two-electron donors. The formulation of such species as "soft" ligands similar to CO, CS, and phosphines, and the description of the interaction with the metal as a metal-to-carbon double bond (σ-donor/π-acceptor interaction) are common features in all textbooks of organometallic chemistry. Theoretically possible cumulene ligands, $C{=}(C{=})_nCR_2$, which are well known in organoruthenium chemistry, have not yet been described in connection with the ^5LFe fragment.

This book contains all the compounds mentioned in "Organoiron Compounds" 16a of the Gmelin Handbook of Inorganic and Organometallic Chemistry and their most important spectroscopic data; more detailed information concerning preparation and references may be obtained from the original version.

1 Carbene Complexes with the $^5LFe{=}C$ Moiety

According to the nature of the $^5L(CO)_2Fe$ fragment as a 17 electron fragment, the emphasis of carbene compounds is on cationic species, which form the majority of complexes. Neutral compounds are formed if one group R formally contains a negatively charged ligand as well in the series of $^5L(^2D)XFe{=}CR_2$ complexes; negatively charged carbene complexes are restricted to one compound, discussed in Sect. 1.3.

If ^2D is isocyanide the cations are described separately in Sect. 1.5.

The cations are arranged first in the order of increasing CO substitution at the iron atom, with the cations of the type $[^5L(^2D)_2Fe{=}CR_2]^+$, $[^5L(^2D)(CO)Fe{=}CR_2]^+$, and $[^5L(CO)_2Fe{=}CR_2]^+$ forming the main sections. Within each section the second criterion of arrangement is the content of a "heteroatom" at the carbene carbon atom. Heteroatoms are considered as being all the atoms except H and C, leading to subsections with the carbene ligands CR_2 (no heteroatom), CR(ER) (one heteroatom), and $C(ER)_2$ (two heteroatoms). ER represents the heteroatom E (mostly O, S, or N, and less frequently F, Cl, or Se), along with further ligands necessary to achieve the noble gas shell.

In order to give a general review of carbene complexes with the $^5LFe{=}CR(ER)$ group, we have also included compounds containing one or two iron atoms in the substituents R or ER.

Apart from the compounds described here, additional compounds with the $^5LFe{=}C$ moiety, which belong to other criteria of arrangement according to the Gmelin system, appear elsewere.

Thus, regardless of the character of the iron-carbon bonds, all ring systems that are ferracycles with a C–Fe=C unit ($^5LFe^2L$ compounds with a bidentate ^2L ligand), are summarized in "Organoiron Compounds" B 16b, 1990, p. 64 and Table 16 on pp. 117–132.

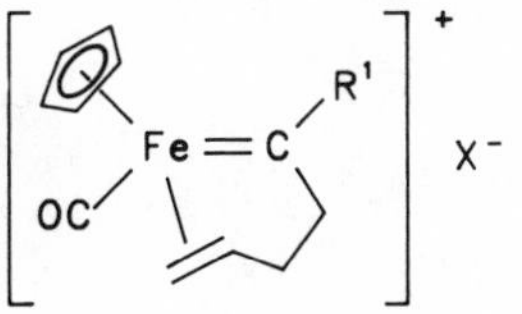

$[\ldots Fe{=}C(R^1)\ldots]^+ X^-$ $[\ldots]^+ [PF_6]^-$ $[\ldots Fe{=}C(OCH_3)O\ldots; H_2C{=}CH_2]^+$

I 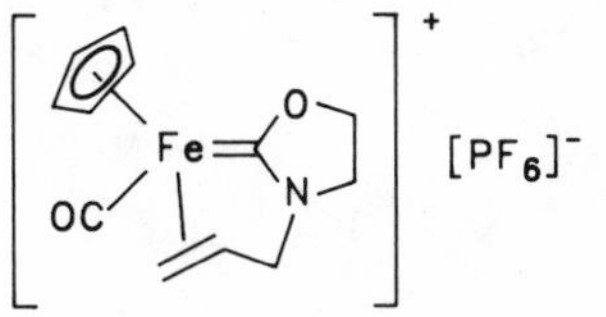 II 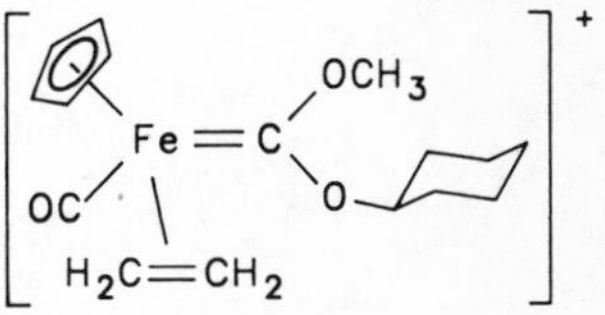 III

Cationic compounds of the types I and II, in which the carbene ligand is part of a ^{3}L ligand (e.g. a ligand which is bonded with three carbon atoms to the metal), are described in "Organoiron Compounds" B 17 1990, p. 167; for the carbene complex III, see ibid p. 138.

Thiocarbonyls such as $[FpC{\equiv}S]^+$ and its derivatives, in which a ^{5}LFe=C moiety can also be formally formulated, have been described separately in "Organoiron Compounds" B 15, pp. 266–279.

1.1 Cationic Carbene Complexes with one Carbene Ligand Bonded to Fe

General remarks. The compounds belong to the electron-rich "Fischer" carbenes, in contrast to the electron-deficient "Schrock" carbene complexes and the carbene ligand as a two electron donating species (singlet carbene) coordinates at a 16-electron transition metal fragment to give a saturated 18-electron complex, mainly with low valent metals of groups 6 to 12. In general, the carbene carbon atom in the Fischer-type compounds are electrophilic and reactions with Lewis bases are possible. Schrock-type carbene complexes are formed with highly oxidized group 4 and 5 metals, the carbene carbon atom is nucleophilic, similar to that in ylides, and the compounds do not always obey the 18-electron rule.

The compounds can be formulated as cationic carbene complexes (Ia) or as Fp-substituted ($Fp = C_5H_5(CO)_2Fe$) carbenium ions as represented by the resonance formula Ib. If the group R is attached to the carbon atom with a heteroatom such as O, S, or N (or others containing filled *p* or *d* orbitals for back donation), a contribution of the resonance formulas Ic and Id occurs. Thus, the cation can be stabilized by π donation from the heteroatoms and from a partial double bond between Fe and the carbene carbon atom. Because of the general shortening of the Fe–C bond with respect to normal Fe–C σ bonds (e.g., in $Fp{-}CH_3$) and the planar arrangement of the ligands at the carbene carbon atom (sp^2 hybridization), the compounds in this section are described with a formal iron-carbon double bond represented by Formula Ia, regardless of the real nature of the bond.

$$\underset{a}{\overset{+}{Fp}{=}C(R)R} \longleftrightarrow \underset{b}{Fp{-}\overset{+}{C}(R)R} \longleftrightarrow \underset{c}{Fp{-}C({=}R^+)R} \longleftrightarrow \underset{d}{Fp{-}C(R){=}R^+}$$

I

The electron density at the iron atom can be increased by the replacement of one or both CO groups of Fp by stronger electron-donor ligands such as phosphines; the result is a stabilizing effect on the cation. Replacement of the ^{5}L ligand C_5H_5 by more bulky ligands shows the same trend.

A similar stabilizing effect, compared with the introduction of an electron-donating heteroatom at the carbene ligand, is shown in complexes in which the fragments CpFe(CO)(^{2}D), Fp, or a substituted Fp group are attached at a cationic $4n+2$ π-electron system, such as the cyclopropenylium ($n = 0$) or the tropylium ($n = 1$) cation. In these compounds, resonance structures with a Fe=C double bond can be formulated as depicted in Formula II and III, respectively (^{2}D = CO, PR$_3$).

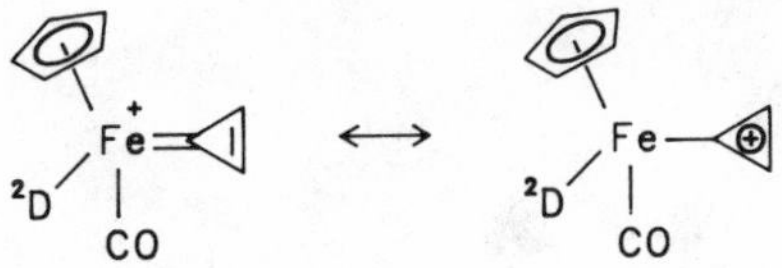

II

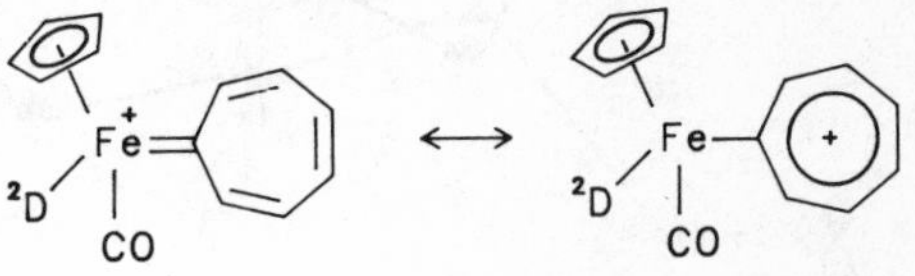

III

Cationic or neutral carbene complexes (adduct formulation IVa, carbenoid formulation IVb) also form if the doubly bonded atom E (E = O, S, or N) in cumulated systems of the general type Fp–(CH=CH)$_n$–CH=E coordinates at various cationic or neutral Lewis acids (LA); compounds of this type are found in Sects. 1.1.3.1 and 1.2.1.

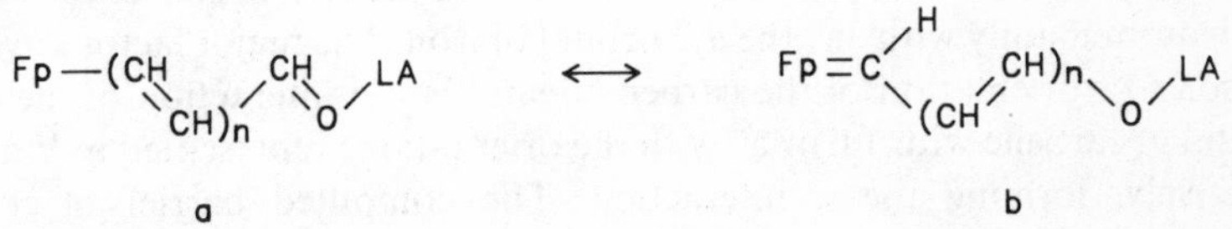

IV

When the carbene ligand is bonded to ^{5}LFe species containing a symmetry plane, such as $C_5H_5(CO)_2Fe$ or $C_5H_5(^2D)_2Fe$ fragments, the orientation can be such that the carbene plane coincides with the symmetry plane of the molecule (Formula V) or is perpendicular to this plane (Formula VI).

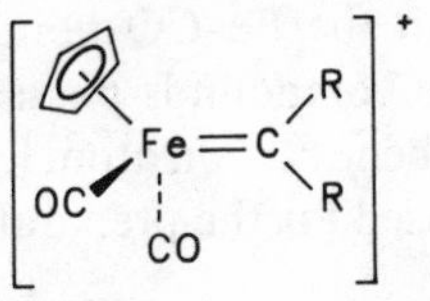

V 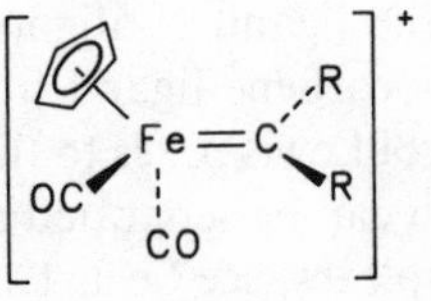 VI 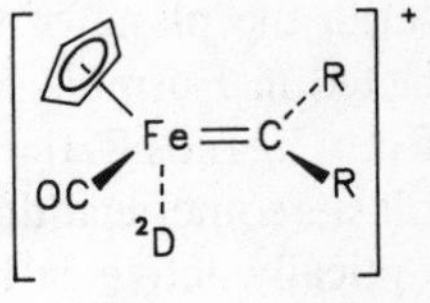 VII

$C_5H_5(CO)_2Fe$ fragment orbitals suitable for σ and π interaction with the carbene ligand have been constructed. The orbital interaction diagram for $[C_5H_5(CO)_2Fe]^+$ and the carbene ligand CH_2 in two different orientations is shown in Fig. 1.

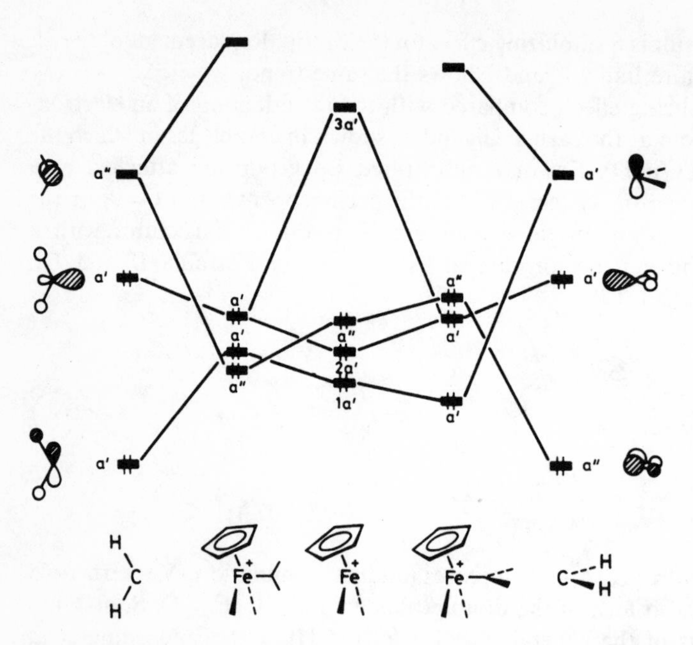

Fig. 1. Orbital interaction diagram for $[C_5H_5(CO)_2Fe]^+$ and a carbene in two different orientations

A σ bond can be formulated by interaction of the σ lone pair of the carbene carbon atom mainly with 3a′, the d_z2 orbital of iron. The major factor governing the choice of orientation of the carbene ligand is the interaction of the empty *p* orbital of carbene with 1a′ or a″ with the orientations represented in V and VI, respectively, forming the π interaction. The computed barrier of rotation between the two orientations is 6.2 kcal/mol, favoring the "upright" conformation V.

The upright orientation is realized in compounds containing the carbene ligands CH_2 or $CHCH_3$, shown by low-temperature NMR spectroscopy. However, with more complex ligands, both orientations are found.

In the cations of the general formula $[Cp(CO)(^2D)Fe{=}CR_2]^+$, the iron atom is surrounded by four different ligands giving rise to "chirality at iron". In general, the plane of the carbene ligand is aligned with the Fe–CO bond as depicted in Formula VII. If the carbene ligand in these compounds possesses different ligands R, the addition of Lewis bases to the carbene carbon atom leads to diastereomeric adducts which can be separated and used for the preparation of optically active carbene complexes; see Sect. 1.1.2.

According to the onium character of the carbenium carbon atom, the ^{13}C NMR shifts appear at very low fields, in the range between 220 and 420 ppm. Within one series of $^5LFe{=}C$ compounds, the high field shift increases with increasing number of π-donating heteroatoms.

The chemistry of the less electron-rich cations $[Fp{=}CR_2]^+$ and $[Cp(CO)(^2D)Fe{=}CR_2]^+$, having carbene ligands without heteroatoms, is mainly governed by the ability to transfer the carbene ligand to an olefinic double bond to form cyclopropanes.

The electrophilic character of the methylene fragment $=CH_2$ decreases when back bonding of the metal increases upon permethylation of the C_5H_5 ring or upon substitution of CO by stronger electron-donor phosphine ligands.

1.1.1 Cationic Complexes Containing the $^5L(^2D)_2Fe{=}C$ or the $^5L(^2D\text{-}^2D)Fe{=}C$ Moiety

The compounds described in this section are electron-rich carbene complexes, because the presence of ^2D ligands (mostly phosphines) increases the electron density at the iron atom relative to that in the dicarbonyl analogue. The higher electron density increases back-bonding to the carbene ligand, resulting in decreased electrophilicity of the carbene carbon atom and increased stability of the cation. Further stabilization is achieved by introduction of a heteroatom at the carbene ligand; see Sect. 1.1.1.2.

In this series the only cationic complex with two heteroatoms at the carbene carbon atom, $[Cp(CH_3NC)_2Fe{=}C(SCH_3)_2]CF_3CO_2$, is described as an isocyanide containing complex, see Sect. 1.1.4.2.

1.1.1.1 Complexes of the Type $[^5L(^2D\text{-}^2D)Fe{=}CR_2]^+$ with No Heteroatom at the Carbene Carbon Atom

The compounds of the general Formula I contain the chelating ligand $(C_6H_5)_2PCH_2CH_2P(C_6H_5)_2$ abbreviated as dppe. The ^5L ligand can be Cp or $C_5(CH_3)_5$. For theoretical considerations concerning the model cation $[(Cp)(PH_3)_2Fe{=}CH_2]^+$, see Sect. 1.1.

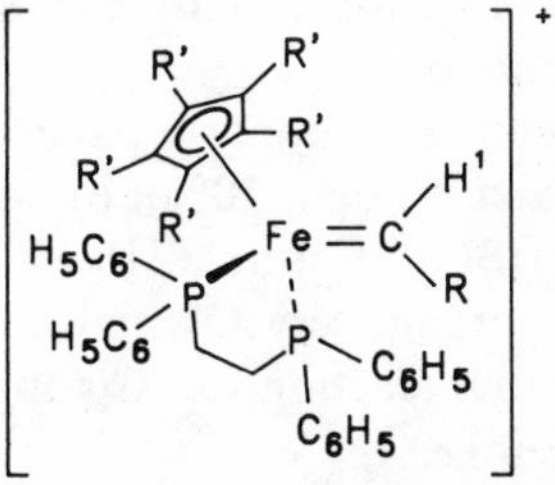

I

a : $R = H^2$; $R' = H$
b : $R = C_4H_9\text{-}t$; $R' = H$
c : $R = H^2$; $R' = CH_3$

General remarks. Variable-temperature NMR spectra of compounds Ia and Ic show hindered rotation about the Fe=C bond at low temperature and an upright orientation of the carbene ligand. The presence of the electron-donating methyl substituents in the corresponding $C_5(CH_3)_5$ complex induces only a weak increase of the Fe=C rotation barrier (Cp, 10.4 ± 0.1 kcal/mol; $C_5(CH_3)_5$, 10.7 ± 0.2 kcal/mol); however, it increases the thermal stability of the complex, which is associated with the absence of electrophilic reactivity. The assignment of the only weakly coupled high-field signal in the ^1H NMR spectrum of Ia at about 17 ppm to the carbene H^1 *syn* to the Cp ring is based on comparison with the benzylidene H in $[Cp(CO)(^2D)Fe{=}CHC_6H_5]^+$ (No. 14 in Table 2 of Sect. 1.1.2.1). In going from Cp to $C_5(CH_3)_5$ a 1.77 ppm upfield shift of H^1 in the ^1H NMR spectrum and a 5.7 ppm upfield shift of the carbene carbon atom in the ^{13}C NMR spectrum are observed.

$[Cp(dppe)Fe{=}CH_2]X$ (Formula Ia, X = CF_3CO_2, CF_3SO_3) forms by treatment of the corresponding $Cp(dppe)FeCH_2OC_2H_5$ complex in CD_2Cl_2 or CD_2Cl_2/SO_2 solution at 78 °C with 2 to 20 equivalents of HX. A homogeneous, dark red solution of the cation is obtained. The cation is stable in this solution at 25 °C for about 1 to 2 h. The formation of the cation was also indicated by the production of ethene and by the transfer of methylene to cyclohexene when $Cp(dppe)FeCH_2OCH_3$ was treated with HBF_4 in acetone/cyclohexene.

Spectroscopic data show that the H-C-H plane of the carbene ligand is perpendicular to the plane of the cyclopentadienyl ligand at low temperature. The ^1H NMR spectrum in CD_2Cl_2 at -90 °C (in ppm) indicates nonequivalency of the methylene protons and exhibits two one-proton resonances for H^1 and H^2 (*syn* and *anti* to the Cp ring): 13.89 (t, H^2; J(P,H) = 14 Hz), 17.29 (s, br, H^1). Above -80 °C the signals of the methylene protons begin to broaden, coalesce at -40 °C, and sharpen at 25 °C to a two-proton triplet (J(P,H) = 7 Hz) centered midway between the two low-temperature signals. The coalescence temperature is independent of the acid concentration and of change in acidity. ^{13}C NMR (same conditions): 317.5 ppm ($CH_2{=}Fe$; J(P,C) = 32, J(H,C) = 140 Hz). From the variable-temperature ^1H NMR experiments the free energy of activation, $\Delta G^{\ddagger}$, for rotation about the iron–carbene carbon bond is estimated to be 10.4 ± 0.1 kcal/mol.

The complex converts olefins into cyclopropanes (ethyl vinyl ether (98%), hex-1-ene (30%), and cyclohexene (about 10%)) by the transfer of methylene.

$[Cp(dppe)Fe{=}CH(C_4H_9\text{-}t)]BF_4$ (Formula Ib, X = BF_4) is obtained in about 60% crude yield as orange crystals mixed with black $[Cp(dppe)Fe(CH_3COCH_3)]BF_4$ from the reaction of $Cp(dppe)FeCH{=}C(CH_3)_2$ with $[(CH_3)_3O]BF_4$ in CH_2Cl_2/acetone.

The ^1H NMR spectrum (probably at room temperature) in CD_2Cl_2 exhibits the following important signals (in ppm): 0.71 (s, C_4H_9), 13.68 (t, CH=Fe; J(P,H) = 14.0 Hz). ^{13}C NMR spectrum in CD_2Cl_2: 28.8 (s, CH_3), 63.0 (s, $C(CH_3)_3$), 88.4 (Cp), 359.5 (t, CH=Fe; J(P,C) = 26.4).

The reaction of the complex with excess $NaBF_4$ in THF gives a 2:3 mixture of the hydride, Cp(dppe)FeH, and the neopentyl complex, $Cp(dppe)FeCH_2C(CH_3)_3$, (8% yield).

[$C_5(CH_3)_5$(dppe)Fe=CH_2]BF_4 (Formula Ic, X = BF_4) is obtained by treatment of a solution of $C_5(CH_3)_5$(dppe)$FeCH_2OCH_3$ in anhydrous diethyl ether with $HBF_4/O(C_2H_5)_2$ at $-90\,°C$ (95% yield). The complex is air-stable and decomposes at 160 °C without melting. It is slightly soluble in THF, and readily soluble in acetonitrile and CH_2Cl_2, and crystallizes from the latter as red-brown microcrystals containing solvent molecules.

The ^1H NMR spectrum in CD_2Cl_2 at $-80\,°C$ shows signals of the carbene ligand (in ppm) at 13.85 (t, H^2; J(P,H) = 13), 15.58 (t, H^1; J(P,H) = 1 Hz). The H^1 and H^2 signals coalesce at $-28\,°C$ to give a single triplet at 14.86; J(P,H) = 7 Hz. ^{13}C NMR spectrum in CD_2Cl_2 at 25 °C (in ppm): 311.8 (tt, C=Fe; J(H,C) = 150, J(P,C) = 24 Hz). The ^{31}P NMR in CD_2Cl_2 at 25 °C shows a signal at 93.3 ppm. The free energy of activation, $\Delta G^{\ddagger}$, for rotation about the iron–carbene carbon bond was estimated to be 10.7 ± 0.2 kcal/mol.

The cyclic voltammogram of the cation in CH_2Cl_2 solution (0.1 M [$N(C_4H_9$-$n)_4$]PF_6 at a platinum electrode vs SCE) indicates a reversible Fe^{II}/Fe^{I} system with an anodic wave (E° = -0.170 V). Activated by electron-transfer catalysis (ETC), probably with intermediate formation of the 19-electron $C_5(CH_3)_5$(dppe)Fe=CH_2 species, rapid removal of the carbene ligand in the presence of excess CH_3CN occurred, to give [$C_5(CH_3)_5$(dppe)Fe(CH_3CN)]$^+$ (90% yield), which was also obtained from treatment of the starting cation in refluxing acetonitrile (3 h, quantitative yield). Under similar ETC conditions rapid methylene exchange also occurred in the presence of $P(CH_3)_3$ or CO. The cation exhibits no electrophilic reactivity and forms no cyclopropanes with olefins and no adducts with nitrogen or phosphorus nucleophiles. However, a rapid electrochemically-induced cyclopropanation of styrene takes place under ETC conditions (coulombic efficiency = 100), using 10 milliequivalents of current or a redox reagent such as cobaltocene. Reduction with a stoichiometric amount of a reducing agent, such as cobaltocene in CH_2Cl_2, produces $C_5(CH_3)_5$(dppe)FeCl in 85% yield by Cl abstraction from the solvent.

1.1.1.2 Complexes [$^5L(^2D)_2$Fe=C(R)ER]$^+$ and [Cp(^2D-^2D)Fe=C(R)ER]$^+$ with One Heteroatom at the Carbene Carbon Atom

Two compounds of the type [$^5L(^2D)_2$Fe=C(R)OR]$^+$ and three compounds of the type [Cp(^2D-^2D)Fe=C(R)OR]$^+$ (Formula I) with the chelating ligand

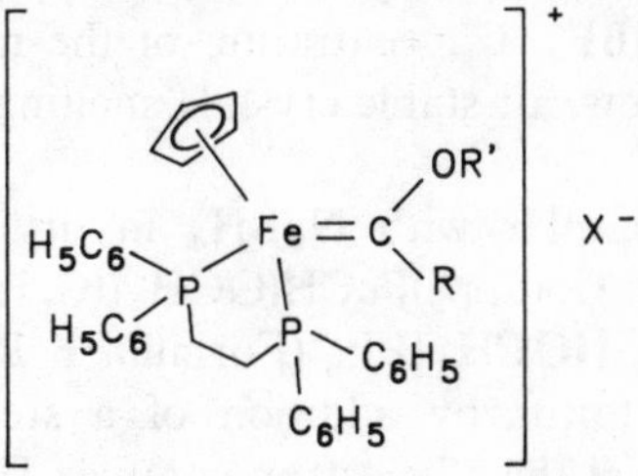

I

$(C_6H_5)_2PCH_2CH_2P(C_6H_5)_2$, abbreviated as dppe, are described. Nothing is reported about the orientation of the carbene plane (upright or perpendicular to the symmetry plane). In the case of the upright orientation the OR group can be arranged in a *cis* or *trans* position with respect to the Cp ring.

[Cp{P(CH$_3$)$_3$}$_2$Fe=C(CH$_3$)OCH$_3$]X (X = FSO_3, Cl, [Cp(CO)$_3$W]). The Cl^- salt is obtained by the reaction of the vinyl complex [Cp{P(CH$_3$)$_3$}$_2$FeC(=CH$_2$)OCH$_3$ with CH_3COCl. The FSO_3^- or $[Cp(CO)_3W]^-$ salts are prepared by the addition of the corresponding HX acid at − 78 or 25 °C in pentane or benzene as yellow crystals in 82 or 75% yield, respectively. Both compounds melt at 132 °C (dec.).

^1H NMR spectrum of the FSO_3^- salt: 2.76 (s, $CH_3C{=}$), 4.06 (s, CH_3O). ^{31}P NMR: 31.90. All spectra are recorded in CD_3CN and given in ppm. ^1H NMR spectrum of the $[Cp(CO)_3W]^-$ salt: 2.87 (s, CH_3C), 4.39 (s, CH_3O), 4.79 (t, CpFe; $^3J(P,H) = 1.50$), 5.15 (s, CpW). ^{31}P NMR 30.76. The spectra (in ppm) are recorded in CD_3NO_2 solution.

Thermal treatment of the $[Cp(CO)_3W]^-$ salt in boiling benzene generates the starting material along with $P(CH_3)_3$, $[Cp(CO)_3W]_2$, $Cp(CO)_3WCH_3$, Cp{P(CH$_3$)$_3$}$_2$FeCOCH$_3$, and Cp(CO)P(CH$_3$)$_3$FeCH$_3$.

[Cp{P(CH$_3$)$_3$}$_2$Fe=C(CH$_2$CS$_2$CH$_3$)OCH$_3$]I forms as brown needles in 68% yield by treatment of a suspension of the neutral carbene complex Cp{P(CH$_3$)$_3$}$_2$Fe=C(CH$_2$CS$_2$)OCH$_3$ (Sect. 1.2.2) in benzene with CH_3I for 20 h at room temperature; m.p. 91 °C (dec.).

Two signals for the Cp, $P(CH_3)_3$, and CH_3O units in the ^1H NMR spectrum are explained with the presence of two rotamers in solution. ^1H NMR spectrum (in ppm) in CD_3CN: 1.56/1.57 (m's, CH_3P; J(P,H) = 9.10), 2.17 (s, CH_2), 2.68 (s, CH_3S), 4.12 (s, CH_3O), 4.82/4.83 (t's, Cp; $^3J(P,H) = 1.55$). ^{31}P NMR: 27.08 ppm.

[Cp(dppe)Fe=C(C$_6$H$_5$)OH]BF$_4$ (Formula I, R′ = H, R = C_6H_5) is obtained by addition of HBF_4 (40% in H_2O) to a solution of the acyl complex Cp(dppe)FeCOC$_6$H$_5$ in CH_3OH. Concentration of the mixture after 10 min, followed by addition of H_2O, precipitates yellow crystals; m.p. 148 to 150 °C (dec.) (81% yield).

Deprotonation with formation of the starting material is achieved with excess $KHCO_3$ in ethanol.

[Cp(dppe)Fe=C(C$_6$H$_5$)OCH$_3$]BF$_4$ (Formula I, R′ = CH_3, R = C_6H_5) is prepared by methylation of the acyl complex, Cp(dppe)FeCOC$_6$H$_5$, in dry CH_2Cl_2 with [O(CH$_3$)$_3$]BF$_4$. Concentration of the mixture and addition of dry ether precipitate yellow, air stable crystals, melting at 172 to 175 °C (dec.); yield 67%.

Reduction of the complex with $NaBH_4$ in dry ethanol results in the formation of the complex Cp(dppe)FeCH(OCH$_3$)(C$_6$H$_5$) in 68% yield.

[C$_5$(CH$_3$)$_5$(dppe)Fe=CHOCH$_3$]PF$_6$ (Formula I, R′ = CH_3, R = H; Cp is replaced by $C_5(CH_3)_5$) forms by addition of a stoichiometric amount of KOC_4H_9-t to a solution of the 17-electron complex II (obtained from IV and [C(C$_6$H$_5$)$_3$]PF$_6$ or [Cp$_2$Fe]PF$_6$) in THF solution at − 80 °C (1 h, 45% yield) along with equivalent amounts of C$_5$(CH$_3$)$_5$(dppe)FeCH$_2$OCH$_3$ (Formula IV) and C_4H_9OH. In the first step, II is deprotonated to give the unstable neutral 19-electron carbene complex, $C_5(CH_3)_5(dppe)Fe{=}CHOCH_3$ (Formula III), which is converted into the title complex by an immediate electron transfer to II, resulting in IV.

The complex exhibits the following NMR signals in CD_2Cl_2 solution (in ppm): ^1H NMR: 3.23 (s, CH_3O), 12.31 (t, CH=Fe, $^3J(P,H) = 3$ Hz). ^{13}C NMR: 68.5 (CH_3O, $J(H,C) = 145$ Hz), 97.6 (s, C_5), 306.2 (dt, CH=Fe, $J(H,C) = 142$, $J(P,C) = 57$ Hz). ^{31}P NMR: 106.4.

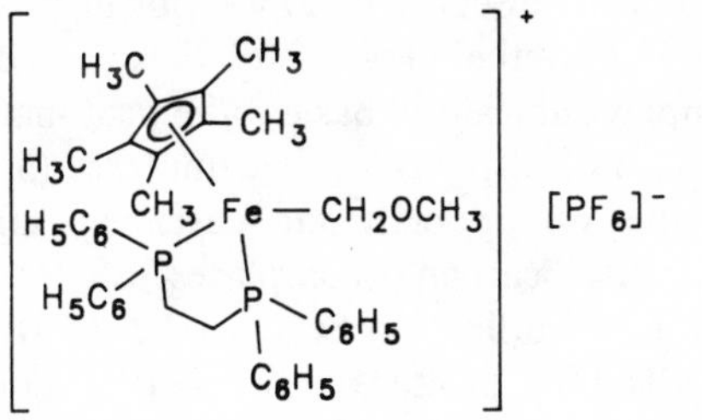

$[PF_6]^-$

II

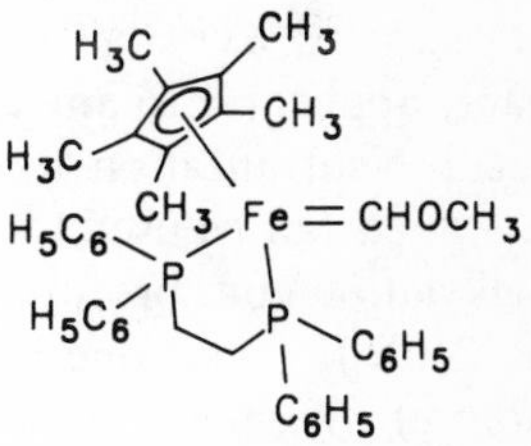

III

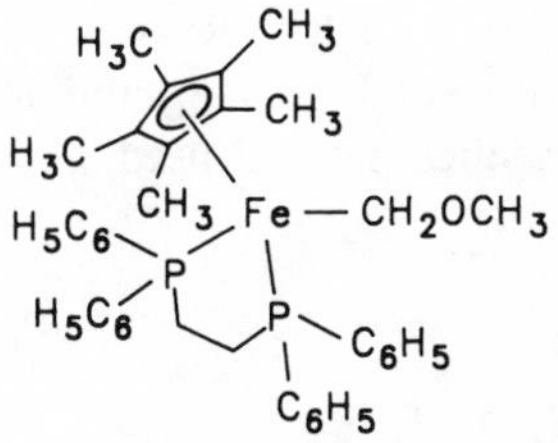

IV

1.1.2 Cationic Complexes Containing the $^5L(CO)(^2D)Fe{=}C$ Moiety

The cations studied in this series contain four different ligands at the iron atom in a pseudotetrahedral arrangement, being therefore chiral-at-iron (Formula I). The compounds are used as chiral auxiliaries to carry out diastereo- and enantioselective reactions. If the carbene carbon atom has two different substituents it is prochiral and the addition of nucleophiles at the carbene carbon atom gives rise to the formation of diastereoisomers, but diastereoselectivity is achieved since one face of the carbene ligand normally is more shielded than the other one; see Sects. 1.1.2.1 and 1.1.2.2.

Optically pure cations have been prepared by two routes in the series $[Fe{=}CRR']^+$ with no heteroatom and in the series $[Fe{=}CRER']^+$ with one heteroatom at the carbene carbon atom (Fe represents the appropriate Cp(CO)(^2D)Fe fragment). Starting from the racemic acyl complex II, protonation with

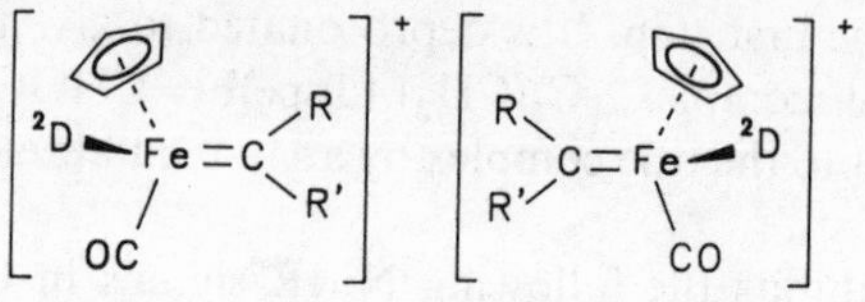

(*R*)-I (*S*)-I

either (1*S*)-(+)- or (1*R*)-(−)-10-camphorsulfonic acid gives a 1 : 1 mixture of the diastereomeric hydroxycarbene salts (see Sect. 1.1.2.2) which can be separated by successive crystallization in CH_2Cl_2-ether at − 25 °C. This procedure is probably applicable to any acyl complex sufficiently basic to be protonated by this acid. Neutralization recovers the optically pure acyl complexes (*S*)-II or (*R*)-II, which can be used for the synthesis of optically pure carbene complexes via alkylation or protonation to give carbene complexes of the type $[Fe{=}CR(ER')]^+$ with one heteroatom, as outlined in Sect. 1.1.2.2. $NaBH_4$ reduction of these compounds in CH_3OH produces the neutral complex Fe–CHR(ER′) with retention of the configuration at the iron atom serving as starting materials for the $[Fe{=}CRR']^+$ carbene compounds in Sect. 1.1.2.1. The other route starts from substitution of $Cp(CO)_2FeCOCH_3$ by one enantiomer of the phosphine ligand $P(C_6H_5)_2R^*$ (R^* = (*S*)-2-methylbutyl), generating the diastereomeric acyl complexes III ($S_{Fe}S_C$ and $R_{Fe}S_C$), which can be separated chromatographically on silica gel and used similarly as starting materials.

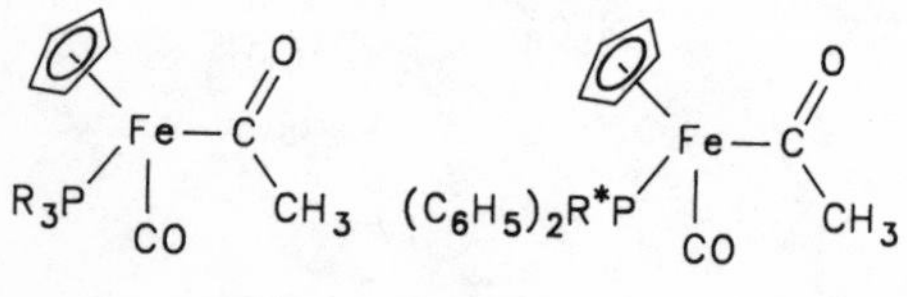

(*S*)-II, (*R*)-II (*RS*)-III, (*SS*)-III

For better understanding, the first letter in *SS*,*RR*,*RS*, or *SR* designates the chirality of the iron center and the second letter the chirality of the carbon center. The same assignment is given for the resulting adducts of nucleophiles at the prochiral carbene ligands of $[Fe{=}CRR]^+$ compounds.

The *R* and *S* designation for chiral iron centers is based on the priority sequence Cp > PR_3 > CO > COR > CH(OR)R′ > =CHR.

Other chiral acyl complexes as possible precursors for the synthesis of carbene complexes are of the type $Cp(CO)(P(C_6H_5)_2NR'R^*)FeCOR$ (R^* = (*S*)-$CH(CH_3)C_6H_5$).

In general these cations are more stable than the unsubstituted derivatives, $[Cp(CO)_2Fe{=}CRR']^+$, described in Sect. 1.1.3.1. The stability also increases with the introduction of heteroatoms at the carbene carbon atom in the series $[Fe{=}CRR']^+ < [Fe{=}CRER']^+ < [Fe{=}CE'R(ER')]^+$.

In this series the carbene plane is aligned with the CO–Fe bond, which leads to synclinal (*syn*) and anticlinal (*anti*) orientations of the carbene ligands in the case of the corresponding $[Fe{=}CHR]^+$ compounds.

1.1.2.1 Complexes $[^5L(CO)(^2D)Fe{=}CRR']^+$ with No Heteroatom at the Carbene Carbon Atom

This section comprises cationic carbene complexes of the general type $[Cp(CO)(^2D)Fe{=}CRR']^+$, in which *R* and R′ are represented by *H*, alkyl, aryl, or vinyl; in Nos. 24 and 25 the carbene carbon atom is incorporated in a cycloheptatriene ring, forming an aromatic π-system. Except the two species described at the end of the section, ^5L is generally C_5H_5, abbreviated as Cp.

The compounds Nos. 10, 11, and 12 to 16 can also be viewed as $^5L(CO)(^2D)$ Fe-substituted allyl cations.

The cation $\mathbf{[Cp(CO)(P(C_6H_5)_3)Fe{=}CHC(CH_3)_2CH_2CH_2OH]^+}$ (Formula II) was proposed as an intermediate in the acid-induced, stereoselective epimerization of the *RR,SS* diastereomer Ia into *RS,SR* diastereomer Ib. The intermediate cation $[Cp(CO)(P(C_6H_5)_3)Fe{=}CH_2]^+$ may also play a role in the reactions of $Cp(P(C_6H_5)_3)FeCH_2Cl$ with various nucleophiles.

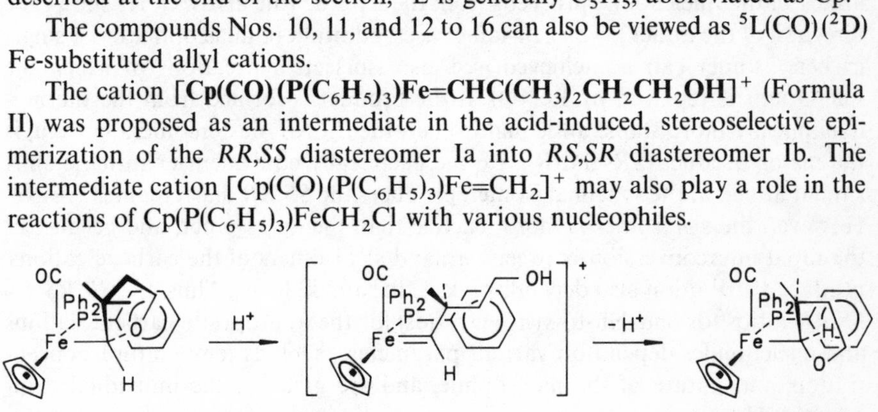

Ia —H⁺→ II —−H⁺→ Ib

Conformation. In the cations $[Cp(CO)(^2D)Fe{=}CRR']^+$ normally the carbene plane is aligned with the Fe–CO bond and the presence of a carbene ligand with one bulky substituent of the type =CHR (R = CH_3, C_6H_5) gives rise to a synclinal or anticlinal orientation of the planar ligand (Formula III, *syn* and *anti*, respectively). The anticlinal conformation is energetically more favored on steric grounds, with the more bulky group R at the CO side. For R = CH_3, both isomers can be observed by low-temperature ^1H NMR. Benzylidene complexes (R = C_6H_5) show no temperature-dependent ^1H NMR spectra and a high anticlinal:synclinal ratio (> 30:1) is inferred.

The rate of synclinal to anticlinal interconversion is first order; for equilibrium constants, K_{eq}, or $\Delta G^{\ddagger}$ values for interconverions, see the individual compounds.

(*syn*) (*anti*)

III

Reaction with nucleophiles. Addition of nucleophiles, Nu^-, to give compounds of the type Cp(CO)(^{2}D)FeCHRNu (Formula IV) proceeds stereospecifically because the phosphine ligand shields one face of the carbene ligand and the attack of the nucleophile proceeds from the Cp side (the arrow in III shows the direction of the attack). Thus, in some cases, addition of nucleophiles to a single carbene isomer can be achieved (see also isoelectronic cations in which the FeCO unit is replaced by ReNO). If the product is formed from the thermodynamically more stable anticlinal isomer, the *SR,RS* diastereomer, *anti*-Nu, is the major product (*SR* and *RS* are the chemically equivalent enantiomers). A similar attack at the synclinal isomer, produces the *SS,RR* diastereomer *syn*-Nu. However, the *syn* isomer is more reactive than the *anti* isomer, and because of the rapid interconversion between *anti* and *syn* isomers of the carbene cations, product distribution also depends on k_a [Nu] and k_s [Nu]. Thus, the *SR,RS*-to-*SS,RR* ratios (or *anti*-Nu-to-*syn*-Nu ratios) for the appropriate carbene cations and nucleophiles depend on various parameters, such as temperature, concentration, and nature of the nucleophile, and are given at the individual compounds. The *anti* Nu:*syn*-Nu ratio can be discussed in terms of the Curtin-Hammett-Winstein-Holness equation with the following boundary conditions: the ratio corresponds approximately to K_{eq} if k_a [Nu], k_s [NU] $\gg K_{a,s}$ and to $K_{eq} \cdot k_a/k_s$ if $k_{a,s}$, $k_{s,a} \gg k_a$ [Nu], k_s [Nu].

Concerning the resulting addition compounds the first letter designates the chirality of the iron center and the second letter the chirality of the carbon center.

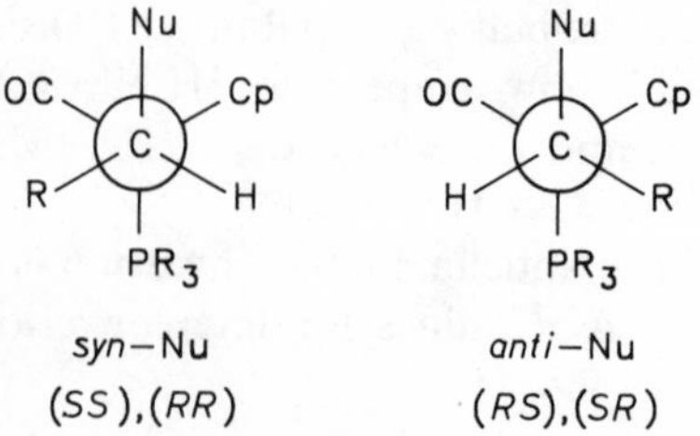

syn–Nu (*SS*),(*RR*) *anti*–Nu (*RS*),(*SR*)

IV

Cyclopropanation reactions. These reactions were carried out by generating the cation in situ at low temperature according to method Ib in the presence of an appropriate olefin (excess), followed by warming the mixture to room temperature as shown in Table 1. With optically pure carbene complexes ethylidene transfer proceeds with high enantioselectivity with ee values (enantiomeric excess) greater than 90%. The fact that the *SS* and *RS* cations (Table 1, Nos. 7 and 8) give cyclopropanes of opposite configuration in almost identical optical purities indicates that the chirality at the iron atom is primarily responsible for asymmetric induction and that the phosphine chirality plays little or no role. Methylene transfer proceeds with less stereoselectivity (low ee's) than ethylidene transfer (high ee values); benzylidene transfer occurs with moderate to high enantioselectivity.

The compounds in Table 1 correspond to the carbene complexes Nos. 2 to 6, 11, 17 and 18 in Table 2. Similar cyclopropanation reactions were conducted with the cation $[Cp(CO)(P(C_6H_5)_3)Fe{=}CHC_6H_5]^+$ and various alkenes.

Reaction Nos. 1 to 14 in Table 1 describe cyclopropanations with optically pure carbene cations and Nos. 15 and 16, reactions with racemic (at iron) compounds. The numbering of chiral carbon atoms of the resulting cyclopropane derivatives is given in Formula V.

For the mechanism of the cyclopropanation reaction with styrene, the attack of the weakly nucleophilic olefin at the more reactive synclinal isomer followed by reverse-side closure (attack of the electrophilic center developing at C_τ on the Fe–C_α bond) is favored over attack at the major anticlinal isomer with front-side closure. In compounds with a high anticlinal : synclinal ratio the attack may also occur at the anticlinal isomer with the same reverse-side closure leading to a reduction of the enantiomeric excess (ee).

Preparation. Because of the low thermal stability of these compounds, solutions have mostly been prepared at low temperature and used immediately for further reactions. The compounds are obtained according to the following general methods:

Method I Abstraction of $[OR']^-$ from the corresponding ether, $Cp(CO)(^2D)FeCHROR'$.
- a. With HX (X = BF_4, PF_6, $OOCCF_3$, or OSO_2CF_3)
- b. With $(CH_3)_3SiOSO_2CF_3$

Method II Protonation at low temperature of the appropriate vinyl complex, $Cp(CO)(^2D)FeCR{=}CHR'$, with the acids HBF_4, HPF_6, or HSO_3CF_3.

With the exception of Nos. 24 and 25, these cations are unstable at room temperature and were prepared and characterized only at low temperatures. Arylcarbene complexes have half-lives on the order of days.

General remarks. The Cp ligand appears as a singlet or doublet (J(P,H) = 1 to 2 Hz) in the region between 4.80 to 5.50 ppm in the ^1H NMR spectrum and from 89 to 93 ppm in the ^{13}C NMR spectrum. The IR spectrum exhibits one strong ν(CO) absorption between 1950 and 2025 cm^{-1}. Table 2 contains selected spectroscopic data.

V

Table 1. Cyclopropanation of Various Alkenes with $[Cp(CO)(^2D)Fe{=}CHR]^+$ Compounds. The mole ratio corresponds to the alkene:carbene ratio

No.	alkene (mole ratio)	cyclopropane	*cis*:*trans* (overall yield)
$[Cp(CO)(P(C_6H_5)_3)Fe{=}CH_2]^+$ (prepared from (−)-VI, Method Ia)			
1	E-C_6H_5–CH=CH–CH_3	CH_3, H_5C_6 (1S,2S)	(9%)
$[Cp(CO)(P(C_6H_5)_3)Fe{=}CH_2]^+$ (prepared from (+)-IV, Method Ia)			
2	E-C_6H_5–CH=CH–CH_3	H_3C, C_6H_5 (1*R*,2*R*)	(9%)
(*R*)-$[Cp(CO)(P(CH_3)_3)Fe{=}CHCH_3]^+$			
3	CH_2=CH–$COOC_2H_5$ (10:1)	O_2CCH_3, H_3C; H_3C, O_2CCH_3	1:1.6 (50 to 60%)

(*S*)-[Cp(CO)(P(CH$_3$)$_3$)Fe=CHCH$_3$]$^+$

4	CH_2=CH–$COOC_2H_5$ (10:1)		1:1.6 (50 to 60%)

(*R*)-[Cp(CO)(P(C$_2$H$_5$)$_3$)Fe=CHCH$_3$]$^+$ (acyl precursor ee = 76%)

5	CH_2=CH–$COOC_2H_5$ (10:1)	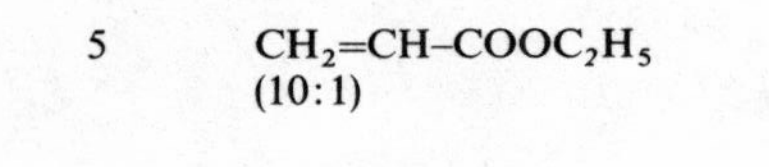 	1:1.8 (27 to 35%)

(*S*)-[Cp(CO)(P(C$_2$H$_5$)$_3$Fe=CHCH$_3$]$^+$ (acyl precursor ee = 77%)

6	CH_2=CH–$COOC_2H_5$ (10:1)	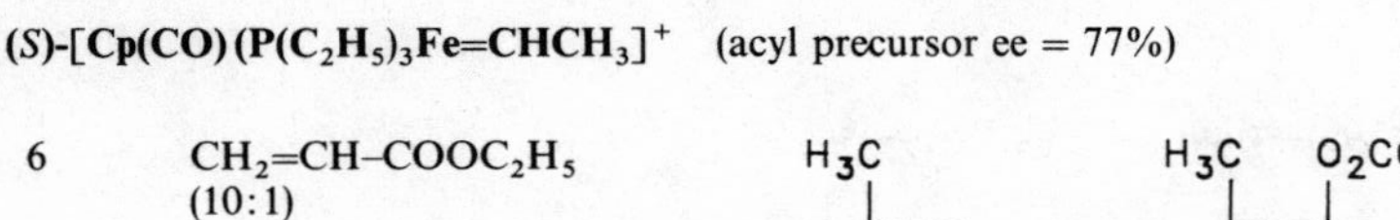 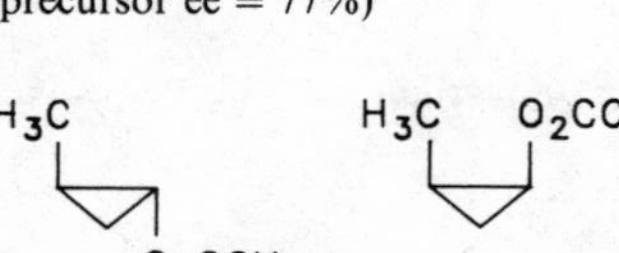	1:1.8 (27 to 35%)

Table 1. Continued

No.	alkene (mole ratio)	cyclopropane	cis:trans (overall yield)
(SS)-[Cp(CO)(P(C_6H_5)$_2$R)Fe=CHCH$_3$]$^+$ (R = (S)-2-methylbutyl)			
7	C_6H_5–CH=CH_2 (8:1)	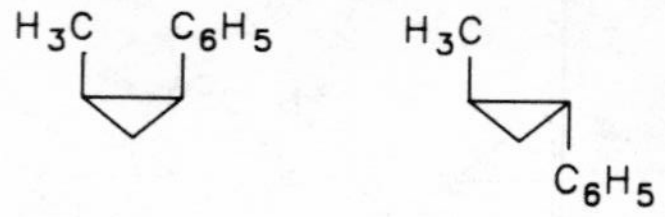 H_3C, C_6H_5 cis-(1R,2S); H_3C, C_6H_5 trans-(1R,2R)	1:3.5 (75%)
(RS)-[Cp(CO)(P(C_6H_5)$_2$R)Fe=CHCH$_3$]$^+$ (R = (S)-2-methylbutyl)			
8	C_6H_5–CH=CH_2 (8:1)	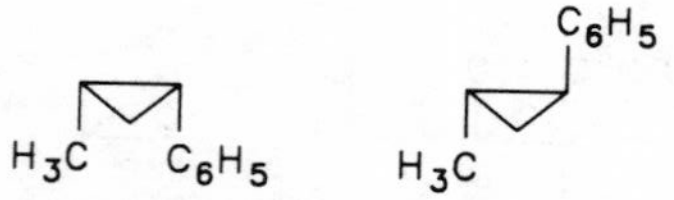 H_3C, C_6H_5 cis-(1S,2R); C_6H_5, H_3C trans-(1S,2S)	1:4.0 (75%)
(R)-[Cp(CO)(P(C_2H_5)$_3$)Fe=CHC$_6$H$_5$]$^+$			
9	CH_3COO–CH–CH_2 (10:1)	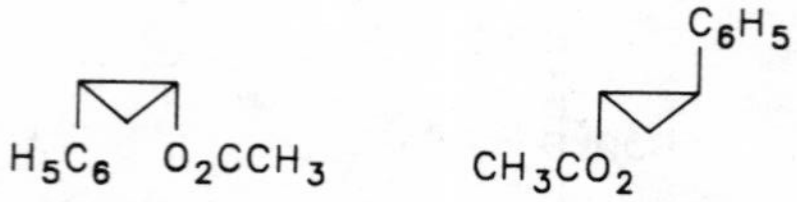 H_5C_6, O_2CCH_3 cis-(1R,2R); C_6H_5, CH_3CO_2 trans-(1S,2R)	4:1 (24%)

(*S*)-[Cp(CO)(P(C_2H_5)$_3$)Fe=CHC_6H_5]$^+$

10	$CH_3COO–CH{=}CH_2$ (10:1)	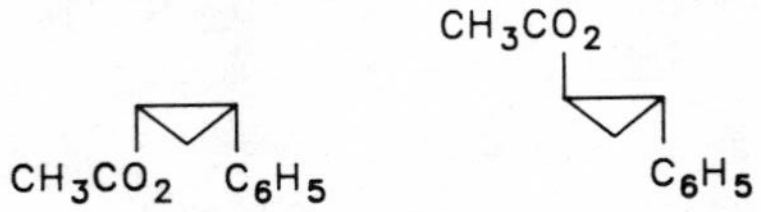 CH_3CO_2, C_6H_5 *cis*-(1*S*,2*S*)	CH_3CO_2, C_6H_5 *trans*-(1*R*,2*S*)	4:1 (21%)

(*RS*)-[Cp(CO)(P(C_6H_5)$_2$R)Fe=CHC_6H_5]$^+$ (R = (*S*)-2-methylbutyl)

11	$CH_3–CH{=}CH_2$ (high excess)	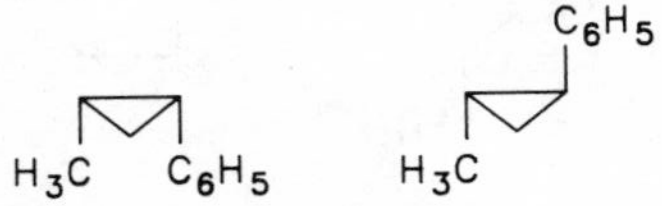 H_3C, C_6H_5 *cis*-(1*S*,2*R*)	C_6H_5, H_3C *trans*-1*S*,2*S*)	2:3 (22%)

Table 1. Continued

No.	alkene (mole ratio)	cyclopropane	*cis*:*trans* (overall yield)
12	$CH_3COO{-}CH{=}CH_2$ (12:1)	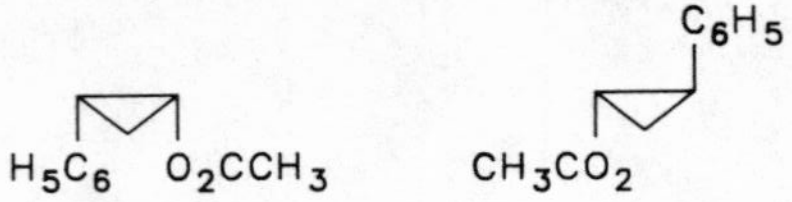 H_5C_6 O_2CCH_3 *cis*-(1*R*,2*R*); CH_3CO_2 C_6H_5 *trans*-(1*S*,2*R*)	4:1 (30%)
(*SS*)-[Cp(CO)(P(C_6H_5)$_2$R)Fe=CHC$_6H_5$]$^+$ (R = (*S*)-2-methylbutyl)			
13	$CH_3{-}CH{=}CH_2$ (high excess)	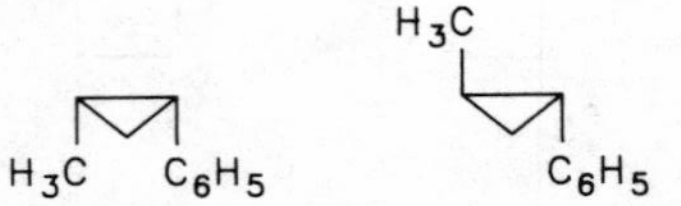 H_3C C_6H_5 *cis*-1*S*,2*R*); H_3C C_6H_5 *trans*-(1*R*,2*R*)	2:3 (21%)
14	$CH_3COO{-}CH{=}CH_2$ (12:1)	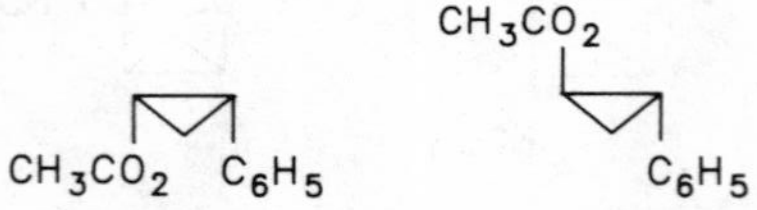 CH_3CO_2 C_6H_5 *cis*-(1*S*,2*S*); CH_3CO_2 C_6H_5 *trans*-(1*R*,2*S*)	4:1 (30)

$[Cp(CO)(P(C_6H_5)_3)Fe{=}CHCH_3]^+$

15	$C_6H_5{-}CH{=}CH_2$	 H_5C_6–(cyclopropane)–CH_3	1:3 (good yields)

$[Cp(CO)(P(OCH_3)_3)Fe{=}CHCH{=}C(CH_3)_2]^+$

16	$C_6H_5{-}CH{=}CH_2$ (10:1)	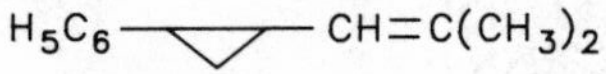 H_5C_6–(cyclopropane)–$CH{=}C(CH_3)_2$	5:1 (15%)

Table 2. Cationic Carbene Compounds of the Type $[Cp(CO)(^2D)Fe{=}CRR']^+$ Fe represents the corresponding fragment $Cp(CO)(^2D)Fe$. An asterisk indicates further information at the end of the table

No.	Fe=CRR ^{2}D	anion, method of preparation (yield) properties and remarks
*1	Fe=CH_2 THF	Cl^- salt
*2	Fe=CH_2 $P(C_6H_5)_3$	Cl^- salt BF_4^- salt, Ia
*3	Fe=$CHCH_3$ $P(CH_3)_3$	$SO_3CF_3^-$ salt, Ib ^{1}H NMR (CD_2Cl_2 at − 114 °C): 15.92 (CH=Fe, *syn*), 17.46 (CH=Fe, *anti*) BF_4^- salt, Ia ^{1}H NMR (CD_2Cl_2): 17.26 (dq, CH=Fe, J(P,H) = 1.4, J(P,H) = 7.47)
*4	Fe=$CHCH_3$ $P(C_2H_5)_3$	$SO_3CF_3^-$ salt, Ib ^{1}H NMR (CD_2Cl_2 at − 114 °C): 15.69 (CH=Fe, *syn*), 17.65 (CH=Fe, *anti*) BF_4^- salt, Ia ^{1}H NMR (CD_2Cl_2): 16.99 (q, CH=Fe, J(H,H) = 7.5)
*5	Fe=$CHCH_3$ $P(C_6H_5)_2R$ R=(*S*)-2-methylbutyl	$SO_3CF_3^-$ salt, Ib ^{1}H NMR (CD_2Cl, at − 78 °C): 17.27 for the *SS* and 17.42 for the *RS* diastereomer (CH =Fe)
*6	Fe=$CHCH_3^-$ $P(C_6H_5)_3$	$SO_3CF_3^-$ salt, Ib ^{1}H NMR: 2.84 (d, CH_3, J = 7.8), 17.94 (q, CH, J = 7.8) in CD_2Cl_2 at − 78 °C or 0 °C; 17.05 (CH=Fe, synclinal), 18.24 (CH=Fe, anticlinal) in CD_2Cl_2-SO_2ClF at − 126 °C ^{13}C NMR (CD_2Cl_2 at 0 °C): 51.0 (CH_3), 213.4 (d, CO, J(P,C) = 28.7 Hz), 380.0 (d, C=Fe, J(P,C) = 25.1) PF_6^- salt, Ia, II; yellow crystals at − 80 °C ^{1}H NMR (CF_3COOH): 2.94 (d, CH_3, J(H,H) = 8), 17.85 (q, CH=Fe, J = 8)
*7	Fe=$CHCH_3$ $P(OC_6H_5)_3$	PF_6^- salt, Ia, II; yellow crystals at − 80 °C
*8	Fe=CHC_2H_5 $P(C_6H_5)_3$	$SO_3CF_3^-$ salt, Ib ^{1}H NMR (CD_2Cl_2 at − 78 °C): 0.90 (s, br, CH_3), 2.96 (s, br, CH_2), 17.52 (s, br, CH=Fe) ^{13}C NMR (CD_2Cl_2 at − 78 °C): 48.3 (CH_2), 213 (d, CO, J(P,C) = 28.6), 383.2 (d, C=Fe, J(P,C) = 25.1)

No.	Compound	Data
9	$Fe{=}CHC_3H_7$-*i* $P(C_6H_5)_3$	$SO_3CF_3^-$ salt, Ib ^{1}H NMR (CD_2Cl_2 at − 78 °C): 17.8 (CH=Fe) not isolated; only tentative evidence for the existence was obtained
*10	$Fe{=}CHCH{=}CHC_2H_5$ $P(OCH_3)_3$	$CF_3SO_3^-$ salt, II ^{1}H NMR (CD_2Cl_2 at − 80 °C): 16.75 (s, CH=Fe)
*11	$Fe{=}CHCH{=}C(CH_3)_2$ $P(OCH_3)_3$	$CF_3SO_3^-$ salt, II (60%) ^{1}H NMR (CD_2Cl_2 at − 45 °C): 7.97 (d, CH=Fe, J = 13.8), 16.03 (d, CH=Fe, J = 13.8, J(P,H) = 1.7) ^{13}C NMR (CD_2Cl_2) 212.7 (d, CO, J(P,C) = 46.9), 314.3 (d, C=Fe; J(P,C) = 33.5)
*12	$Fe{=}CHCH{=}C(CH_3)OC_2H_5$ $P(C_6H_5)_3$	PF_6^- salt; red crystals, m.p. 135 to 137 °C ^{1}H NMR ($CDCl_3$): 2.05 (CH_3C), 7.00 (m, CH=Fe), 14.92 (m, CH=Fe; J(P,H) = 4 to 5, J(H,H) = 14.5) ^{13}C NMR: 296.7 (C=Fe) BF_4^- salt; red-orange crystals
*13	$Fe{=}CHCH{=}C(C_6H_4CH_3\text{-}4)OC_2H_5$ $P(C_6H_5)_3$	PF_6^- salt; m.p. 177 to 180 °C (dec.) ^{1}H NMR ($CDCl_3$): 7.67 (m, CH=), 13.84 (m, CH=Fe; J(P,H) = 4 to 5, J(H,H) = 14.6) ^{13}C NMR: 296.7 (C=Fe) BF_4^- salt; red-orange crystals
*14	$Fe{=}CHCH{=}C(CH_3)NHC_6H_5$ $P(C_6H_5)_3$	BF_4^- salt; mixture of isomers A, B; m.p. 196 to 198 °C (dec.) ^{1}H NMR (CD_3SOCD_3, A/B): 12.20/12.60 (m, CH=Fe; J(P,H) = 4 to 5, J(H,H) = 15.2)
*15	$Fe{=}CHCH{=}C(CH_3)NHC_6H_{11}$-*c* $P(C_6H_5)_3$	BF_4^- salt; mixture of isomers A, B; m.p. 204 to 206 °C ^{1}H NMR (CD_3SOCD_3, A/B): 11.70/12.20 (m, CH=Fe; J(P,H) = 4 to 5, J(H,H) = 16.0)
*16	$Fe{=}CHCH{=}C(CH_3)N(C_2H_5)_2$ $P(C_6H_5)_3$	BF_4^- salt; m.p. 196 to 198 °C (dec.) ^{1}H NMR (CD_3SOCD_3): 12.27 (m, CH=Fe; J(P,H) = 4 to 5, J(H,H) = 15)
*17	$Fe{=}CHC_6H_5$ $P(C_2H_5)_3$	$CF_3SO_3^-$ salt, Ib; racemic, (*R*) and (*S*) ^{1}H NMR: 17.04 (br, CH=Fe) in CD_2Cl_2 at 25 °C, − 71 °C, and in CD_2Cl_2-SO_2ClF at − 125 °C ^{13}C NMR (CD_2Cl_2, − 70 °C): 216 (d, CO, J(P,C) = 30.1), 333.15 (d, C=Fe, J(P,C) = 23.1) $CF_3CO_2^-$ salt, Ia
*18	$Fe{=}CHC_6H_5$ $P(C_6H_5)_2R$ R=(*S*)-2-methylbutyl	$CF_3SO_3^-$ salt, Ib; dark red solution of a mixture of (*SS*) and (*RS*) diastereomers ^{1}H NMR (CD_2Cl_2 at − 71 °C): 16.9 (d, CH=Fe, J(P,H) = 9.9) ^{13}C NMR (CD_2Cl_2 at − 71 °C): 341.0 (d, C=Fe, J(P,C) = 24)

Table 2. Continued

No.	Fe=CRR ^2D	anion, method of preparation (yield) properties and remarks
*19	$Fe{=}CHC_6H_5$ $P(C_6H_5)_3$	$CF_3CO_2^-$ salt, Ia; maroon, deep red CD_2Cl_2 solution ^1H NMR (CF_3COOH at 0 °C): 17.43 (d, CH=Fe, J(P,H) = 1.12) ^{13}C NMR (CF_3COOH at 0 °C): 215.4 (d, CO, J(P,C) = 29), 341.2 (dd, C=Fe, ^1J(H,C) = 136, J(P,C) = 21) PF_6^- salt; maroon AsF_6^- salt; burgundy $CF_3SO_3^-$ salt, Ia, Ib ^1H NMR (CD_2Cl_2-SO_2ClF at − 125 °C): 17.48 (s, CH=Fe) ^{13}C NMR (CD_2Cl_2 at − 20 °C): 215.5 (CO), 342.0 (C=Fe, J(P,C) = 23)
*20	$Fe{=}CHC_6H_4CH_3$-4 $P(C_6H_5)_3$	$CF_3SO_3^-$ salt, Ia, Ib ^1H NMR (CD_2Cl_2 at 0 °C): 16.93 (s, CH=Fe) ^{13}C NMR (CD_2Cl_2 at 0 °C): 215.7 (CO), 336.4 (C=Fe, J(P,C) = 22)
*21	$Fe{=}CHC_6H_4F$-4 $P(C_6H_5)_3$	$CF_3SO_3^-$ salt, Ia, Ib ^1H NMR (CD_2Cl_2 at − 40 °C): 17.20 (s, CH=Fe) ^{13}C NMR (CD_2Cl_2 at − 40 °C): 215.5 (CO), 334.3 (C=Fe, J(P,C) = 20)
*22	$Fe{=}CHC_6H_4OCH_3$-4 $P(C_6H_5)_3$	$CF_3SO_3^-$ salt, Ia, Ib ^1H NMR (CD_2Cl_2 at − 60 °C): 16.62 (s, CH=Fe); ^{13}C NMR (CD_2Cl_2 at − 20 °C): 216.4 (CO), 323.6 (C=Fe, J(P,C) = 23)
*23	$Fe{=}C(CH_3)_2$ $P(C_6H_5)_3$	BF_4^- salt, II (93%); yellow, m.p 140 °C (dec.) ^1H NMR (in $CDCl_3$ or CD_2Cl_2): 3.13 (s, CH_3) ^{13}C NMR (CD_2Cl_2 at 0 °C): 59.4 (s, CH_3), 214.2 (d, CO, J(P,C) = 29), 407.5 (s, br, C=Fe) $CF_3SO_3^-$ salt, II (83%); yellow, m.p 100 °C (dec). ^1H NMR (CD_2Cl_2): 3.14 (s, CH_3) ^{13}C NMR (CD_2Cl_2 at − 20 °C): 59.1 (CH_3), 213.9 (d, CO, J(P,C) = 26), 406.5 (d, C=Fe, J(P,C) = 18)

*24

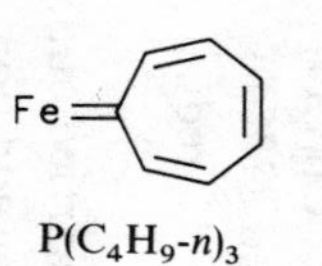

$P(C_4H_9\text{-}n)_3$

PF_6^- salt; purple plates
^1H NMR (CD_3COCD_3): 7.0 to 8.0 (m, 4 H), 9.6 (d, 2 H)
^{13}C NMR (CD_3COCD_3): 217.5 (d, CO, J(P,C) = 27.5), 278.8 (d, C=Fe, J(P,C)=22.5)

*25

Fe=

$P(C_4H_9\text{-}n)_3$

PF_6^- salt; blue-purple solid
^1H NMR (CD_3COCD_3): 7.30 (d, 2 H), 7.70 to 8.00 (AA'BB', 4 H), 9.10 (d, 2 H)
^{13}C NMR (CD_3COCD_3 at − 10 °C): 129.3 (C-6, 7), 133.6 (C-5, 8), 135.5 (C-4, 9), 139.4 (C-3, 10), 157.6 (C-2, 11)

Further Information

$[Cp(CO)C_4H_8OFe{=}CH_2]Cl$ (Table 2, No. **1**) was believed to be an intermediate in the reduction of coordinated CO in $Mg[Cp(CO)_2Fe]_2$ by HCl in THF at $-78\,^\circ C$ to give CH_4 and C_2H_6.

$[Cp(CO)(P(C_6H_5)_3)Fe{=}CH_2]X$ (Table **2**, No. **2**, X = Cl, BF_4). An approximately 1:1:1 mixture of unreacted starting complex, $[Cp(CO)(P(C_6H_5)_3)Fe{=}CHOCH_3]^+$, and $Cp(CO)(P(C_6H_5)_3)FeCH_3$ was obtained when 3 equivalents of $Cp(CO)(P(C_6H_5)_3)FeCH_2OCH_3$ were added slowly to 1 equivalent of HBF_4 etherate in THF at $-78\,^\circ C$ (inverse Method Ia). This indicates that the initially formed title complex is a good hydride abstractor. The freshly prepared BF_4^- salt reacts with LiC_6H_5 in THF at $-78\,^\circ C$ to give $Cp(CO)(P(C_6H_5)_3)$-$FeCH_2C_6H_5$. Cyclopropanation reactions were carried out by generation of the two enantiomers of the cation according to Method Ia from either (+)- or (−)-$Cp(CO)(P(C_6H_5)_3)FeCH_2OC_{10}H_{19}$ (Formula VI) in the presence of *trans*-1-phenylpropene. The cation was proposed to be an intermediate in the insertion of SO_2 into the C–O bond of $Cp(CO)(P(C_6H_5)_3)FeCH_2OCH_3$ to form $Cp(CO)(P(C_6H_5)_3)FeCH_2SO_2OCH_3$, or in the conversion of the menthyl ether VI into $Cp(CO)(P(C_6H_5)_3)FeCH_2Cl$ by anhydrous HCl at 0 °C in ether.

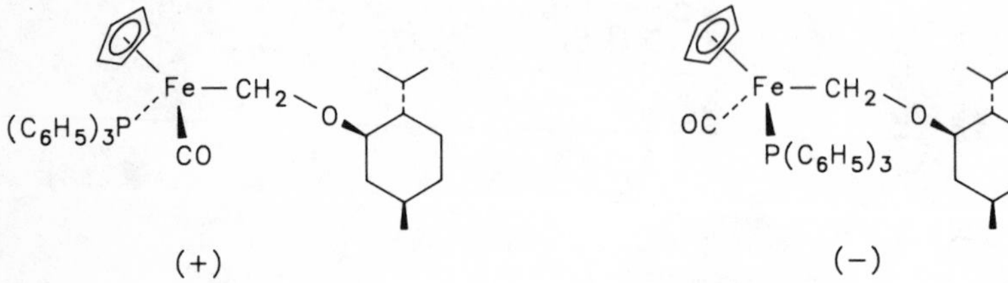

(+) (−)

VI

$[Cp(CO)(P(CH_3)_3)Fe{=}CHCH_3]X$ (Table **2**, No. **3**, X = SO_3CF_3, BF_4). The equilibrium constant for the anticlinal (*anti*) to synclinal (*syn*) interconversion (see Formula III) for X = SO_3CF_3 at $-104\,^\circ C$ was estimated to be $K = a/s = 4.6$; extrapolated to $-30\,^\circ C$, $K = 2.8$. The activation energies are $\Delta G^{\ddagger}_{AS} = 9.3$ and $\Delta G_{AS} = 8.8$ kcal/mol. Optically pure cations (S_{Fe} or R_{Fe}) were obtained starting from the optically pure acyl complex $Cp(CO)$-$P(CH_3)_3FeCOCH_3$ with an enantiomeric excess of 77% for the (*R*) and 87% for the (*S*) enantiomer following the reaction sequence: alkylation at 25 °C, reduction with $NaBH_4/NaOCH_3/CH_3OH$, and Method Ib.

Reactions with the nucleophiles $NaSC_6H_5$ and $KSCOCH_3$ were carried out at $-30\,^\circ C$ by quenching in a stirred methanol solution, resulting in the addition compounds IV (Nu = SC_6H_5, $SCOCH_3$; $PR_3 = P(CH_3)_3$). At a given Nu^- concentration the *syn*-Nu:*anti*-Nu ratios was estimated. For the weaker nucleophile $[SCOCH_3]^-$, a higher *syn*-Nu:*anti*-Nu ratio was obtained than for the stronger nucleophile $[SC_6H_5]^-$; see also general remarks. The enantiomeric S_{Fe} and R_{Fe} cations undergo enantioselective ethylidene transfer with vinyl acetate to give a mixture of *cis* and *trans* cyclopropanes, as shown in Table 1, with optical yields between 89 and 93%. The cations were prepared in situ at $-78\,^{\circ}C$ in CH_2Cl_2 according to method Ib in the presence of the vinyl acetate.

$[Cp(CO)(P(C_2H_5)_3)Fe{=}CHCH_3]X$ (Table **2**, No. **4**, $X = SO_3CF_3$, BF_4). The equilibrium constant for the anticlinal (*anti*) to synclinal (*syn*) interconversion for $X = SO_3CF_3$ at $-104\,^{\circ}C$ was estimated to be $K = anti/syn = 10.9$; extrapolated to $-30\,^{\circ}C$, $K = 5.2$. The activation energies are $\Delta G^{\ddagger}_{anti,syn} = 9.6$ and $\Delta G^{\ddagger}_{syn,anti} = 8.8$ kcal/mol. The optically pure cations are obtained similarly as described for No. 3 starting with the acyl complexes (*R*)- and (*S*)-$Cp(CO)(P(C_2H_5)_3)FeCOCH_3$ with ee = 76% and 77%, respectively.

Addition of the nucleophiles $NaSC_6H_5$ and $KSCOCH_3$ under various conditions was carried out at $-30\,^{\circ}C$ by quenching in a stirred methanol solution, forming the adducts $Cp(CO)(PR_3)FeCHCH_3$-Nu (Formula IV, Nu = SC_6H_5, $SCOCH_3$; $PR_3 = P(C_2H_5)_3$). With $NaSC_6H_5$ the *anti*-Nu:*syn*-Nu ratios were found to be between 1:1.1 and 1:6.0, and the less nucleophilic $KSCOCH_3$ exhibits values between 1:3.0 and 1:13. Low concentrations of the nucleophile favor the *S*-Nu adduct; see general remarks. The chiral S_{Fe} and R_{Fe} cations (Formula I, $^2D = P(C_2H_5)_3$ in 1.1.2) undergo enantioselective ethylidene transfer with vinyl acetate to give the corresponding cyclopropanes, as outlined in Table 1, in 27 to 35% yields with optical yields between 89 and 97%.

$[Cp(CO)(P(C_6H_5)_2C_5H_{11})Fe{=}CHCH_3]SO_3CF_3$ (Table **2**, No. **5**). The reaction of the enantiomeric cations $R_{Fe}S_C$ and $S_{Fe}S_C$ (Formula VII and VIII, R = (*S*)-2-methylbutyl; prepared in situ according to Method Ib from the corresponding *SS* and *RS* methyl ether) with styrene leads to the enantioselective synthesis of the cyclopropanes with optical yields between 84 and 90%, as depicted in Table 1. A mechanism involving attack of the nucleophile, styrene, at the anticlinal isomer is assumed.

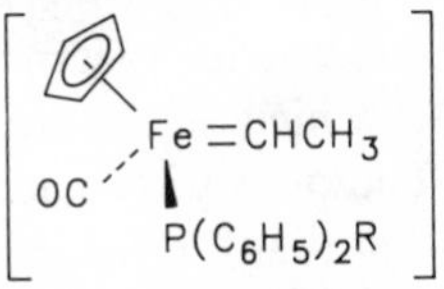

$[Cp(OC)Fe{=}CHCH_3(P(C_6H_5)_2R)]^+$ (*RS*) VII

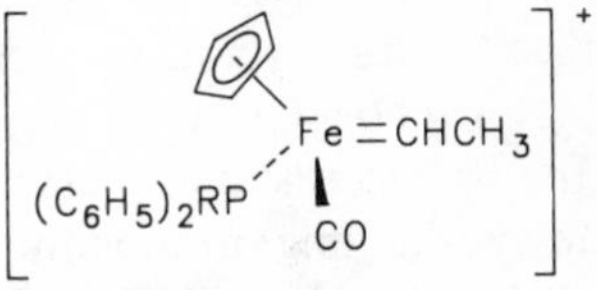

$[Cp((C_6H_5)_2RP)Fe{=}CHCH_3(CO)]^+$ (*SS*) VIII

$[Cp(CO)(P(C_6H_5)_3)Fe{=}CHCH_3]X$ (Table **2**, No. **6**; $X = CF_3SO_3$, PF_6). The cation is moderately stable in CH_2Cl_2 solution at $25\,^{\circ}C$ with $t_{1/2} = 3$ h; decomposition gives the ethylene complex $[Cp(CO)(P(C_6H_5)_3Fe(\eta^2\text{-}CH_2{=}CH_2)]PF_6$ in about 10% yield along with an uncharacterized, probably binuclear material. The formation of cyclopropanes upon decomposition was also suggested. For the thermal decomposition, see also No. 8. For the synclinal-to-anticlinal interconversion, the values $\Delta G^{\ddagger}_{anti,syn} = 7.8$ and $\Delta G^{\ddagger}_{syn,anti} = 7.3$ kcal/mol have been estimated; K_{eq} for *anti*/*syn* = 4.6 at $-104\,^{\circ}C$.

The PF_6^- salt adds $P(C_6H_5)_3$ to give the ylide complex IX ($^2D = P(C_6H_5)_3$, $R = CH_3$, $X = PF_6$). With the base $NC_2H_5(C_3H_7\text{-}i)_2$ the starting material (Method II) is recovered. Reactions with the nucleophiles $NaSC_6H_5$ and $KSCOCH_3$ were carried out at $-30\,°C$ by quenching into a stirred methanol solution with result of addition compounds IV (Nu = SC_6H_5, $SCOCH_3$; $PR_3 = P(C_6H_5)_3$). At a given Nu^- concentration the *syn*-Nu:*anti*-Nu ratios were estimated; the *syn*-Nu is the major product. For the weaker and less reactive nucleophile $[SCOCH_3]^-$, a higher *syn*-Nu:*anti*-Nu ratio was obtained than for the stronger nucleophile $[SC_6H_5]^-$.

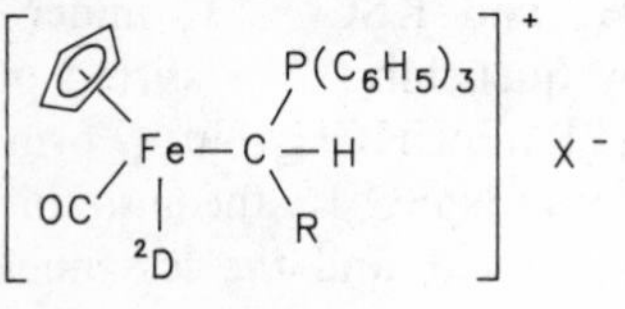

$P(C_6H_5)_3$
Fe—C—H
OC
^2D
R
X^-

IX

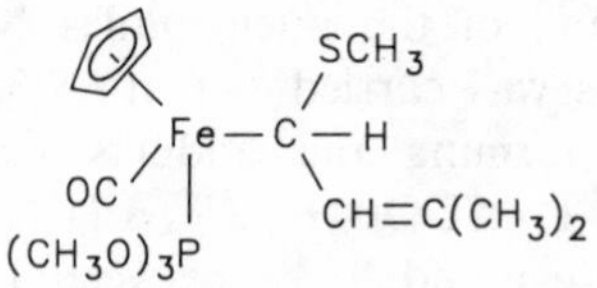

SCH_3
Fe—C—H
OC
$CH{=}C(CH_3)_2$
$(CH_3O)_3P$

X

$[Cp(CO)(P(OC_6H_5)_3)Fe{=}CHCH_3]PF_6$ (Table **2**, No. **7**) is unstable at room temperature and decomposes at 25 °C with hydride migration to give the cationic olefin complex $[Cp(CO)\,(P(OC_6H_5)_3)\,Fe(\eta^2\text{-}CH_2{=}CH_2)]PF_6$.

Similarly to No. 6, addition of $P(C_6H_5)_3$ at $-80\,°C$ generates the ylide complex IX ($^2D = P(OC_6H_5)_3$, $R = CH_3$, $X = PF_6$) with 56% yield, but with the base $NC_2H_5(C_3H_7\text{-}i)_2$ the starting vinyl complex is recovered. Reduction with $Li[BH(C_2H_5)_3]$ at $-80\,°C$ produces $Cp(CO)(P(OC_6H_5)_3)FeC_2H_5$ in 81% yield.

$[Cp(CO)(P(C_6H_5)_3)Fe{=}CHC_2H_5]CF_3SO_3$ (Table **2**, No. **8**). The cation is supposed to be an intermediate by the epimerization of the diastereoisomers $Cp(CO)(P(C_6H_5)_3)FeCHC_2H_5OCH_3$ via reversible loss of $[OCH_3]^-$. It undergoes hydride migration even at $-40\,°C$ ($t_{1/2}$ about 1 h) to give first the thermodynamically less stable, kinetic diastereomeric olefin complexes $[Cp(CO)(P(C_6H_5)_3)Fe(\eta^2\text{-}CH_2CHCH_3)]CF_3SO_3$, which isomerize at higher temperature to the more stable diastereomer. The activation energy for the intramolecular hydride migration is much less than that for No. 6.

$[Cp(CO)(P(OCH_3)_3)Fe{=}CHCH{=}CHC_2H_5]CF_3SO_3$ (Table **2**, No. **10**). An alternative structure of the cation, resulting from protonation at the 2-position (method II) to give $[Cp(CO)(P(OCH_3)_3)Fe{=}CHCH_2CH{=}CHCH_3]^+$, could not be excluded.

$[Cp(CO)(P(OCH_3)_3)Fe{=}CHCH{=}C(CH_3)_2]X$ (Table **2**, No. **11**, $X = BF_4$, CF_3SO_3). This complex (probably as the BF_4^- salt) was trapped as the ylide complex IX ($^2D = P(OCH_3)_3$, $R = CH{=}C(CH_3)_2$, $X = BF_4$) by addition of $P(C_6H_5)_3$ to a CH_2Cl_2 solution of the complex at $-78\,°C$. Addition of $NaSCH_3$ in a similar manner produced a 4:1 mixture of the diastereoisomers of complex X in 52% yield; for cyclopropanation reactions, see general remarks.

$[Cp(CO)(P(C_6H_5)_3)Fe{=}CHCH{=}C(R)OC_2H_5]X$ (Table **2**, No. **12**, $R = CH_3$; No. **13**, $R = C_6H_4CH_3\text{-}4$; $X = BF_4$, PF_6). The BF_4^- salt is obtained by alkylation of the corresponding complex $Cp(CO)(P(C_6H_5)_3)FeCH{=}CHCOR$ with $[O(C_2H_5)_3]BF_4$. The PF_6^- salt is obtained by addition of a methanolic NH_4PF_6 solution to the crude and dried BF_4^- salt; 80 and 75% yield.

Compound No. 12 can be converted with various amines into the complexes 14 to 16.

$[Cp(CO)(P(C_6H_5)_3)Fe{=}CHCH{=}C(CH_3)NRR']BF_4$ (Table **2**, No. **14**, R = H, R′ = C_6H_5; No. **15**, R = H, R′ = C_6H_{11}-*c*; No. **16**, R = R′ = CH_3) are obtained by addition of the appropriate amine RR′NH to a solution of No. 12 in CH_2Cl_2 at −78 °C. The complexes are obtained as mixtures of isomers (4:6, No. 14) in 70, 55, and 55% yields, respectively.

$N(C_2H_5)_3$ in CH_2Cl_2 solution or ethanolic NaOH deprotonates No. 14 to give $Cp(CO)(P(C_6H_5)_3)FeCH{=}CHC({=}NC_6H_5)CH_3$ (reverse reaction with HBF_4). No. 15 gives No. 12 when treated with ethanolic NaOH.

$[Cp(CO)(P(C_2H_5)_3)Fe{=}CHC_6H_5]X$ (Table **2**, No. **17**, X = CF_3CO_2, CF_3SO_3). The corresponding cation forms also by addition of H^+ to a solution of $Cp(CO)(P(C_2H_5)_3)FeCHC_6H_5(OCH_3)$ in CD_3OD. The deuteriobenzylidene cation, $[Cp(CO)(P(C_2H_5)_3)Fe{=}CDC_6H_5]^+$, is obtained according to Method Ia from $Cp(CO)(P(C_2H_5)_3)FeCD(OCH_3)C_6H_5$ as starting material. A deep red solution of the $CF_3SO_3^-$ salt at −78 °C is obtained with Method Ib. Starting with the *RR* and *SR* diastereomeric precurors $Cp(CO)(P(C_2H_5)_3)FeCH(C_6H_5)OCH_3$, solutions of the *R* and *S* enantiomers are obtained according to Method Ib. At −125 °C a high anticlinal:synclinal equilibrium ratio (> ca. 30) was found with activation energies for the interconversion of less than 7 kcal/mol.

Methoxide addition was carried out at 0 °C with various concentrations of $NaOCH_3$ or $NaOCD_3$ in CH_3OH or CH_3OD, respectively. The resulting diastereomer ratios of the adduct (Formula IV) were concentration-dependent, and excess of the *SR,RS* diastereomer (*anti*-Nu) was formed; see general remarks. A similar H^-/D^- addition to the cation and the α-D cation with $LiHB(C_2H_5)_3$ or $LiDB(C_2H_5)_3$, respectively, was carried out in THF at −45 °C according to Scheme I. The diastereomer XI arising from the addition of the nucleophile at the anticlinal isomer was always the major product with a 30:1 ratio (*SS,RR* in the case of D^- addition and *RS,SR* in the case of H^- addition).

$[Cp(CO)(P(C_6H_5)_2C_5H_{11})Fe{=}CHC_6H_5]CF_3SO_3$ (Table **2**, No. **18**). A deep red solution of a mixture of the *RS*/*SS* diastereomers was obtained in CD_2Cl_2 at −78 °C according to the procedure outlined in Method Ib using a racemic mixture (*SSS*/*SSR* and *RSS*/*RSR*; sequence of chirality centers: Fe, C at phosphine, C–Fe) of the carbene precursor $Cp(CO)(P(C_6H_5)_2C_5H_{11})$ FeCH$(C_6H_5)OCH_3$. For in situ cyclopropanation the diastereomers were separately prepared starting from the *SSS*/*SSR* or *RSS*/*RSR* precursors, respectively.

For the cyclopropanation of propene and vinyl acetate, a mechanism involving reaction of the synclinal isomer followed by backside closure has been proposed.

$[Cp(CO)(P(C_6H_5)_3)Fe{=}CHC_6H_5]X$ (Table **2**, No. **19**; X = PF_6, AsF_6, CF_3SO_3, CF_3CO_2). The PF_6^- and AsF_6^- salts were obtained in a way similar to Method I, but using the corresponding $[C(C_6H_5)_3]X$ salts as abstracting agent.

XI

The PF_6^- salt shows 50% decomposition after 60 h at 25 °C. At − 125 °C a high anticlinal:synclinal equilibrium ratio (>ca. 30) was found with an activation energy of interconversion of <7 kcal/mol. For correlations of CH=Fe ^{1}H and ^{13}C NMR shifts with Hammett σ^+ constants, see also Nos. 20 to 22.

Methoxide addition was carried out at 0 °C with various concentrations of $NaOCH_3$ or $NaOCD_3$ in CH_3OH or CH_3OD, respectively. The resulting diastereomer ratios of the adducts (Formula IV) were concentration-dependent. At low methoxide concentration the major product arises from the more reactive synisomer whereas higher methoxide concentrations give excess of the *SR*,*RS* diastereomer resulting from addition at the *anti* isomer. Irradiation of the cation (as the $CF_3CO_2^-$ salt) in CD_2Cl_2 at − 78 °C results in the disappearance of the benzylidene ^{1}H NMR resonance signal; no photolytically induced synclinal/anticlinal conversion is found as it is with the isoelectronic neutral Re complex (FeCO unit replaced by the ReNO unit).

[Cp(CO)(P(C_6H_5)$_3$)Fe=CHC$_6$H$_4$R-4)]CF$_3$SO$_3$ (Table 2, Nos. **20** to **22**, R = CH_3, F, OCH_3) exhibit barriers of aryl rotation, $\Delta G^{\ddagger}$, of 9.0, 9.0, and 10.9 kcal/mol, respectively. The aryl ring in solution is coplanar with the plane of the carbene ligand, as shown by low temperature (0 to − 40 °C) ^{1}H NMR and ^{13}C NMR spectroscopy. The ^{1}H and ^{13}C chemical shifts of CH=Fe show an approximately linear correlation with Hammett σ^+ constants (including No. 19).

[Cp(CO)(P(C_6H_5)$_3$)Fe=C(CH$_3$)$_2$]X (Table 2, No. **23**, X = BF_4, CF_3SO_3). Both salts are stable as solids but decompose in CD_2Cl_2 solution; the $CF_3SO_3^-$

salt decomposes faster (half-life of 15 min at 40 °C) than the BF_4^- salt, which decomposes slowly at room temperature and rapidly at 88 °C.

Addition of pyridine to the cation in ether at room temperature regenerates the starting vinyl complex, $Cp(CO)(P(C_6H_5)_3)FeC(CH_3){=}CH_2$, in 78% yield. No cyclopropanation reaction is observed with excess isobutylene at 88 °C in CD_2Cl_2 solution (3 h).

$[Cp(CO)(P(C_4H_9\text{-}n)_3)Fe{=}C_7H_6]PF_6$ and **$[Cp(CO)(P(C_4H_9\text{-}n)_3)Fe{=}C_{11}H_8]PF_6$** (Table **2**, Nos. **24** and **25**). The complexes were obtained by H^- abstraction from the cycloheptatrienes XII to XIV (No. 24) and XV (No. 25) by dropwise addition of $[C(C_6H_5)_3]PF_6$ in CH_2Cl_2 solution at − 78 °C. The air-sensitive compounds are isolated in 54 and 45% yields, respectively. No. 25 is less stable than No. 24 and decomposes at room temperature.

The barriers of rotation about the Fe=C axis have been estimated by variable temperature ^1H NMR to be $\Delta G^\ddagger = 9.6 \pm 0.2$ kcal/mol (coalescence temperature − 84 °C) for No. 24 and $\Delta G^\ddagger = 10.4 \pm 0.2$ kcal/mol (coalescence temperature − 68 °C) for No. 25.

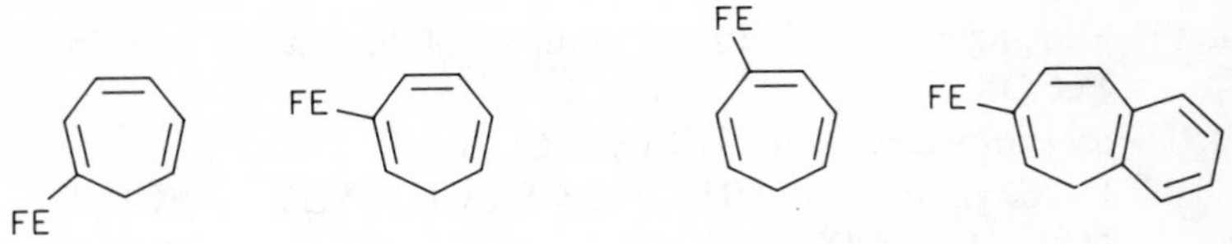

FE FE FE FE

XII XIII XIV XV

Compounds with the ^5L ligand $C_5(CH_3)_5$

$[C_5(CH_3)_5(CO)(P(C_6H_5)_3Fe{=}CH_2]X$ ($X = CF_3CO_2$, SO_3CF_3) are prepared by treatment of a CD_2Cl_2 solution of the neutral $Cp^*(CO)(P(C_6H_5)_3)FeCH_2OCH_3$ at − 90 °C with the appropriate acid HX.

^1H NMR spectrum at − 85 °C in CD_2Cl_2 gives signals for the non-equivalent methylene protons at 15.10 and 16.67 ppm (s's, br, $CH_2{=}Fe$). They coalesce at − 45 °C giving a single broad signal at 15.95 ppm at − 20 °C. The free energy of activation, ΔG^*, for the Fe–C bond rotation is calculated to be 44.35 kJ/mol. ^{13}C NMR (in CD_2Cl_2 at − 80 °C): 215.3 (d, CO, J(P,C) = 31.5 Hz), 351.2 ppm (d, CH_2, J(P,C) = 23.6 Hz).

The complex (as the $CF_3SO_3^-$ salt) is stable up to − 10 °C.

$[C_5(CH_3)_5(CO)P(CH_3)_3Fe{=}C(CH_3)_2]BF_4$ has been prepared according to preparation Method II from the corresponding vinyl complex in ether at − 78 °C as a red microcrystalline powder; m.p. 190 °C (dec.).

^1H NMR (CH_2Cl_2): 3.32 ppm (s, CH_3). ^{13}C NMR (CD_2Cl_2): 59.0 (CH_3 of $(CH_3)_2C$), 215.8 (d, CO; J(P,C) = 27), 399.5 ppm (d, C=Fe; J(P,C) = 25). ^{31}P NMR ($CDCl_3$): 22.96 ppm. IR (CH_2Cl_2): ν (CO) 1965 cm^{-1}.

The complex reacts with $Na[C_5(CH_3)_5(CO)_2Fe]$ to give the starting material and $Cp^*(CO)_2FeH$.

1.1.2.2 Cations of the Type $[{}^5L(CO)({}^2D)Fe{=}C(R)ER']^+$ with One Heteroatom at the Carbene Carbon Atom

The compounds in this section with one heteroatom at the carbene carbon atom are more stable than those of the preceding section. The heteroatoms are Cl (1 compound), O (66 compounds), S (4 compounds), and N (17 compounds). Complex No. 85 in Table 3 contains a chelating ligand of the type ${}^1L\text{-}{}^2D$ in which ${}^1L$ represents the carbene group. Only one complex in this series with ${}^5L$ other than C_5H_5 (Cp) is known. It contains the bulky $C_5(CH_3)_5$ ligand and is described at the end of this section. The compounds are prepared according to the following general methods:

Method I Starting complex is vinylidene complex of the general type $[Cp(CO)({}^2D)Fe{=}C{=}CR_2]X$.
a. Addition of H_2S, HCl, HSO_3CF_3.
b. Addition of HOR, HSR, or HNR_2.
c. Cycloaddition of $RHC{=}CNCH_3$ or the 2-thiazoline I via the intermediate II.

Method II Starting complex is an acyl complex of the general type $Cp(CO)({}^2D)FeCOR$.
a. Protonation with HX (X = Cl, Br, BF_4, SO_3F).
b. Alkylation with $CH_3OSO_2CF_3$, CH_3OSO_2F, $[O(CH_3)_3]BF_4$, or $[O(C_2H_5)_3]X$ (X = BF_4, PF_6).

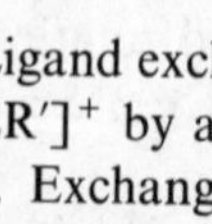

I

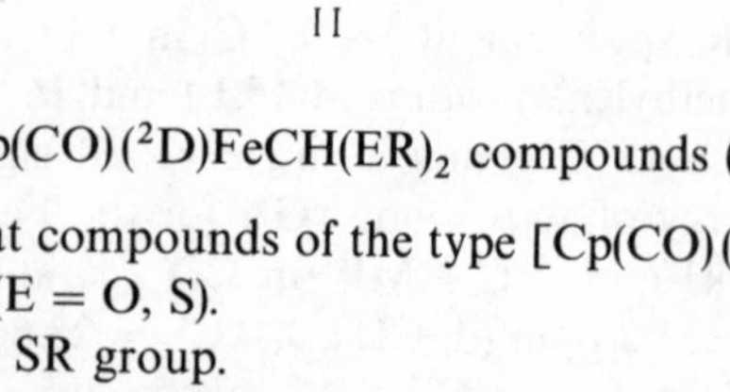

II

Method III Protonation of $Cp(CO)({}^2D)FeCH(ER)_2$ compounds (E = O, S).

Method IV Ligand exchange at compounds of the type $[Cp(CO)({}^2D)Fe{=}C(R)ER']^+$ by amines (E = O, S).
a. Exchange of an SR group.
b. Exchange of an OR group.

General remarks. The cations are chiral at iron. Separation of enantiomers can be achieved starting with racemic acyl compounds of the type $Cp(CO)({}^2D)FeC(O)CH_3$ which are protonated with either (1*S*)-(+)-or (1*R*)-(−)-10-camphorsulfonic acid (method IIa) to form diastereomeric carbene salts *RS,SS* or *SR,RR*, respectively. These can be separated by successive crystallization. Neutralization recovers the *S* or *R* acyl isomer which can be alkylated (method IIb) to the optically pure carbene compounds. The other route starts from the racemic carbene complex $[Cp(CO)(P(C_6H_5)_3)Fe{=}C(CH_3)OCH_3]BF_4$ by reaction with (*S*)-(−)-α-phenylethylamine as described in method IVb.

The barrier of rotation about the FeC double bond is less than that of compounds with no heteroatom and is estimated to be less than 7 kal/mol, accompanied by a decrease of reactivity due to diminished electrophilicity of the carbene carbon atom.

According to the alignment of the carbene plane with the Fe–CO bond (see also Sect. 1.1.2.1), the heteroatom can occupy a position *cis* or *trans* to the oxygen atom of CO as depicted in Formula III and IV (shown for the *S* enantiomer). In the case of E = O and $^2D = P(C_6H_5)_3$ the "*anti* (O,O)" conformation is preferred over the "*syn* (O,O)" conformation. All reactions with nucleophiles are predicted to occur exclusively above this plane from the *anti*-(O,O) conformation as depicted by the arrow in Formula VI rather than between this plane and the parallel plane formed by one C_6H_5 ring of $P(C_6H_5)_3$; see also Formula XII and XIII on p. 54. A similar *anti*-(O,N) conformation is realized in the molecular structure of No. 76.

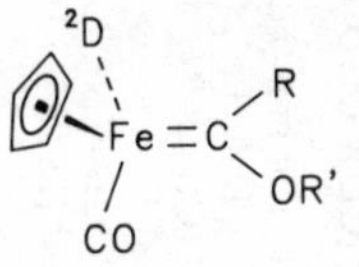

syn O,O

III

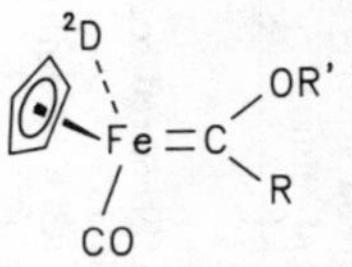

anti O,O

IV

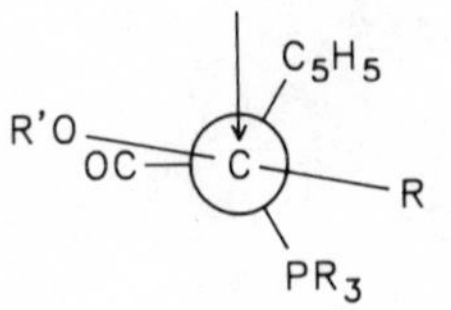

V

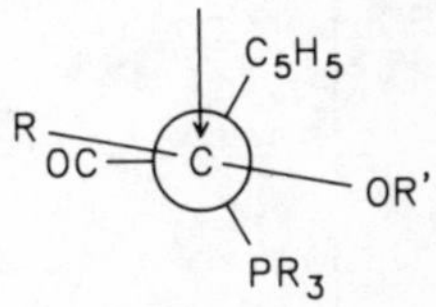

VI

Addition of the nucleophile H^- at the carbene carbon atom to generate the pairs of enantiomers *RS,SR* and *RR,SS* is only observed with E = O and results in ethers of the general type $Cp(CO)(^2D)FeCH(R)OR'$. The stereoselectivity of H^- addition at some racemic carbene cations was studied. Thus, high stereoselectivity (15:1) was observed with the carbene cation $[Cp(CO)(P(C_6H_5)_3)Fe{=}C(C_2H_5)OCH_3]^+$ (No. 35) and hydride donors such as $LiAlH_4$ or $NaBH_4$ in THF. Only low stereoselectivity occurred with $[Cp(CO)(P(C_6H_5)_3)Fe{=}C(CH_3)OCH_3]^+$ (No. 14; 1:1) or $[Cp(CO)(^2D)Fe{=}C(C_6H_5)OCH_3]^+$ ($^2D = P(C_6H_5)_3$, $P(C_2H_5)_3$ Nos. 58 and 57, respectively (1:1 to 1:1.8 at various temperatures and concentrations of $NaBH_4$)), using the reducing system $CH_3OH/NaOCH_3/NaBH_4$. H^- addition to the optically pure cation No. 59 (*RS* and *SS* isomers) to give *RSR,RSS* and *SSR,SSS* diastereomers with a slight

Table 3. Cationic Carbene Compounds of the Type $[Cp(CO)(^2D)Fe{=}C(R)ER]^+$. An asterisk indicates further information at the end of the table

No.	Fe=C(R)ER moiety ^{2}D ligand	method of preparation (yield) properties and remarks
the heteroatom is Cl		
1	$Fe{=}C(CH_3)Cl$ $P(C_6H_5)_3$	BF_4^- salt, Ia; brick-red solid, m.p. 85 °C (dec.) $CF_3SO_3^-$ salt, Ia ^{1}H NMR ($CDCl_3$, both salts): 2.55 (s, CH_3) ^{13}C NMR ($CDCl_3$, $CF_3SO_3^-$ salt): 13.49 (CH_3), 344.50 (C=Fe)
the heteroatom is O		
*2	$Fe{=}C(H)OCH_3$ $P(C_6H_5)_3$	PF_6^- salt; yellowish solid ^{1}H NMR: 13.25 (s, CH) in CD_3NO_2, CD_2Cl_2 BF_4^- salt, III (84%)
*3	$Fe{=}C(H)OC_2H_5$ $P(C_6H_5)_3$	PF_6^- salt ^{1}H NMR (CD_3COCD_3): 13.67 (d, CH; J(P,H) = 1.5)
*4	$Fe{=}C(CH_3)$ OH $P(CH_3)_3$	$SO_3CF_3^-$ salt, IIa; m.p. 112 °C ^{1}H NMR (CD_3CN): 2.87 (s, CH_3C) ^{13}C NMR (CD_3CCN): 48.8 (q, CH_3, J(H,C) = 128.90), 215.9 (d, CO; J(P,C) = 29.33), 335.6 (s, C=Fe) ^{31}P NMR (CD_3CN): 32.5 (1*S*)-(+)- and (1*R*)-(−)-10-camphorsulfonate, IIa
*5	$Fe{=}C(CH_3)OH$ $P(C_2H_5)_3$	(1*S*)-(+)- and (1*R*)-(−)-10-camphorsulfonate, IIa
*6	$Fe{=}C(CH_3)OH$ $P(C_6H_{11})_3$	BF_4^- salt, IIa; yellow, m.p. 120 °C (dec.) ^{1}H NMR (SO_2): 3.17 (s, CH_3) Cl^- salt, IIa; stable product
*7	$Fe{=}C(CH_3)OH$ $P(CH_3)_2C_6H_5$	BF_4^- salt, IIa

No.	Compound	Data
*8	Fe=C(CH$_3$)OH P(C$_6$H$_5$)$_3$	Cl$^-$ salt, IIa; creamy yellow powder Br$^-$ salt, IIa; creamy yellow powder ^{1}H NMR (SO$_2$): 2.59 (s, CH$_3$) BF$_4^-$ salt, IIa
9	Fe=C(CH$_3$)OD P(C$_6$H$_5$)$_3$	Cl$^-$ or Br$^-$ salt, IIa removal of HX on standing in vacuum
10	Fe=C(CH$_3$)OH P(OCH$_3$)$_3$	SO$_3$CF$_3^-$ salt, IIa is stable for 24 h in CH$_2$Cl$_2$ solution; deprotonation with N(C$_2$H$_5$)$_3$
*11	Fe=C(CH$_3$)OCH$_3$ P(CH$_3$)$_3$	SO$_3$F$^-$ salt, IIa; yellow, m.p. 154 °C (dec.) ^{1}H NMR (CD$_3$CN): 2.92 (s, CH$_3$C), 4.30 (s, CH$_3$O) ^{31}P NMR (CD$_3$CN): 33.54 SO$_3$CF$_3^-$ salt, IIb; (*R*)- and (*S*)-enantiomers
12	Fe=C(CH$_3$)OCH$_3$ P(C$_2$H$_5$)$_3$	SO$_3$CF$_3^-$ salt, IIb not isolated; see further information for No. 11
*13	Fe=C(CH$_3$)OCH$_3$ P(C$_6$H$_{11}$)$_3$	BF$_4^-$ salt, IIb; yellow, m.p. 170 °C (dec.) ^{1}H NMR (CD$_3$COCD$_3$): 3.29 (s, CH$_3$C), 4.45 (s, CH$_3$O)
*14	Fe=C(CH$_3$)OCH$_3$ P(C$_6$H$_5$)$_3$	SO$_3$F$^-$ salt, IIb (90%); yellow, m.p. 147 °C (dec.) ^{1}H NMR (CD$_3$CN): 2.90 (s, CH$_3$C), 4.06 (s, CH$_3$O) ^{31}P NMR (CD$_3$CN): 65.44 PF$_6^-$ salt; dark yellow, m.p. 179 to 181 °C ^{1}H NMR 2.34 (s, CH$_3$C), 4.15 (s, CH$_3$O) in CD$_3$CN, similar values in CF$_3$COOH, CD$_3$COCD$_3$ BF$_4^-$ salt, Ib (70%), IIb; yellow crystals, m.p. 164 °C ^{1}H NMR (CDCl$_3$): 2.84 (s, CH$_3$C), 3.98 (s, CH$_3$O) CF$_3$SO$_3^-$ salt, Ib, IIb
*15	Fe=C(CH$_3$)OCH$_3$ P(OCH$_3$)$_3$	PF$_6^-$ salt; yellow crystals ^{1}H NMR (CF$_3$COOH): 2.94 (s, CH$_3$), 3.82 (d, CH$_3$OP; J(P,H) = 12), 4.39 (s, CH$_3$O),

Table 3. Continued

No.	Fe=C(R)ER moiety ^2D ligand	method of preparation (yield) properties and remarks
*16	$Fe{=}C(CH_3)OCH_3$ $P(OC_6H_5)_3$	PF_6^- salt; yellow powder ^1H NMR (CF_3COOH or $CDCl_3$?): 2.86 (s, CH_3), 4.22 (s, CH_3O)
17	$Fe{=}C(CH_3)OC_2H_5$ $P(C_6H_{11})_3$	BF_4^- salt, IIb; m.p. 175 °C (dec.) ^1H NMR (CD_3COCD_3): 1.53 (t, CH_3; J = 6.9), 3.15 (s, CH_3C), 4.53 (m, CH_2)
18	$Fe{=}C(CH_3)OC_3H_7$-*i* $P(C_6H_5)_3$	BF_4^- salt, Ib (72%); yellow crystals, m.p. 173 °C ^1H NMR ($CDCl_3$): 1.05, 1.40 (d's, CH_3 of C_3H_7; J = 6), 2.95 (s, CH_3C), 3.40 (m, CH); diastereotopic methyl groups
*19	$Fe{=}C(CH_3)OC_2H_5$ $P(C_6H_5)_3$	BF_4^- salt, Ib (65%); yellow, m.p. 108 °C (dec.) ^1H NMR (CD_3Cl): 1.40 (t, CH_3; = 7), 2.83 (s, CH_3C), 4.25 (m, CH_2O) PF_6^- salt; bright yellow powder ^1H NMR (CF_3COOH): 2.79 (s, CH_3), 4.17 (m, CH_2) ^{13}C NMR (CH_2Cl_2): 14.8 (CH_3 of C_2H_5), 46.1 (CH_3), 75.2 (CH_2), 216.4 (d, CO; J(P,C) = 23), 342.0 (d, C=Fe; J(P,C) = 27)
*20	$Fe{=}C(CH_3)OCH_2CH{=}CH_2$ $P(C_6H_5)_3$	BF_4^- salt, Ib (61%); dark yellow, m.p. 130 °C ^1H NMR ($CDCl_3$): 2.88 (s, CH_3), 4.65 (d, CH_2O), 5.31, 5.51 (m's, $CH_2{=}$), 5.77 (m, $CH{=}$)
*21	$Fe{=}C(CH_3)OCH_2CH{=}CH_2$ $P(OCH_3)_3$	BF_4^- salt, Ib ^1H NMR ($CDCl_3$): 3.05 (s, CH_3), 5.18 (s, CH_2O)
*22	$Fe{=}C(CH_3)OCH_2C{\equiv}CH$ $P(OCH_3)_3$	BF_4^- salt, Ib
*23	$Fe{=}C(CH_3)OCH_2CH{=}CHCH_3$ $P(OCH_3)_3$	BF_4^- salt, Ib
*24	$Fe{=}C(CH_3)OCH(CH_3)CH{=}CH_2$ $P(OCH_3)_3$	BF_4^- salt, Ib
*25	$Fe{=}C(CH_3)OCH_2FeCp(CO)_2$ $P(C_6H_5)_3$	PF_6^- salt

No.	Compound	Properties
*26	$Fe{=}C(CH_3)OCH_2MoCp(CO)_3$ $P(C_6H_5)_3$	PF_6^- salt; yellow powder ^1H NMR (CD_3COCD_3): 2.68 (s, CH_3), 4.82 (s, br, CH_2), 5.01 (d, CpFe; J(P,H) = 1.5), 5.56 (s, CpMo) ^{13}C NMR (CH_2Cl_2): 45.04 (CH_3), 72.48 (CH_2), 87.76 (CpFe), 94.00 (CpMo), 228.08 (COMo-*trans*), 237.28 (COMo-*cis*) IR (CH_2Cl_2): 1979 (Fe), 1955, 2036 (Mo)
*27	$Fe{=}C(CH_3)OSO_2CF_3$ $P(C_6H_{11})_3$	$CF_3SO_3^-$ salt
*28	$Fe{=}C(CH_3)OSO_2CF_3$ $P(CH_3)_2C_6H_5$	$CF_3SO_3^-$ salt ^1H NMR ($CDCl_3$): 2.73 (s, CH_3)
*29	$Fe{=}C(CH_3)OSO_2CF_3$ $P(C_6H_5)_3$	$CF_3SO_3^-$ salt ^1H NMR ($CDCl_3$): 2.38 (s, CH_3
*30	$Fe{=}C(CH_3)OSi(CH_3)_3$ $P(C_6H_5)_3$	$SO_3CF_3^-$ salt; m.p. 50 °C (dec.) ^1H NMR (CD_3CN): 2.58 (s, CH_3C)
*31	$Fe{=}C(CH_3)OFeCp(CO)_2$ $P(C_6H_5)_3$	BF_4^- salt; red purple solid PF_6^- salt ^1H NMR (CD_3COCD_3 of PF_6^- salt); 2.63 (s, CH_3), 5.92 (s, Fp), 4.60 (d, Cp; J(P,H) = 1.5)
*32	$Fe{=}C(CH_3)OMoCp(CO)_3$ $P(C_6H_5)_3$	PF_6^- salt; red powder ^1H NMR (CD_3COCD_3): 2.41 (s, CH_3), 4.75 (d, CpFe; J(P,H) = 1.5), 6.13 (s, CpMo)
33	$Fe{=}C(CH_2D)OCH_3$ $P(C_6H_5)_3$	BF_4^- salt, Ib (75%); pale yellow, m.p. 108 °C ^1H NMR ($CDCl_3$): 2.83 (t, CH_2D; J(D,H) = 2), 3.98 (s, CH_3O)

Table 3. Continued

No.	$Fe{=}C(R)ER$ moiety ^2D ligand	method of preparation (yield) properties and remarks
*34	$Fe{=}C(C_2H_5)OCH_3$ $P(CH_3)_3$	SO_3^-F salt, IIb; yellow, m.p. 165 °C (dec.) ^1H NMR (CD_3CN): 1.17 (t, CH_3C), 3.21 (dq), CH_2; J(P,H) = 1.58), 4.18 (s, CH_3O) ^{31}P NMR (CD_3CN): 40.53 I^- salt
*35	$Fe{=}C(C_2H_5)OCH_3$ $P(C_6H_5)_3$	SO_3F^- salt ^1H NMR (CD_3CN): 1.08 (t, CH_3), 3.47 (dq, CH_2; J(P,H) = 1.45), 4.18 (s, CH_3O) ^{31}P NMR (CD_3CN): 64.17 BF_4^- salt, Ib, IIb $SO_3CF_3^-$, salt; Ib
*36	$Fe{=}C(C_2H_5)OCH_2CH{=}CH_2$ $P(OCH_3)_3$	BF_4^- salt, Ib
*37	$Fe{=}C(C_2H_5)OCH_2CH{=}CHCH_3$ $P(OCH_3)_3$	BF_4^- salt, Ib
38	$Fe{=}C(C_3H_7\text{-}i)OH$ $P(C_6H_5)_3$	BF_4^- salt, IIa; yellow oil addition of $O(SO_2CF_3)_2$ gives the vinylidene cation $[Cp(CO)(P(C_6H_5)_3)Fe{=}C{=}C(CH_3)_2]^+$
*39	$Fe{=}C(C_3H_7\text{-}i)OCH_3$ $P(CH_3)_3$	SO_3F^- salt, IIb; yellow, m.p. 143 °C (dec.) ^1H NMR (CD_3CN): 1.05, 1.10 (d's, CH_3C), 3.83 (hept, CH), 4.60 (s, CH_3O) ^{31}P NMR (CD_3CN): 34.23 I^- salt
*40	$Fe{=}C(C_3H_7\text{-}i)OCH_3$ $P(C_6H_5)_3$	FSO_3^- salt ^1H NMR (CD_3CN): 0.63, 0.67 (d's, CH_3C), 3.91 (hept, CH), 4.60 (s, CH_3O) ^{31}P NMR (CD_3CN): 64.70 $SO_3CF_3^-$ salt, IIb

No.	Compound	Remarks
*41	$Fe{=}C(C_3H_7\text{-}i)OCH_2CH{=}CH_2$ $P(OCH_3)_3$	$SO_3CF_3^-$ salt, Ib
*42	$Fe{=}C(C_3H_7\text{-}i)OSO_2CF_3$ $P(C_6H_5)_3$	$CF_3SO_3^-$ salt ^1H NMR ($CDCl_3$: 0.21, 1.11 (d's, CH_3), 3.75 (hept, CH)
43	$Fe{=}C(C_4H_9\text{-}n)OCH_3$ $P(C_6H_5)_3$	BF_4^- salt for the deprotonation, see No. 35
44	$Fe{=}C(CH_2CH_2C_6H_5)OC_2H_5$ $P(C_6H_5)_3$	BF_4^- salt, IIb ^1H NMR (CD_3COCD_3); 1.53 (t, CH_3), 2.7 to 3.4 (m, CH_2CH_2), 4.73 (m, CH_2O) 22% $[Cp(CO)_2(P(C_6H_5)_3)Fe]^+$ as by-product
45	$Fe{=}C(CH_2CH_2Fc)OC_2H_5$ $P(C_6H_5)_3$ Fc = $-C_5H_4FeCp$	BF_4^- salt, IIb ^1H NMR (CD_3COCD_3): 1.53 (t, CH_3), 2.8 to 3.2 (m, CH_2CH_2), 4.26 (s, Cp of Fc), 4.0 to 4.8 (m, CH_2O), C_5H_4), 5.13 (d, Cp) 30% $[Cp(CO)_2(P(C_6H_5)_3)Fe]^+$ as by-product
*46	$Fe{=}C(CH_2CH_2CH{=}CH_2)OC_2H_5$ $P(C_6H_5)_3$	BF_4^- salt, IIb; yellow
*47	$Fe{=}C(CH_2CH_2C{\equiv}CH)OC_2H_5$ $P(C_6H_5)_3$	BF_4^- salt, IIb; yellow
*48	$Fe{=}C(CH{=}CHSi(CH_3)_3)OCH_3$ $P(C_6H_5)_3$	BF_4^- salt, IIb (47%); red powder ^1H NMR (?): 4.25 (s, CH_3O), 6.26 (d, CHSi) 6.67 (d, CHC) ^{13}C NMR (CD_3COCD_3): 67.60 (CH_3O), 150.05 (CHSi), 151.79 (CH), 217.45 (d, CO; J(P,C) = 24), 333.58 (d, C=Fe; J(P,C) = 24.4)
*49	$Fe{=}C(CH_2C_6H_5)OCH_3$ $P(C_6H_5)_3$	BF_4^- salt, Ib; m.p. 154.0 to 154.7 °C ^1H NMR (CD_3COCD_3); 4.00, 4.61 (AB m's, CH_2), 4.22 (s, CH_3O)
*50	$Fe{=}C(CH_2OCH_3)OH$ $P(OCH_3)_3$	BF_4^- salt, IIa

Table 3. Continued

No.	Fe=C(R)ER moiety ²D ligand	method of preparation (yield) properties and remarks
*51	Fe=C(CH_2OCH_3)OCH_3 P(C_6H_5)$_3$	PF_6^- salt, IIb (70%); yellow solid ^{1}H NMR (CD_3COCD_3): 3.91 (s, br, CH_2), 4.17 (s, CH_3OC=Fe)
*52	Fe=C(CH_2OCH_3)OC_2H_5 P(C_6H_5)$_3$	PF_6^- salt, IIb (70%); yellow solid ^{1}H NMR (CD_3COCD_3): 3.96 (s, br, CH_2), 4.37 (m, CH_2 of C_2H_5)
53	Fe=C(CH_2OCH_3)OC_2H_5 P(OCH_3)$_3$	PF_6^- salt, IIb (86%); reddish-brown gum ^{1}H NMR (CD_3COCD_3): 4.44 (s, CH_2C), 4.92 (q, CH_2 of C_2H_5) starting complex regenerated with I^-
*54	Fe=C(CH_2OCH_3)OMoCp(CO)$_3$ P(C_6H_5)$_3$	
*55	Fe=C(CH_2OCH_3)OMoCp(CO)$_3$ P(OCH_3)$_3$	
56	Fe=C($CH_2OC_2H_5$)OC_2H_5 P(C_6H_5)$_3$	PF_6^- salt, IIb (70%); yellow solid ^{1}H NMR (CD_3COCD_3): 4.06 (s, br, CH_2), 4.36 (q, CH_2 of C_2H_5OC) for reactions, see Nos. 51 and 52
*57	Fe=C(C_6H_5)OCH_3 P(C_2H_5)$_3$	$CF_3SO_3^-$ salt, IIb; dark red oil ^{1}H NMR (CD_2Cl_2): 3.48 (s, CH_3O) ^{13}C NMR (CD_2Cl_2): 61.3 (CH_3O), 216.5 (CO; J(P,C) = 28.8), 331.6 (C=Fe; J(P,C) = 21.2)
*58	Fe=C(C_6H_5)OCH_3 P(C_6H_5)$_3$	$CF_3SO_3^-$ salt, IIb; dark red oil ^{1}H NMR (CD_2Cl_2): 3.39 (s, CH_3O) ^{13}C NMR (CD_2Cl_2): 60.8 (CH_3O), 217.0 (d, CO; J(P,C) = 28.7), 333.6 (C=Fe; J(P,C) = 24.7)

No.	Compound	Properties
*59	$Fe{=}C(C_6H_5)OCH_3$ $P(C_6H_5)_2R$ R = (*S*)-2-methylbutyl	$CF_3SO_3^-$ salt, IIb; *SS* and *RS* diastereomers ^1H NMR (*SS* diastereomer in CD_2Cl_2): 4.01 (s, CH_3O), (*RS* diastereomer in CD_2Cl_2): 3.99 (s, CH_3O) ^{13}C NMR (*SS* diastereomer in CD_2Cl_2): 62.5 (CH_3O), (*RS* diastereomer in CD_2Cl_2): 62.6 (CH_3O), 215.6 (d, CO; J(P,C) = 27.5), 332.0 (d, C=Fe; J(P,C) = 24.5)
*60	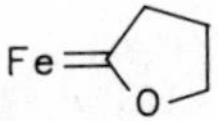 Fe= O $P(C_6H_5)_3$	Br^- salt; yellow PF_6^- salt; yellow, m.p. 223 to 227 °C (dec.) ^1H NMR (CD_3COCD_3): 3.45 (q, CH_2C), 4.69 (t, CH_2O) $B(C_6H_5)_4^-$ salt; yellow needles, m.p. 182 to 185 °C ^1H NMR (CD_3COCD_3): 3.74 (t, CH_2C), 4.63 (br, CH_2O)
*61	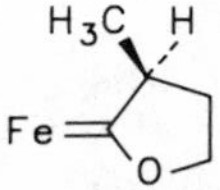 H_3C H Fe= O $P(C_6H_5)_3$	I^- salt ^1H NMR: 1.33 (d, CH_3)
*62	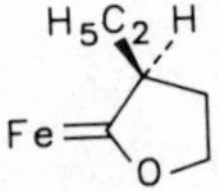 H_5C_2 H Fe= O $P(C_6H_5)_3$	I^- salt
*63	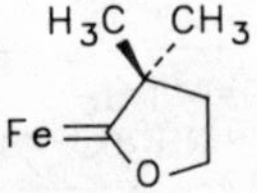 H_3C CH_3 Fe= O $P(C_6H_5)_3$	I^- salt ^1H NMR: 0.62, 1.45 (s's, CH_3)

Table 3. Continued

No.	Fe=C(R)ER moiety ^{2}D ligand	method of preparation (yield) properties and remarks
*64	H_3C C_2H_5 Fe= O $P(C_6H_5)_3$	I^- salt ^{1}H NMR: 0.38 (s, CH_3 isomer A), 1.26 (s, CH_3, isomer B); see Formula XIII
the heteroatom is S		
65	Fe=C(H)SCH_3 $P(C_6H_5)_3$	$CF_3SO_3^-$ salt, III (71%); yellow oil ^{1}H NMR (CD_3CN): 2.99 (d, CH_3S; J(P,H) = 0.73), 14.94 (CH)
66	Fe=C(H)SCH_3 $P(OC_6H_5)_3$	$CF_3SO_3^-$ salt, III (72%); yellow oil ^{1}H NMR (CD_3CN): 2.96 (s, CH_3S), 14.92 (CH) ^{13}C NMR (CD_3CN): 34.6 (CH_3S), 212.2 (d, CO; J(P,C) = 39.07), 320.6 (d, C Fe; J(P,C) = 33.21)
*67	Fe=C(CH_3)SH $P(C_6H_5)_3$	BF_4^- salt, Ia (65%); orange, m.p. 104 °C ^{1}H NMR ($CDCl_3$): 2.60 (s, CH_3), 4.32 (s, br, SH)
68	Fe=C(CH_3) SCH_3 $P(C_6H_5)_3$	BF_4^- salt, Ib (79%); yellow, m.p. 141 °C (dec.) ^{1}H NMR ($CDCl_3$): 2.61 (s, CH_3C), 3.19 (s, CH_3S)
the heteroatom is N		
*69	Fe=C(H)NHC_6H_{11}-*c* $NCCH_3$	$CF_3SO_3^-$ salt; golden crystals ^{1}H NMR ($CDCl_3$): 11.62 (s, CH), 11.84 (s, NH) ^{13}C NMR ($CDCl_3$): 216.1 (CO), 243.9 (C=Fe)

No.	compound	method of preparation (yield) properties and remarks
70	$Fe{=}C(H)N(C_2H_5)_2$ $P(OC_6H_5)_3$	$CF_3SO_3^-$ salt, IVa (56%); bright yellow crystals ^1H NMR (CD_3CN): 3.89, 4.11 (q's, CH_2), 11.36 (d, CH=Fe; J(P,H) = 5.13) ^{13}C NMR (CD_3CN): 13.9, 14.8 (CH_3), 52.1, 60.1 (CH_2N), 216.6 (d, CO; J(P,C) = 41.02), 239.0 (d, C=Fe; J(P,C) = 39.07)
*71	$Fe{=}C(CH_3)NHCH_3$ $P(C_6H_5)_3$	BF_4^- salt, Ib (75%); orange, m.p. 137 °C ^1H NMR ($CDCl_3$): 2.14 (s, CH_3C), 2.98 (d, CH_3N), 9.90 (s, br, HN)
72	$Fe{=}C(CH_3)NHCH_2C_6H_5$ $P(C_6H_5)_3$	BF_4^- salt, Ib (74%), IVb; yellow, m.p. 203 °C ^1H NMR ($CDCl_3$): 2.34 (s, CH_3), 4.67 (d, CH_2), 10.81 (s, br, HN)
*73	$Fe{=}C(CH_3)NHCH(CH_3)C_6H_5$ $P(C_6H_5)_3$	BF_4^- salt, IVb; mixture of diastereomers isomer B (more soluble) m.p. 186 °C, isomer (A less soluble) m.p. 195 °C (both dec.) ^1H NMR (CD_3COCD_3, B/A): 2.92/2.85 (s's, $CH_3C{=}Fe$)
*74	$Fe{=}C(CH_3)N(CH_3)_2$ $P(C_6H_5)_3$	BF_4^- salt, Ib (77%); orange, m.p. 176 °C ^1H NMR ($CDCl_3$): 2.37 (s, CH_3C), 3.34, 3.94 (s's, CH_3N)
*75	H_3C CH_3 Fe= (ring) N ~ C_6H_5 N–CH_3 $P(C_6H_5)_3$	BF_4^- salt, Ic (52%); golden foam, 3:1 mixture of diastereomers A and B ^1H NMR ($CDCl_3$, isomer A/B): 0.67/0.54 (s's, CH_3C), 1.41/1.05 (s's, CH_3C), 3.23/3.30 (s's, CH_3N) $CF_3SO_3^-$ salt, Ic (58%); golden oil
*76	H_3C CH_3 Fe= (ring) N ~ C_6H_5 N–CH_3 $P(OCH_3)_3$	$CF_3SO_3^-$ salt, Ic (36 to 72%) 8:5 mixture of diastereomers A and B, m.p. of B 213 to 214 °C (dec.) ^1H NMR ($CDCl_3$, A/B): 0.71/0.78 (s's, CH_3C), 1.42/1.33 (s's, CH_3C), 3.38/3.41 (s's, CH_3N) ^{13}C NMR ($CDCl_3$, A/B): 60.5/60.2 (CH_3N), 215.6 (d, CO; J(P,C) = 43), 275.0/277.2 (d's, C=Fe; J(P,C) = 36/35)

Table 3. Continued

No.	Fe=C(R)ER moiety [2]D ligand	method of preparation (yield) properties and remarks
*77	Fe=C(CMe₂)–CH(∼$C_6H_4CH_3$–3)–N(CH_3) (four-membered ring: H_3C, CH_3 on C; N–CH_3) $P(C_6H_5)_3$	BF_4^- salt, Ic (31%); 4:3 mixture of diastereomers A and B ^{1}H NMR ($CDCl_3$, C, isomers A/B): 0.67/0.52, (s's, CH_3C), 1.36/1.00 (s's, CH_3C), 3.20/3.28 (s's, CH_3N) ^{13}C NMR ($CDCl_3$): 62.0, 65.9 (CH_3N), 218.4 (CO), 278.9 (C=Fe)
*78	Fe=C(CMe₂)–CH(∼$C_6H_4CH_3$–4)–N(CH_3) (four-membered ring: H_3C, CH_3 on C; N–CH_3) $P(C_6H_5)_3$	BF_4^- salt, Ic (38%); oil, 4:3 mixture of diastereomers A and B ^{1}H NMR ($CDCl_3$, isomers A/B): 0.68/0.54 (s's, CH_3C), 1.37/0.99 (s's, CH_3C), 3.20/3.28 (s's, CH_3N)
*79	Fe=C(CMe₂)–CH(H)(–CH=CH–C_6H_5)–N(CH_3) (four-membered ring: H_3C, CH_3 on C; N–CH_3; C=C with H, H, C_6H_5) $P(C_6H_5)_3$	BF_4^- salt, Ic (33%); (*E*) configuration, 4:1 mixture of diastereomers A and B ^{1}H NMR ($CDCl_3$, A/B): 0.75/0.87 (s's, CH_3C), 1.27/0.98 (s's, CH_3C), 3.1/2.75 (s's, CH_3N)

*80

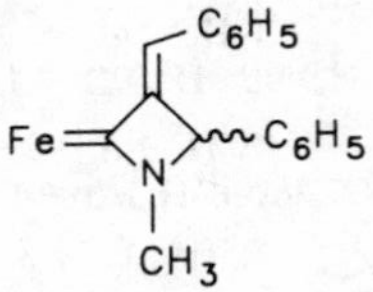

$P(C_6H_5)_3$

BF_4^- salt, Ic (46%); golden foam (E), mixture of diastereomers
^1H NMR ($CDCl_3$): 3.50 (s, CH_3)

*81

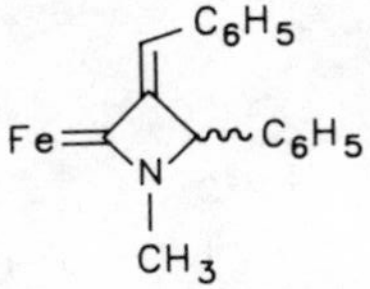

$P(OCH_3)_3$

$CF_3SO_3^-$ salt, Ic (11%); (*E*) configuration, 2:1 mixture of diastereomers A and B
^1H NMR ($CDCl_3$, A/B): 3.33/3.36 (s's, CH_3N)
^{13}C NMR ($CDCl_3$): 216.4 (d, CO; J(P,C) = 40)
BF_4^- salt, Ic

*82

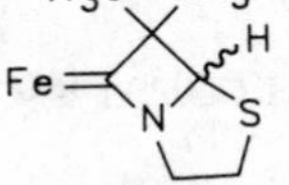

$P(OCH_3)_3$

$CF_3SO_3^-$ salt, Ic (90%); 15:1 mixture of diastereomers A and B
^1H NMR ($CDCl_3$, major isomer A): 1.13, 1.42, (s's, CH_3)
^{13}C NMR ($CDCl_3$, major isomer A): 214.9 (d, CO; J(P,C) = 42), 284.9 (m, C=Fe)
BF_4^- salt, Ic (14%); tan solid

*83

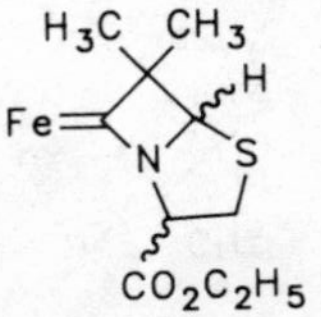

$P(OCH_3)_3$

$CF_3SO_3^-$ salt, Ic (72%); orange oil, 8:1 mixture of diastereomers A and B
^1H NMR ($CDCl_3$, A/B): 1.16/1.21 (s, CH_3), 1.46/1.43 (s, CH_3)
^{13}C NMR ($CDCl_3$, major isomer): 214.4 (d, CO; J(P,C) = 42), 294.0 (d, C=Fe; J(P,C) = 37)

Table 3. Continued

No.	Fe=C(R)ER moiety ^{2}D ligand	method of preparation (yield) properties and remarks
*84	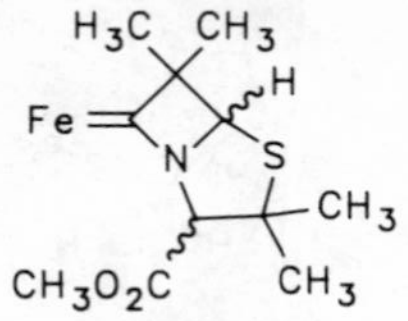 $P(OCH_3)_3$	$CF_3SO_3^-$ salt, Ic (58%); 6:1 mixture of diastereomers A and B ^{13}C NMR ($CDCl_3$): 214.7 (d, CO; J(P,C) = 42), 289.8 (d, C=Fe; J(P,C) = 36)
compounds of the type $[Cp(CO)Fe{=}CRE\text{-}^{2}D]^+$		
*85	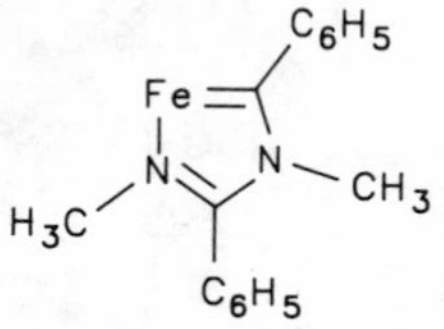	PF_6^- salt, dark red-brown ^{1}H NMR (CD_3COCD_3): 3.33, 3.43 (s's, CH_3N) ^{13}C NMR (CD_3COCD_3): 41.8, 48.8 (CH_3N), 171.3 (C=N), 213.2 (CO), 292.9 (C=Fe)
Supplement		
the heteroatom is O		
86	$Fe{=}C(CH_3)ORe(NO)(CO)Cp$ $P(C_6H_5)_3$	
87	$Fe{=}C(CH_3)OW(CO)_3Cp$ $P(C_6H_5)_3$	
88	$Fe{=}C(C_3H_7\text{-}i)OCH_3$ $P(OCH_3)_3$	$SO_3CF_3^-$ salt, Ib (96%); red oil ^{1}H NMR ($CDCl_3$): 1.01, 1.07 (d's, CH_3; J = 6.0, 6.8), 3.82 (m, CH), 4.72 (s, CH_3O)

stereoselectivity is described. With the nucleophile $C_6H_5^-$ (as LiC_6H_5) the formation of only one diastereomer was observed.

The low stereoselectivity by the addition of H^- is discussed in terms of rapid interconversion between *anti*-(O,O) and the more reactive *syn*-(O,O) conformations (attack of the nucleophile shown by the arrow in V), as outlined in Sect. 1.1.2.1.

Al compounds show one strong ν(CO) absorption in the IR spectrum in the range between 1950 to 2000 cm^{-1} for the $Cp(CO)(^2D)Fe$ fragment.

Further Information

$[Cp(CO)(P(C_6H_5)_3)Fe{=}C(H)OCH_3]X$ (Table **3**, No. **2**; X = PF_6, BF_4). The PF_6^- salt is obtained in 80 to 90% yield by hydride abstraction from $Cp(CO)(P(C_6H_5)_3)FeCH_2OCH_3$ with $[C(C_6H_5)_3]PF_6$ in CH_2Cl_2 solution. $[CpMo(CO)_3]PF_6$ (29% yield) or $[Cp(CO)(P(C_6H_5)_3)Fe{=}CH_2]BF_4$ can also serve as H^- abstractor; see also the quantitative generation of a 1:1:1 mixture of $Cp(CO)(P(C_6H_5)_3)FeCH_2OCH_3$, $[Cp(CO)(P(C_6H_5)_3)Fe{=}CHOCH_3]^+$ and $Cp(CO)(P(C_6H_5)_3)FeCH_3$ when 3 equivalents of $Cp(CO)(P(C_6H_5)_3)FeCH_2OCH_3$ are slowly treated with 1 equivalent of HBF_4 etherate in THF at − 78 °C. The cation is stable as a solution in dry CH_3COCH_3, CH_2Cl_2, nitromethane, and trifluoracetic acid.

Reaction with excess $[P(CH_3)(C_6H_5)_3]I$ in CH_2Cl_2 at room temperature (2 h) converts the complex into an equimolar mixture of $[CpFe(CO)_2(P(C_6H_5)_3)]^+$ and $Cp(CO)(P(C_6H_5)_3)FeCH_2OCH_3$ with elimination of CH_3I. A mechanism via an intermediate formyl complex followed by hydride transfer to the starting complex is discussed. A similar dealkylation occurs with Cp(dppe)FeH (dppe = $(C_6H_5)_2PCH_2CH_2P(C_6H_5)_2$) producing the same product mixture. With $[P(CH_3)(C_6H_5)_3]BH_4$ the methyl complex $Cp(CO)(P(C_6H_5)_3)FeCH_3$ is quantitatively formed. Addition of $[CH_3O]^-$ produces the acetal complex $Cp(CO)(P(C_6H_5)_3)FeCH(OCH_3)_2$ in 99% yield in a reversible reaction. Similarly, OCD_3^- at − 78 °C in CH_2Cl_2 gives a 9:1 mixture of the diastereomers of $Cp(CO)(P(C_6H_5)_3)FeCH(OCH_3)(OCD_3)$. With LiC_6H_5 one single diastereomer of $Cp(CO)(P(C_6H_5)_3)FeCH(OCH_3)C_6H_5$ is obtained. Similarly, with LiC_2H_5 in CH_2Cl_2 at − 78 °C the resulting ratio of the diastereomers is about 30:1.

$[Cp(CO)(P(C_6H_5)_3)Fe{=}C(H)OC_2H_5]PF_6$ (Table **3**, No. **3**) is prepared similarly (80 to 90%) and behaves in solution as No. 2.

It is less reactive toward excess I^- than No. 2 (half-life of at least 12 h) but gives an analogous product mixture when it is allowed to react with $[P(CH_3)(C_6H_5)_3]I$.

$[Cp(CO)(P(CH_3)_3)Fe{=}C(CH_3)OH]SO_3CF_3$ (Table **3**, No. **4**). For the (1*S*)-(+)- or (1*R*)-(−)-10-camphorsulfonate, see procedure outlined for No. 5.

It is deprotonated by $CH_2P(CH_3)_3$ to give the starting acyl complex and $[P(CH_3)_4]SO_3CF_3$; see No. 5.

[Cp(CO)(P(C_2H_5)$_3$)Fe=C(CH_3)OH]$SO_3C_{10}H_{15}O$ (Table **3**, No. **5**). 1:1 mixtures of the diastereomeric carbene compounds are obtained according to Method IIa either with (1*S*)-(+)- or (1*R*)-(−)-10-camphorsulfonic acid in CH_2Cl_2 solution. The resulting diastereomers can be separated by successive crystallization from CH_2Cl_2-ether at −25 °C. Neutralization generates the optically pure neutral acyl complexes with ee values of 76 to 99%.

[Cp(CO)(P(C_6H_{11})$_3$)Fe=C(CH_3)OH]BF_4 and **[Cp(CO)(P(CH_3)$_2C_6H_5$)Fe=C(CH_3)OH]BF_4** (Table **3**, Nos. **6** and **7**). Subsequent addition of $O(SO_2CF_3)_2$ to the oily compounds obtained according to Method IIa in ether effects precipitation of the corresponding vinylidene compounds, [Cp(CO)(^{2}D)Fe=C=CH_2]BF_4; see also No. 8. No. 6 reacts with CH_2N_2 in ether to give No. 13 along with the acetyl complex Cp(CO)(P(C_6H_{11})$_3$)FeC(O)CH_3.

[Cp(CO)(P(C_6H_5)$_3$)Fe=C(CH_3)OH]X (Table **3**, No. **8**, X = Cl, Br, BF_4). The compounds (X = Cl, Br) lose HX on prolonged standing in vacuum or washing with petroleum ether, forming the starting acetyl complex; see also No. 9.

For reaction of the BF_4^- salt with $O(SO_2CF_3)_2$, see Nos. 6 and 7.

[Cp(CO)(P(CH_3)$_3$)Fe=C(CH_3)OCH]X (Table **3**, No. **11**, X = SO_3F, I) also forms along with the carbene complexes Nos. 34 and 39 as a 1.7:1.0:1.1 mixture upon methylation of the vinyl complex Cp(CO)(P(CH_3)$_3$)FeC(OCH_3)=CH_2 with CH_3OSO_2F or CH_3I, respectively. A mechanism involving formation of No. 34 in the first step, followed by proton transfer to the vinyl complex to give No. 11 and the propenyl complex, Cp(CO)(P(CH_3)$_3$)FeC(OCH_3)=$CHCH_3$, which is further methylated to give No. 39, is discussed. The (*R*) and (*S*) enantiomers are intermediates upon alkylation of the corresponding (*R*) and (*S*) acyl complexes followed by reduction with $NaBH_4$–$NaOCH_3$-CH_3OH to give the corresponding ethers (*R*)- and (*S*)-Cp(CO)(P(CH_3)$_3$)FeCHCH$_3$OCH$_3$. These ethers are precursors of heteroatom free carbene cations, see Sect. 1.1.2.1. Similar reactions give the intermediate No. 12.

The complex reacts with $(CH_3)_3P{=}CH_2$ in pentane at −40 °C with H^+ abstraction to give Cp(CO)(P(CH_3)$_3$)FeC(OCH_3)=CH_2.

[Cp(CO)(P(C_6H_{11}-*c*)$_3$)Fe=C(CH_3)OCH_3]BF_4 (Table **3**, No. **13**) is also obtained in 60% yield by addition of an excess of an ethereal solution of CH_2N_2 to solid No. 6.

[Cp(CO)(P(C_6H_5)$_3$)Fe=C(CH_3)OCH_3]X (Table **3**, No. **14**; X = SO_3F, PF_6, BF_4, SO_3CF_3). The PF_6^- salt can be prepared nearly quantitatively, similar to Method IIb using [$CH_3C(OCH_3)_2$]PF_6 (generated from [$CH_3C(OCH_3)_3$ and [C(C_6H_5)$_3$]PF_6) as alkylating agent in CH_2Cl_2 (1 h). The PF_6^- salt is also obtained from the SO_3F^- salt in 77% yield by addition of NH_4PF_6 in CH_3OH or acetone solution. The compounds are stable in air. The cation also forms along with Nos. 35 and 40 by the addition of CH_3OSO_2F to the complex Cp(CO)(P(C_6H_5)$_3$)FeC(OCH_3)=CH_2; for the discussion of the mechanism, see No. 11.

A β-H abstraction occurs with $(CH_3)_3P{=}CH_2$ to give the vinyl complex Cp(CO)(P(C_6H_5)$_3$)FeC(OCH_3)=CH_2. Hydride addition by quenching a CH_2Cl_2

solution of the $SO_3CF_3^-$ salt into $CH_3OH/NaOCH_3/BH_4^-$ at $-78\,°C$ generates $Cp(CO)(P(C_6H_5)_3)FeCH(OCH_3)(CH_3)$ in high yields; this is also obtained by the reaction of the PF_6^- salt with $Li[HB(C_2H_5)_3]$ in THF at $-80\,°C$. Selective reduction of the cation to the ethyl complex $Cp(CO)(P(C_6H_5)_3)FeCH_2CH_3$ (77%) is performed by $[CH_3P(C_6H_5)_3]BH_4$. With Cp(dppe)FeH the acetyl complex $Cp(CO)(P(C_6H_5)_3)FeCOCH_3$ is formed.

[Cp(CO)(P(OCH$_3$)$_3$)Fe=C(CH$_3$)OCH$_3$]PF$_6$ (Table **3**, No. **15**) is prepared similarly to Method IIb in 88% yield using $[CH_3C(OCH_3)_2]PF_6$ (generated from $CH_3C(OCH_3)_3$ and $[C(C_6H_5)_3]PF_6$) as alkylating agent in CH_2Cl_2 (0.5 h).

Treatment with $[CH_3P(C_6H_5)_3]BH_4$ in CH_2Cl_2 solution gives the alkyl complex $Cp(CO)(P(OCH_3)_3)FeC_2H_5$ in quantitative yield.

[Cp(CO)(P(OC$_6$H$_5$)$_3$)Fe=C(CH$_3$)OCH$_3$]X (Table **3**, No. **16**; X = SO_3F, PF_6). The PF_6^- salt is obtained by dissolving the oily SO_3F^- salt (from Method IIb) in $CHCl_3$/acetone followed by addition of NH_4PF_6. For the preparation from the acetyl precursor and $CH_3C(OCH_3)_2$, see No. 15.

Reduction with $Li[HB(C_6H_5)_3]$ in THF at $-80\,°C$ gives the alkyl complex $Cp(CO)(P(OC_6H_5)_3)FeCH(OCH_3)CH_3$.

[Cp(CO)(P(C$_6$H$_5$)$_3$)Fe = C(CH$_3$)OC$_2$H$_5$]BF$_4$ (Table **3**, No. **19**) is also obtained by the reaction of the vinylidene precursor with $C_2H_5OCH{=}CH_2$ (Method Ib) with loss of HC≡CH. It is formed (see the reverse reaction below) from the vinyl ether $Cp(CO)(P(C_6H_5)_3)FeC(OC_2H_5){=}CH_2$ and HBF_4/propionic anhydride mixture. The (+)-(*R*)-cation is obtained according to Method IIb (with $[O(C_2H_5)_3]BF_4$) starting from the corresponding (+)-(*R*)-acyl enantiomer.

NaI in THF eliminates C_2H_5I to give $Cp(CO)(P(C_6H_5)_3)FeC(O)CH_3$. With $[C_2H_5O]^-$ in ethanol, proton abstraction occurs to afford the vinyl ether $Cp(CO)(P(C_6H_5)_3)FeC(OC_2H_5){=}CH_2$. Treatment with an equimolar amount of $NaBH_4$ or $NaBD_4$ in ethanol generates approximately equal amounts of $Cp(CO)(P(C_6H_5)_3)FeCR(OC_2H_5)CH_3$ and $Cp(CO)(P(C_6H_5)_3)FeCR_2CH_3$ with R = H or D, respectively. Primary amines and ammonia, but not secondary amines, produce carbimonium salts, and (*S*)-(−)-α-phenylethylamine gives a diastereomeric mixture of the carbene compound No. 73. Benzylamine is reported to give the carbene cation No. 72.

[Cp(CO)(P(C$_6$H$_5$)$_3$)Fe=C(CH$_3$)OCH$_2$CH=CH$_2$]BF$_4$ (Table **3**, No. **20**) is deprotonated with KH in the presence 18-crown-6 (C_6H_6, 0 °C) to the neutral vinyl allylether VIII or in refluxing C_6H_6 (120 h) to the rearranged complex VIII (R-1 = R-2 = R-3 = R-4 = H).

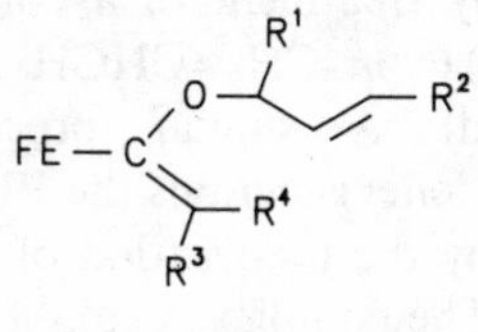

VII 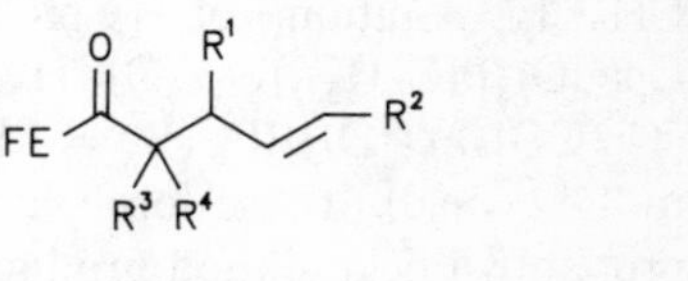 VIII

[Cp(CO)(P(OCH$_3$)$_3$)Fe=C(CH$_3$)OR]BF$_4$ (R = $CH_2CH{=}CH_2$, $CH_2C{\equiv}CH$, $CH_2CH{=}CHCH_3$, $CH(CH_3)CH{=}CH_2$; Table **3**, Nos. **21** to **24**). The compounds Nos. 21, 23 and 24 react similarly to No. 20 to give the corresponding ethers (VII) and under more drastic conditions the rearranged products (VIII) in overall yields between 10 and 60% based on four steps. No. 22 gives at first the corresponding allene, $Cp(CO)(D)FeC(O)CH_2CH{=}C{=}CH_2$, which isomerizes to the corresponding diene, $Cp(CO)(D)FeC(O)CH{=}CHCH{=}CH_2$ (28% overall yield).

[Cp(CO)(P(C$_6$H$_5$)$_3$)Fe=C(CH$_3$)OCH$_2$Fe(CO)$_2$Cp]PF$_6$ (Table **3**, No. **25**) is obtained similar to Method IIb using $[Fp{=}CH_2]PF_6$ as alkylating agent ($Fp{=}Cp(CO)_2Fe$). It could not be obtained pure and decomposes with formation of paramagnetic species.

Treatment of the resulting mixture with $[N(C_4H_9\text{-}n)_4]I$ produces a mixture of $Cp(CO)(P(C_6H_5)_3)FeC(O)CH_3$, $FpCH_2I$, and FpI.

[Cp(CO)(P(C$_6$H$_5$)$_3$)Fe=C(CH$_3$)OCH$_2$Mo(CO)$_3$Cp]PF$_6$ (Table **3**, No. **26**) is similarly obtained as described for No. 25 using $[Cp(CO)_3Mo{=}CH_2]PF_6$ (from $AgPF_6$ and $Cp(CO)_3MoCH_2Cl$ in situ at $-78\,°C$) as the alkylating agent; golden yellow solid (68% yield). The complex is stable as solid but decomposes in solution. In acetone, nitromethane, or CH_2Cl_2 a mixture of No. 32, $Cp(CO)(P(C_6H_5)_3)FeC(O)CH_3$, and $Cp(CO)_3MoCH_3$ is obtained.

Reaction with $[(C_2H_5)_3NCH_2C_6H_5]Cl$ in CH_2Cl_2 produces a mixture of the starting iron acyl complex and $Cp(CO)_3MoCH_2Cl$ (97%). With the reducing agents $[CH_3P(C_6H_5)_3]$ $[BH_3CN]$, $Cp(CO)(P(C_6H_5)_3)FeH$, or $Li[HB(C_2H_5)_3]$, quantitative conversion into the iron acyl complex and $Cp(CO)_3MoCH_3$ is observed; no reaction occurs with $BH_3{\cdot}THF$.

[Cp(CO)(^{2}D)Fe=C(CH$_3$)OSO$_2$CF$_3$]SO$_3$CF$_3$ ($^2D = P(C_6H_{11})_3$, $P(CH_3)_2C_6H_5$, $P(C_6H_5)_3$); (Table **3**, Nos. **27** to **29**) are rapidly obtained as intermediates in CH_2Cl_2 solution by the reaction of $O(SO_2CF_3)_2$ with the appropriate acyl complex $Cp(CO)(^2D)FeC(O)CH_3$ and have been characterized by ^1H NMR and IR spectroscopy.

The compounds slowly eliminate $HOSO_2CF_3$ to give the corresponding vinylidene complexes, $[Cp(CO)(^2D)Fe{=}C{=}CH_2]SO_3CF_3$.

[Cp(CO)(P(C$_6$H$_5$)$_3$)Fe=C(CH$_3$)OSi(CH$_3$)$_3$]SO$_3$CF$_3$ (Table **3**, No. **30**) forms from the reaction of $Cp(CO)(P(C_6H_5)_3)FeC(O)CH_3$ with $(CH_3)_3$ Si OSO_2CF_3 similar to Method IIb.

It transfers the $(CH_3)_3Si$ group to the ylide $(CH_3)_3P{=}CH_2$, resulting in $[(CH_3)_3PCH_2Si(CH_3)_3]SO_3CF_3$ and starting material.

[Cp(CO)(P(C$_6$H$_5$)$_3$)Fe=C(CH$_3$)OFe(CO)$_2$Cp]X (Table **3**, No. **31**, X = BF_4, PF_6). The BF_4^- salt is obtained by treatment of a cold ($-78\,°C$) CH_2Cl_2 solution of $[Cp(CO)(O(C_2H_5)_2)Fe(\eta^2\text{-}CH_2{=}CHCH_3)]BF_4$ with $Cp(CO)(P(C_6H_5)_3)FeC(O)CH_3$ (25% yield). A similar procedure with $[Cp(CO)_2Fe(^2D)]PF_6$ (^2D = THF or isobutylene) generates the PF_6^- salt in 50 to 85% yield). It also forms as by-product by the preparation of No. 25; it is probably a degradation product of No. 25. The complex is soluble in CH_2Cl_2

and insoluble in ether. It forms also from the reaction of $[Cp(CO)_2Fe{=}C(CH_3)OFp]^+$ with the acyl complex $Cp(CO)(P(C_6H_5)_3)FeC(O)CH_3$.

$[Cp(CO)(P(C_6H_5)_3)Fe{=}C(CH_3)OMo(CO)_3Cp]PF_6$ (Table **3**, No. **32**) is obtained similarly to No. 31 from $Cp(CO)(P(C_6H_5)_3FeC(O)CH_3$ and $Cp(CO)_3MoFPF_5$ in 50 to 85% yield. It also forms as a by-product by the preparation of No. 26 (39%) and is one product from the decomposition of No. 26.

$[Cp(CO)(P(CH_3)_3)Fe{=}C(C_2H_5)OCH_3]X$ (Table **3**, No. **34**, X = SO_3F, I). For the formation of the complex along with Nos. 11 and 39, see No. 11. The complex reacts in pentane at −40 °C to give the vinyl complex, $Cp(CO)(P(CH_3)_3)FeC(OCH_3){=}CHCH_3$ (mainly the *Z* isomer) along with $[P(CH_3)_4]SO_3F$.

$[Cp(CO)(P(C_6H_5)_3)FeC(C_2H_5)OCH_3]X$ (Table **3**, No. **35**, X = SO_3F, BF_4, SO_3CF_3). Alkylation of the vinyl complex $Cp(CO)(P(C_6H_5)_3)FeC(OCH_3){=}CH_2$ with CH_3OSO_2F produces a mixture of the title complex (X = SO_3F), No. 14, and No. 40; for the mechanism proposed, see No. 11. An X-ray analysis of the BF_4^- salt shows that the oxygen atoms are in *anti* positions (see also General Remarks) and that one face of the carbene ligand is shielded by the proximate phenyl group of the $P(C_6H_5)_3$ ligand.

The complex can be stereoselectively (> 95%) deprotonated with $NaOCH_3$ or KOC_4H_9-*t* at −78 °C to yield the (*Z*)-configured vinyl complex $Cp(CO)(P(C_6H_5)_3)FeC(OCH_3){=}CHCH_3$; analogous results are obtained with No. 43. Stereoselective reduction with a variety of hydride donors, such as $LiAlH_4$ or $NaBH_4$, occurs to give the two diastereomers of $Cp(CO)(P(C_6H_5)_3)FeCH(OCH_3)C_2H_5$ in the ratios 15:1 ($NaBH_4$, −100 °C) and 12:1 ($NaBH_4$, −78 °C), whereas with a $CH_3OH/NaOCH_3/NaBH_4$ mixture the ratio of the diastereomers together with the vinyl complex is 8.5:1:4.5. With $K[HB(C_4H_9\text{-}n)_3]$ only deprotonation occurs to generate the vinyl complex.

$[Cp(CO)(P(OCH_3)_3)Fe{=}C(C_2H_5)OCH_2CH{=}CHR]BF_4$ (R = H, CH_3; Table **3**, Nos. **36**, **37**) react with KH in C_6H_6 in the presence of 18-crown-6 at 0 °C to give the corresponding vinyl ethers $Cp(CO)_2(^2D)FeC({=}CHCH_3)OCH_2CH{=}CHR$ and in refluxing C_6H_6 to give the rearranged products $Cp(CO)_2(^2D)FeC(O)CH(CH_3)CH(R)CH{=}CH_2$ in 13% yield as a 3:1 mixture of diastereomers for R = H and in 34% yield as a complex mixture of diastereomers for R = CH_3.

$[Cp(CO)(P(CH_3)_3)Fe{=}C(C_3H_7\text{-}i)OCH_3]X$ (Table **3**, No. **39**, X = SO_3F, I). The iodide forms along with the corresponding I^- salts of Nos. 11 and 34 by the addition of CH_3I to the vinyl ether, $Cp(CO)(P(CH_3)_3)\ FeC(OCH_3){=}CH_2$; see No. 11.

The SO_3^- salt is deprotonated by $(CH_3)_3P{=}CH_2$ in pentane at −40 °C to give the vinyl ether $Cp(CO)(P(CH_3)_3)FeC(OCH_3){=}C(CH_3)_2$ and $[P(CH_3)_4]SO_3F$.

$[Cp(CO)(P(C_6H_5)_3)Fe{=}C(C_3H_7\text{-}i)OCH_3]X$ (Table **3**, No. **40**, X = SO_3F, SO_3CF_3). The SO_3F^- salt forms along with Nos. 14 and 35 by the addition of

CH_3OSO_2F to the vinyl ether complex, $Cp(CO)(P(C_6H_5)_3)FeC(OCH_3){=}CH_2$; for the discussion of the mechanism, see No. 11.

Reduction of the $SO_3CF_3^-$ salt with $CH_3OH/NaOCH_3/NaBH_4$ in CH_2Cl_2 at $-78\,°C$ produces the diastereomeric mixture (49:51 ratio) of $Cp(CO)(P(C_6H_5)_3)FeCH(OCH_3)C_3H_7$-*i* in 60% yield.

$[Cp(CO)(P(OCH_3)_3)Fe{=}C(C_3H_7\text{-}i)OCH_2CH{=}CH_2]SO_3CF_3$ (Table **3**, No. **41**) reacts with KH in C_6H_6 in the presence of 18-crown-6 at 0 °C to give the corresponding vinyl ether $Cp(CO)(^2D)FeC({=}C(CH_3)_2)OCH_2CH{=}CH_2$ and in refluxing C_6H_6 (21 h) the rearranged product $Cp(CO)(^2D)FeC(O)C(CH_3)_2CH_2CH{=}CH_2$ in 21% yield.

$[Cp(CO)(P(C_6H_5)_3)Fe{=}C(C_3H_7\text{-}i)OSO_2CF_3]SO_3CF_3$ (Table **3**, No. **42**) is rapidly obtained as an intermediate in CH_2Cl_2 solution by the reaction of $O(SO_2CF_3)_2$ with $Cp(CO)(P(C_6H_5)_3)FeC(O)C_3H_7$-*i*. It slowly eliminates $HOSO_2CF_3$ to give the vinylidene complex, $[Cp(CO)(P(C_6H_5)_3)Fe{=}C{=}C(CH_3)_2]SO_3CF_3$.

$[Cp(CO)(P(C_6H_5)_3)Fe{=}C(CH_2CH_2R)OC_2H_5]BF_4$ ($R = CH_2{=}CH_2$, $C{\equiv}CH$; Table **3**, Nos. **46**, **47**) are reduced immediately after preparation (Method IIb) with $NaBH_4/CH_3OH/NaOCH_3$ in CH_2Cl_2 (0.5 h) to give the corresponding (*E*)-configured compounds $Cp(CO)(^2D)FeCH{=}CHCH_2CH{=}CH_2$ and $Cp(CO)(^2D)FeCH{=}CHCH_2C{\equiv}CH$, respectively, in 96% yields. UV irradiation at 0 °C results in decomposition into unidentifiable products.

$[Cp(CO)(P(C_6H_5)_3)Fe{=}C(CH{=}CHSi(CH_3)_3)OCH_3]BF_4$ (Table **3**, No. **48**, Formula IX) is obtained according to Method IIb from the (*E*)-configured acyl precursor with retention of the (*E*) configuration.

Addition of solid $[S(N(CH_3)_2)_3][Si(CH_3)_3F_2]$ to a THF solution of the complex at $-78\,°C$ produces the vinyl ether X with a diastereomeric ratio >92:8; the major product is supposed to be the (*E*)-*SS*, (*E*)-*RR* enantiomeric pair.

IX: $[Cp(OC)((C_6H_5)_3P)Fe{=}C(OCH_3)CH{=}CHSi(CH_3)_3]^+[BF_4]^-$

X: $Cp(OC)((C_6H_5)_3P)Fe{-}C(OCH_3){=}CH{-}CH(CH_3)Si(CH_3)_2F$

$[Cp(CO)(P(C_6H_5)_3)Fe{=}C(CH_2C_6H_5)OCH_3]BF_4$ (Table **3**, No. **49**) is obtained in 82% yield by the reaction of $Cp(CO)(P(C_6H_5)_3)FeC{\equiv}CC_6H_5$ with $HBF_4{\cdot}O(C_2H_5)_2$ in CH_3OH solution. It can also be prepared in 73% yield by a stepwise addition of HBF_4 giving first the vinylidene cation $[Cp(CO)(P(C_6H_5)_3)Fe{=}C{=}CHC_6H_5]^+$, followed by reaction with CH_3OH (Method Ib).

$[Cp(CO)(P(OCH_3)_3)Fe{=}C(CH_2OCH_3)OH]BF_4$ (Table **3**, No. **50**). Quenching a solution of the complex freshly prepared according to Method IIa with triethylamine regenerates the starting complex with 90% yield. Rearrangement occurs within 2 h to give the cation XI along with small amounts of the cation $[Cp(CO)_2FeP(OCH_3)_3]^+$.

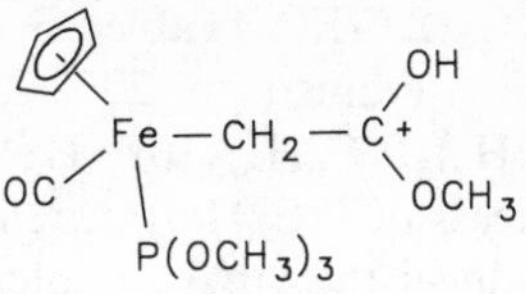

OH
Fe—CH$_2$—C+
OC OCH$_3$
P(OCH$_3$)$_3$

XI

$[Cp(CO)(P(C_6H_5)_3)Fe{=}C(CH_2OCH_3)OR]PF_6$ (R=CH_3, C_2H_5; Table **3**, Nos. **51, 52**) are air stable solids and can be reduced with hydride donors, such as $[CH_3P(C_6H_5)_3]BH_4$ in CH_2Cl_2 to afford $Cp(CO)(P(C_6H_5)_3)FeC_2H_5$ in 69% yield. However, one equivalent of $Li[HB(C_2H_5)_3]$ or $Li[HB(C_4H_9\text{-}s)_3]$ in THF at − 80 °C transforms the compounds into $Cp(CO)(P(C_6H_5)_3)FeCH_2CHO$ in 63% yield; the same results are obtained with No. 56.

$[Cp(CO)(^2D)Fe{=}C(CH_2OCH_3)OMo(CO)_3Cp]^+$ ($^2D = P(C_6H_5)_3$, $P(OCH_3)_3$; Table **3**, Nos. **54**, **55**) probably as the BF_4^- or PF_6^- salts are assumed to be intermediates in the reaction of the corresponding acyl compounds $Cp(CO)(^2D)FeCOCH_2OCH_3$ with $[CpMo(CO)_3]^+$ in CH_2Cl_2 at − 40 °C followed by addition of $[CH_3P(C_6H_5)_3]\ BH_4$ at 0 °C to generate the corresponding ethyl compounds, $Cp(CO)(^2D)FeC_2H_5$, with 33 to 38% yield; see also Nos. 51 and 52.

$[Cp(CO)(^2D)Fe{=}C(C_6H_5)OCH_3]SO_3CF_3$ ($^2D = P(C_2H_5)_3$, $P(C_6H_5)_3$; Table **3**, Nos. **57**, **58**). The barrier to rotation about the Fe–C bond is < 7 kcal/mol. Addition of a CH_2Cl_2 solution of No. 57 to a $CH_3OH/NaOCH_3/NaBH_4$ mixture (BH_4^- to OCH_3^- mol ratio = 5:1 in all cases) at 0 °C produces $Cp(CO)(P(C_2H_5)_3)FeCH(OCH_3)C_6H_5$ in 93% yield. A similar procedure with No. 58 is carried out with a $CD_3OD/NaOCD_3/NaBH_4$ or $NaBD_4$ mixture at − 78 °C, resulting in the analogous H^- or D^- adduct. The diastereomeric selection is low and nearly independent of BH_4^- concentration and temperature; diastereomer ratios, *SS*,*RR* (major product) to *RS*,*SR* (minor product), are determined by ^1H NMR spectroscopy and are listed in a table. A mechanism involving attack of the nucleophile at the *syn* or *anti* isomer (see Formula III to VI and Sect. 1.1.2.1) is discussed. Further reduction, giving the corresponding $Cp(CO)(^2D)FeCH_2C_6H_5$ compounds, is observed with $LiAlH_4$ or $NaBH_4$ in THF at − 78 °C.

$[Cp(CO)(P(C_6H_5)_2C_5H_{11})Fe{=}C(C_6H_5)OCH_3]SO_3CF_3$ (Table **3**, No. **59**, C_5H_{11} = (*S*)-2-methylbutyl). The *RS* and the *SS* diastereomers are obtained according to Method IIb, starting with the corresponding *RS* (> 98:2) and *SS* (> 92:8) diastereomeric acyl complexes.

Addition of a CH_2Cl_2 solution of the complex to a $CH_3OH/NaOCH_3/NaBH_4$ mixture at − 78 °C yields the corresponding $Cp(CO)(^2D)FeCH(OCH_3)C_6H_5$ compounds in 95% yields. Thus, the starting *RS* isomer produces the *RSS*,*RSR* pair of diastereomers, whereas the starting *SS* isomer gives the *SSS,SSR* pair of diastereomers (sequence of chirality: Fe, phosphine ligand, C-α; see Sect. 1.1.2). In both cases one diastereomer is the major product; see General Remarks.

$[Cp(CO)(P(C_6H_5)_3)Fe{=}C_4H_6O]X$ (Table **3**, No. **60**, X = Br, PF_6, $B(C_6H_5)_4$). The Br^- salt is obtained in 29% yield along with the salt $[Cp(CO)(P(C_6H_5)_3)FeCO(CH_2)_3P(C_6H_5)_3]Br$ (19%) by the reaction of $Fp(CH_2)_3Br$ with a slight excess of $P(C_6H_5)_3$ in refluxing CH_3CN for 5 h.

The action of a base produces the vinyl complex XII (R = H) and methylation with CH_3I in the presence of a base gives No. 61 and 63.

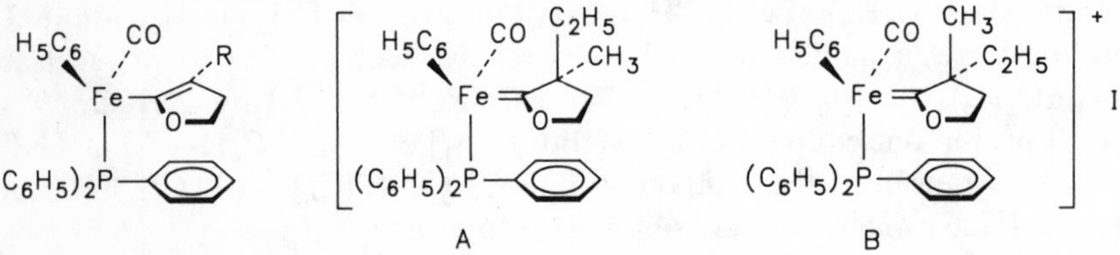

A B

XII XIII

$[Cp(CO)(P(C_6H_5)_3)Fe{=}C_4H_5(R)O]I$ (R = CH_3, C_2H_5; Table **3**, Nos. **61**, **62**). The compounds are obtained by alkylation of the vinyl complex XII (R = H) with CH_3I or C_2H_5I, respectively. The chiral center at C-β forms stereoselectively (>98%), as described for No. 64.

Deprotonation occurs by treatment with a base with formation of the vinyl compounds XII with R = CH_3 or C_2H_5, respectively.

$[Cp(CO)(P(C_6H_5)_3)Fe{=}C_6H_{10}O]I$ (Table **3**, No. **63**) is prepared by treatment of No. 60 with excess diisopropylethylamine (formation of XII, R = H) and an excess CH_3I.

Successive treatment of the complex with OH^- and Br_2 releases 2,2-dimethylbutyrolacetone.

$[Cp(CO)(P(C_6H_5)_3)Fe{=}C_7H_{12}O]I$ (Table **3**, No. **64**, Formula XIII). The diastereomer A is obtained upon ethylation of the vinyl complex XII (R = CH_3) with C_2H_5I and diastereomer B via methylation of XII (R = C_2H_5) with CH_3I. The stereoselective quarternation (>98%) of C-β is achieved by attack of the alkylating agent onto the unhindered face of the alkoxyvinyl ligand in the *anti*-(O,O) conformation of XII; see also General Remarks about the addition of nucleophiles to the carbene carbon atom C-α.

$[Cp(CO)(P(C_6H_5)_3)Fe{=}C(CH_3)SH]BF_4$ (Table **3**, No. **67**) is unstable in solution. Attempted deprotonation with the sterically hindered base $N(C_2H_5)(C_3H_7\text{-}i)_2$ produces an unstable material that was suggested to be the thioacyl complex, $Cp(CO)(P(C_6H_5)_3)FeC(S)CH_3$.

$[Cp(CO)(NCCH_3)Fe{=}C(H)NHC_6H_{11}\text{-}c]SO_3CF_3$ (Table **3**, No. **69**) is obtained by irradiation of a solution of $[Cp(CO)_2Fe{=}C(H)NHC_6H_{11}\text{-}c]SO_3CF_3$ in CH_3CN with UV light at 245 nm for 2 h (85% yield). It is prepared in 93% yield when an acetonitrile solution of $(CH_3)_3NO$ is injected into a CH_2Cl_2 solution of the same starting material.

[Cp(CO)(P(C_6H_5)$_3$)Fe=C(CH_3)NHCH_3]BF_4 (Table **3**, No. **71**) is obtained according to Method Ib along with the corresponding compounds Cp(CO(^{2}D)FeC(O)CH_3 (2%) and Cp(CO(^{2}D)FeC≡CH (11%). The compounds are separated by chromatographing the mixture on deactivated silica gel.

[Cp(CO)(P(C_6H_5)$_3$)Fe=C(CH_3)NHCH(CH_3)C_6H_5]BF_4 (Table **3**, No. **73**). A mixture of diastereomers is obtained according to Method IVb from No. 19 and (*S*)-(−)-α-phenylethylamine. Resolution of the *RS* and *SS* diastereomers is achieved by fractionation from ethanol. The CD spectra of the diastereomers are depicted in Fig. 2. The three Cotton effects in the 500–300 mμ region arise from the chiral Cp(CO)(P(C_6H_5)$_3$)Fe chromophore. The metal center is stable to racemization, and the configuration remains unchanged in refluxing acetone.

[Cp(CO)(P(C_6H_5)$_3$)Fe=C(CH_3)N(CH_3)$_2$]BF_4 (Table **3**, No. **74**). Method Ib produces a mixture with the corresponding compounds Cp(CO)(^{2}D)FeC(O)CH_3(8%), and Cp(CO)(^{2}D)FeC=CH (13%).

XIV XV

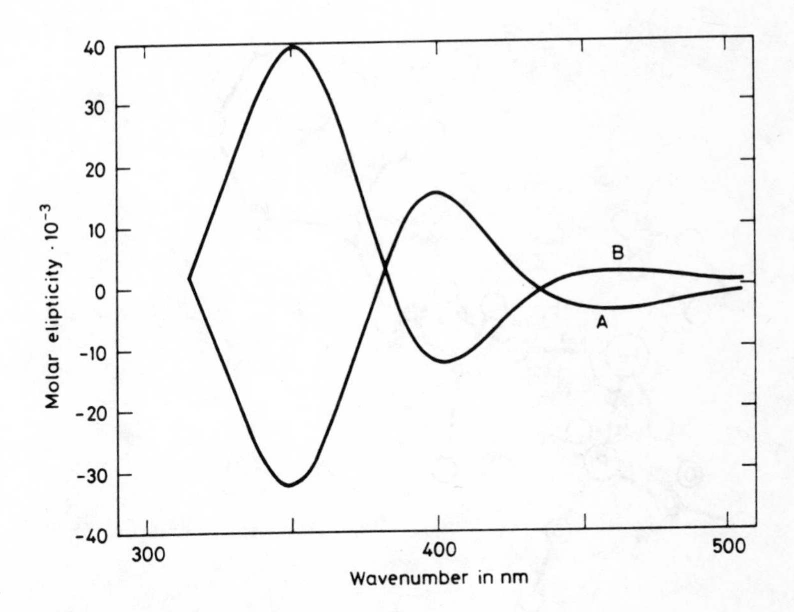

Fig. 2. CD spectra of the diastereomers (A: less soluble; B: more soluble) of (*S*)-[Cp(CO)(P(C_6H_5)$_3$)-Fe=C(CH_3)NHCH(CH_3)C_6H_5]BF_4

$[Cp(CO)(P(C_6H_5)_3)Fe{=}CC_{11}H_{15}N]X$ (Table **3**, No. **75**, X = BF_4, SO_3CF_3; Formula XIV, R = C_6H_5), **$[Cp(CO)(P(OCH_3)_3)Fe{=}CC_{11}H_{15}N]SO_3CF_3$** (Table **3**, No. **76**; Formula XIV, R = C_6H_5), **$[Cp(CO)(P(C_6H_5)_3)Fe{=}CC_{12}H_{17}N]BF_4$** (Table **3**, No. **77**; Formula XIV, R = $C_6H_4CH_3$-3), **$[Cp(CO)(P(C_6H_5)_3)Fe{=}CC_{12}H_{17}N]BF_4$** (Table **3**, No. **78**; Formula XIV, R = $C_6H_4CH_3$-4) and **$[Cp(CO)(P(C_6H_5)_3)Fe{=}CC_{13}H_{17}N]BF_4$** (Table **3**, No. **79**; Formula XIV, R = $CH{=}CHC_6H_5$) are oxidized with C_6H_5IO, O_2, $[N(C_4H_9)_4]NO_2$, or wet Br_2 in CH_2Cl_2 (12 h) to give the corresponding β-lactams XV in 8 to 29% yields (72% with No. 76).

No. 76 crystallizes in the triclinic space group $P\bar{1}$-C_i^1 (No. 2) with the unit cell parameters a = 10.963(2), b = 13.424(2), c = 9.505(2) Å, α = 110.64(2)°, β = 99.76(2) 248, and γ = 85.27(2)°; Z = 2. The molecular structure of the cation is given in Fig. 3. The Fe–$C_{carbene}$ distance lies between that of a formal FeC double bond and a FeC single bond. The azetidinylidene ring is planar and the N atom is arranged *trans* to the CO group.

$[Cp(CO)(P(C_6H_5)_3Fe{=}C_{17}H_{15}N]X$ (Table **3**, No. **80**, X = BF_4; Formula XVI, ^2D = $P(C_6H_5)_3$) and **$[Cp(CO)(P(OCH_3)_3Fe{=}C_{17}H_{15}N]X$** (Table **3**, No. **81**, X = BF_4, SO_3CF_3; Formula XVI, ^2D = $P(OCH_3)_3$) are obtained according to Method Ic using 2 equivalents of the imine.

The compounds react with C_6H_5IO in ethanol to give XVII in 77 and 47% yields, respectively.

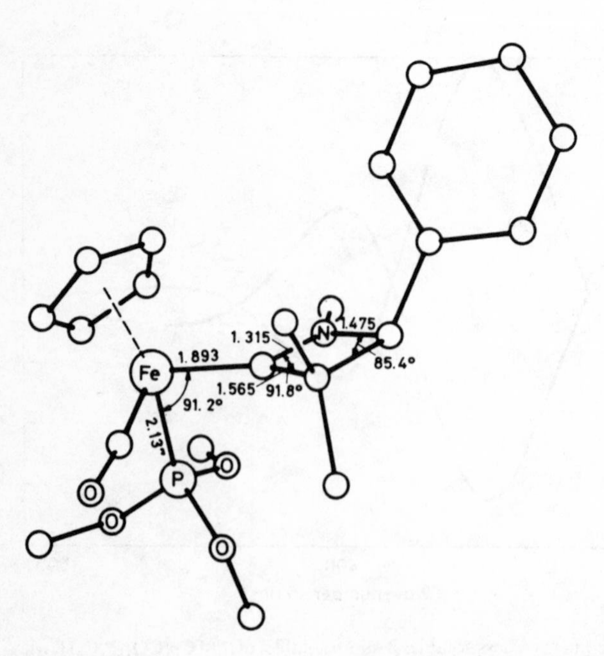

Fig. 3. Molecular structure of the cation *RS,SR*-$[Cp(CO)(P(OCH_3)_3)Fe{=}C_{12}H_{15}N]^+$

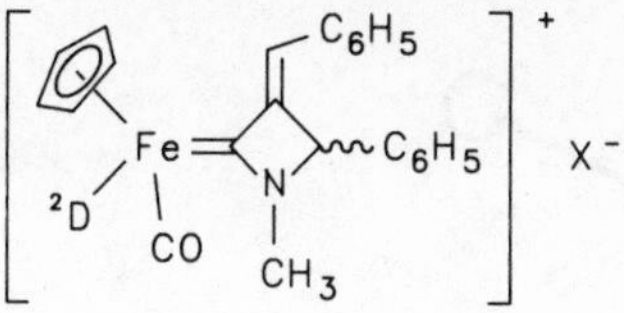

XVI 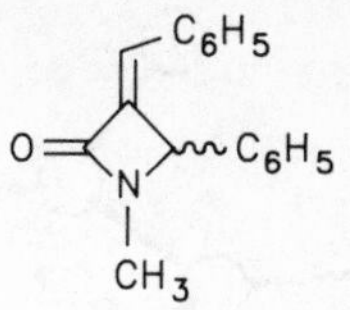 XVII

$[Cp(CO)(P(OCH_3)_3)Fe{=}C_7H_{11}NS]BF_4$ (Table **3**, No. **82**; Formula XVIII, R-1 = R-2 = H), **$[Cp(CO)(P(OCH_3)_3Fe{=}C_{10}H_{15}NO_2S]SO_3CF_3$** (Table **3**, No. **83**; Formula XVIII, R-1 = H, R-2 = $COOC_2H_5$ and **$[Cp(CO)(P(OCH_3)_3)Fe{=}C_{11}H_{17}NO_2S]SO_3CF_3$** (Table **3**, No. **84**; Formula XVIII, R-1 = CH_3, R-2 = $COOCH_3$). The cycloaddition according to Method Ic proceeds with high stereoelectivity controlled solely by the asymmetric iron atom. Thus, the diastereomeric ratios are 15:1 for No. 82, 8:1 for No. 83, and 6:1 for No. 84.

The compounds are oxidized with C_6H_5IO or $[N(C_4H_9\text{-}n)_4]NO_2$ in ethanol at room temperature to give the corresponding compounds XIX (No. 82: 2 d, 36% yield; No. 83: 22 d, 13% yield; No. 84: 4 d, 52% yield) as single isomers.

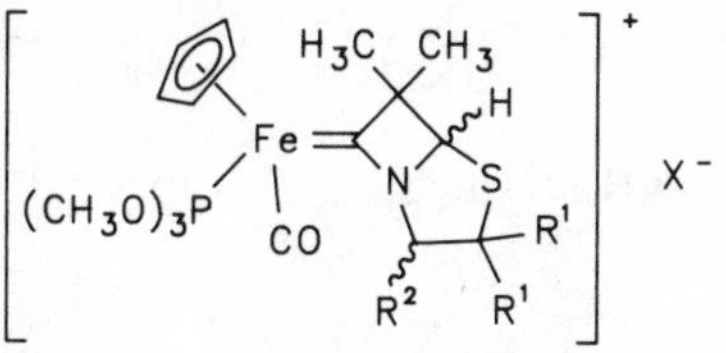

XVIII 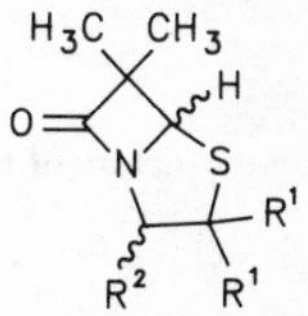 XIX

$[Cp(CO)Fe{=}C_{16}H_{16}N_2]PF_6$ (Table **3**, No. 85, Formula XX). To a solution of Na[Fp] in THF the benzimide chloride $C_6H_5C(Cl){=}NCH_3$ (1:2 mole ratio) is added at ambient temperature. The workup procedure gives some $(Fp)_2$ and small amounts of the neutral complex XXI. Addition of NH_4PF_6 precipitates the title complex in about 40% yield.

The complex is stable in the solid state and in solution. XXI probably forms by H^- addition during the preparation procedure.

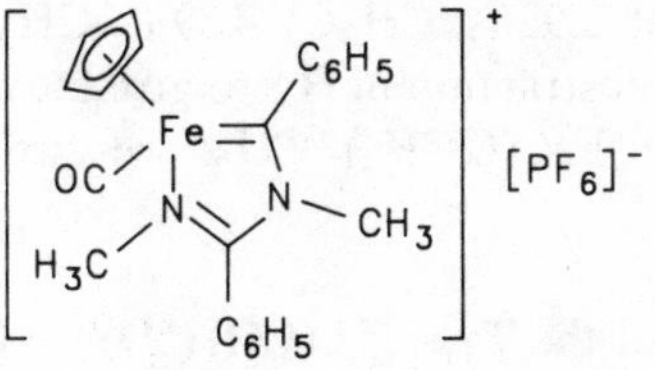

XX 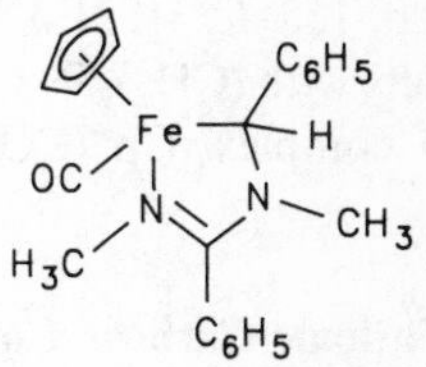 XXI

Compounds with the ^5L ligand $C_5(CH_3)_5$ (Cp*)

$[C_5(CH_3)_5(CO)\{P(CH_3)_3\}Fe{=}CHOCH_3]BF_4$ forms in 49% yield as a yellow powder along with $Cp^*(CO)\{P(CH_3)_3\}FeCH_3$ by treatment of

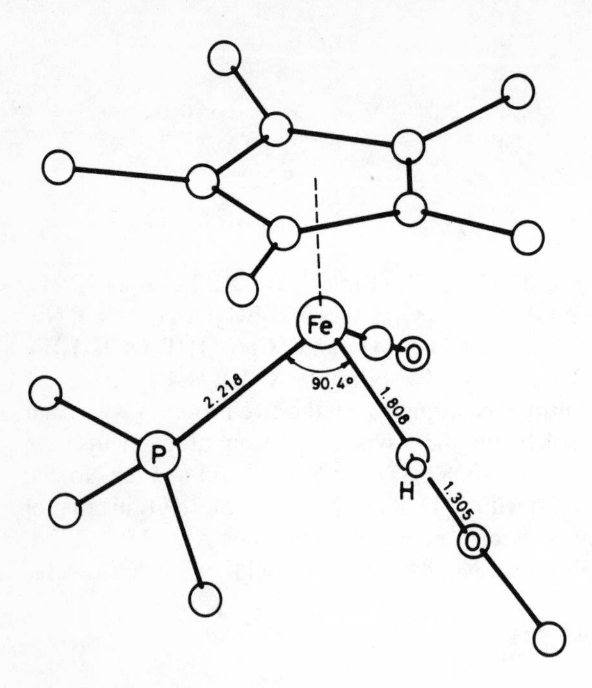

Fig. 4. Molecular structure of the cation $[C_5(CH_3)_5(CO)(P(CH_3)_3)Fe{=}CHOCH_3]^+$

$C_5(CH_3)_5(CO)\{P(CH_3)_3\}FeCH_2OCH_3$ in CH_2Cl_2 solution with $[O(CH_3)_3]BF_4$. The complex melts at 183 °C.

It crystallizes in the space group $P2_1/c$ with a = 12.281(1), b = 13.773(1), c = 12.385(1) Å, β = 90.81(2)°; Z = 4, d_c = 1.351 g/cm^3. The CO group lies within the plane of the carbene ligand, and the OCH_3 group adopts an anticlinal conformation, as shown in Fig. 4.

Conversion into the starting material occurs with NaH in THF (18 h) in 73% yield.

$\mathbf{[C_5(CH_3)_5(CO)(P(CH_3)_3)Fe{=}C(CH_3)OCH_3]SO_3F}$ is obtained analogously to Method IIb from the corresponding acyl complex and CH_3OSO_2F in C_6H_6; m.p. 158 °C. ^{1}H NMR (in ppm): 2.92 (s, CH_3C), 4.39 (s, CH_3O).

It reacts with $(CH_3)_3P{=}CH_2$ with abstraction of H^+ to give the corresponding vinyl complex, $Cp^*(CO)(P(CH_3)_3)FeC(OCH_3){=}CH_2$; see also No. 11 in Table 3.

1.1.2.3 Cationic Carbene Complexes of the Type $[^5L(CO)(^2D)Fe{=}C(ER)_2]^+$ with Two Heteroatoms at the Carbene Carbon Atom

The compounds in this section contain the heteroatoms E = F, Cl, O, S, or N (n = 0 to 2) at the carbene carbon atom with additional ligands R to achieve the rare gas shell. In four compounds the ER_n groups at each carbene carbon atom are different. No cation with the ^{5}L ligand other than Cp is described.

Compounds in which the ^2D ligand is represented by isonitriles are described in Sect. 1.1.4; see also carbene compounds with $^2D = {}^2L = \eta^2$-alkene in Section 1.

The compounds listed in Table 4 are obtained by the following general methods:

Method I: Irradiation of the starting complex $[Fp{=}C(ER)_2]^+$ in CH_3CN solution.

Method II: Ligand replacement starting from $[Cp(CO)(CH_3CN)Fe{=}C(ER)_2]^+$ obtained by method I.

Method III: Reaction of the corresponding cation $[Fp(^2D)]^+$ with oxirane or aziridine in the presence of Br^-.

Method IV: From $[Cp(CO)(PR_3)FeCNR]X$ and H_2NCH_3.

General remarks. The substituents R at the carbene compounds with E = O, S, NH are arranged in the carbene plane as shown in Formula I with one R in *syn* and one R in *anti* position to the iron atom. If the R groups at both heteroatoms are different, isomers can be distinguished as shown for compounds Nos. 23 to 25 with different isomer population; see Formula V.

$$\left[Cp(OC)(^2D)Fe{=}C(E{-}R)(E{-}R) \right]^+$$

I

The R groups in *syn* and *anti* positions with E = S rapidly interconvert and are equivalent according to the NMR spectrum at room temperature, but become nonequivalent as the temperature is lowered. Coalescence temperature, T_C, is in the range of − 28 to − 71 °C corresponding to the barrier to rotation about the $C_{carbene}$–S bonds with free energies of activation values, $\Delta G^\ddagger$, varying from 12.2 to 10.1 kcal/mol. This barrier to rotation is sensitive to changes in electronic density at the iron atom and decreases in the series of ^2D ligands $CH_3CN > CH_3NC > P(C_6H_5)_3 > As(C_6H_5)_3 > P(OCH_3)_3 > P(OC_6H_5)_3 > Sb(C_6H_5)_3 > C_5H_5N$; see individual compounds. In the case of E = NH, higher rotation barriers corresponding to higher coalescence temperatures (about 70 °C) are observed with $\Delta G^\ddagger$ values of about 17 kcal/mol.

The chirality at iron because of the four different ligands causes the CH_2 protons of the carbene ligand of No. 8 to appear as multiplets when they become nonequivalent; no separation of enantiomers or diastereomers is mentioned. In complex No. 9, the ^2D ligand $S(CH_3)_2$ becomes diastereotopic at low temperature.

Table 4. Cationic Carbene Compounds of the Type $[Cp(CO)(^2D)Fe{=}C(ER)_2]^+$. An asterisk indicates further information at the end of the table

No.	$Fe{=}C(ER)_2$ ^2D ligand	properties and remarks method of preparation (yield)
heteroatoms E are F, F		
*1	$Fe{=}CF_2$ $P(C_6H_5)_3$	BF_4^- salt; yellow crystals ^{13}C NMR (CD_2Cl_2): 95 (Cp), 216 (CO) ^{19}F NMR (CD_2Cl_2): 164 (s, CF_2) IR (Nujol mull): 2024; ν (CF) 1182, 1198
heteroatoms E are F, Cl		
*2	$Fe{=}CFCl$ $P(C_6H_5)_3$	identified spectroscopically ^{19}F NMR (CD_2Cl_2): 207 (CFCl)
heteroatoms E are Cl, Cl		
*3	$Fe{=}CCl_2$ $P(C_6H_5)_3$	identified spectroscopically
heteroatoms E are O, O		
*4	$Fe{=}\overline{C(OCH_2CH_2O)}$ $PCH_3(C_6H_5)_2$	PF_6^- salt; III (86%) ^1H NMR (CD_3CN): 4.21, 4.50 (m's, CH_2O) ^{31}P NMR (CD_3CN): 55.65
*5	$Fe{=}\overline{C(OCH_2CH_2O)}$ $P(C_6H_5)_3$	PF_6^- salt, III (84%); light yellow crystals, m.p. 210 °C ^1H NMR (CD_3CN): 4.20 (m, CH_2O) ^{13}C NMR (CD_3CN): 72.46 (CH_2O), 217.20 (d, CO), 254.73 (d, C=Fe)

No.	Compound	Data
heteroatoms E are O, S		
6	Fe=C(OCH$_3$)SCH$_3$ NCCH$_3$	CF$_3$SO$_3^-$ salt, I (34%); m.p. 81 to 83 °C ^{1}H NMR (CD$_3$COCD$_3$): 2.63 (s, CH$_3$S), 4.89 (s, CH$_3$O) ^{13}C NMR (CD$_3$COCD$_3$): 19.9 (CH$_3$S), 69.9 (CH$_3$O), 215.9 (CO), 313.0 (C=Fe)
heteroatoms E are O, N		
7	Fe=C(OCH$_3$)NH$_2$ NCCH$_3$	CF$_3$SO$_3^-$ salt; I; red oil ^{1}H NMR (CD$_3$COCD$_3$): 4.08 (s, CH$_3$O)
8	Fe= (ring: O, N–H) P(C$_6$H$_5$)$_3$	PF$_6^-$ salt, III (70%); yellow flaky cyrstals, mp. 228 °C ^{1}H NMR (CD$_3$CN): 3.30 (m, CH$_2$N), 4.00 (m, CH$_2$O), 8.66 (NH) ^{13}C NMR (CD$_3$CN): 45.41 (CH$_2$N), 72.20 (CH$_2$O), 217.55 (CO), 234.20 (d, C=Fe)
heteroatoms E are S, S		
*9	Fe=C(SCH$_3$)$_2$ S(CH$_3$)$_2$	PF$_6^-$ salt; dark red crystals ^{1}H NMR (CDCl$_3$): 3.20 (s, CH$_3$SC) ^{13}C NMR (CDCl$_3$ at − 30 °C): 22.62, 24.12 (CH$_3$ of S(CH$_3$)$_2$), 28.96 (CH$_3$S of carbene), 215.52 (CO), 318.53 (C=Fe)
*10	Fe=C(SCH$_3$)$_2$ NCCH$_3$	CF$_3$SO$_3^-$ salt, I (88%); red oil PF$_6^-$ salt; red crystals, m.p. 121 to 124 °C ^{1}H NMR (CD$_3$COCD$_3$ or CD$_3$CN): 3.23 (s, CH$_3$S) ^{13}C NMR (CD$_3$COCD$_3$ or C$_6$D$_6$): 28.1 (CH$_3$S), 216.4 (CO), 321.2 (C=Fe)
11	Fe=C(SCH$_3$)$_2$ NC$_5$H$_5$	PF$_6^-$ salt, II (67%); dark red, m.p. 112 to 115 °C (dec.) ^{1}H NMR (CD$_3$COCD$_3$): 2.99 (s, CH$_3$S) ^{13}C NMR (CD$_3$CN): 28.0 (CH$_3$S), 218.6 (CO), 319.6 (C=Fe)
*12	Fe=C(SCH$_3$)$_2$ P(OCH$_3$)$_3$	PF$_6^-$ salt, II (49%); orange, m.p. 195 °C (dec.) ^{1}H NMR (CD$_3$COCD$_3$): 3.19 (s, CH$_3$S) ^{13}C NMR (CD$_3$CN) 29.5 (CH$_3$S), 215.3 (d, CO; J(P,C) = 46.9), 314.3 (d, C=Fe; J(P,C) = 29.3)

Table 4. Continued

No.	$Fe{=}C(ER)_2$ ^2D ligand	properties and remarks method of preparation (yield)
*13	$Fe{=}C(SCH_3)_2$ $P(OC_6H_5)_3$	PF_6^- salt, II (78%); dark yellow, m.p. 133 to 135 °C ^1H NMR (CD_3COCD_3): 3.07 (s, CH_3S) ^{13}C NMR (CD_3CN): 29.9 (CH_3S), 214.9 (d, CO; J(P,C) = 43.0), 310.6 (d, C=Fe; J(P,C) = 29.3)
*14	$Fe{=}C(SCH_3)_2$ $PCH_3(C_6H_5)_2$	PF_6^- salt; not isolated pure ^1H NMR (CD_3CN): 2.81 (s, CH_3S)
15	$Fe{=}C(SCH_3)_2$ $P(C_6H_5)_3$	PF_6^- salt, II (58%); orange, m.p. 205 °C (dec.) ^1H NMR (CD_3COCD_3): 3.06 (s, CH_3S) ^{13}C NMR (CD_3CN): 29.7 (CH_3S), 218.0 (d, CO; J(P,C) = 29.3), 317.7 (d, C=Fe; J(P,C) = 17.6)
*16	$Fe{=}C(SCH_3)_2$ $As(C_6H_5)_3$	PF_6^- salt, II (73%); m.p. 180 to 184 °C (dec.) ^1H NMR (CD_3COCD_3): 3.09 (s, CH_3S) ^{13}C NMR (CD_3CN): 29.9 (CH_3S), 217.4 (CO), 317.5 (C=Fe)
17	$Fe{=}C(SCH_3)_2$ $Sb(C_6H_5)_3$	PF_6^- salt, II (74%); red, m.p. 165 °C (dec.) ^1H NMR (CD_3COCD_3): 3.12 (s, CH_3S) ^{13}C NMR (CD_3CN): 30.2 (CH_3S), 216.4 (CO), 316.1 (C=Fe)
*18	$Fe{=}\overline{CSCH_2CH_2S}$ $NCCH_3$	PF_6^- salt, I (64%); dark red crystals ^1H NMR (CD_2Cl_2): 3.14 (s, CH_2S) ^{13}C NMR (CD_2Cl_2): 46.43 (CH_2S), 214.49 (CO), 307.07 (C=Fe)
*19	$Fe{=}\overline{CSCH_2CH_2S}$ $PCH_3(C_6H_5)_2$	PF_6^- salt ^1H NMR (CD_2Cl_2): 3.36, 3.72 (m's, CH_2S)

No.	compound	method of preparation (yield) properties and remarks
*20	Fe=C (1,3-dithian-2-ylidene, ring $S-CH_2-CH_2-CH_2-S$) $NCCH_3$	PF_6^- salt, I (89%); dark red, m.p. 72 to 73 °C ^1H NMR (CD_2Cl_2): 2.50 (m, 2 CH_2), 3.20 (m, CH_2S) ^{13}C NMR (CD_2Cl_2): 19.22 (CH_2), 37.85 (CH_2S), 215.27 (CO), 300.56 (C=Fe)
heteroatoms E are N, N		
*21	$Fe{=}C(NHCH_3)_2$ $P(CH_3)_3$	I^- salt, IV; m.p. 218 °C (dec.) ^1H NMR ($CDCl_3$): 2.92 (d, CH_3N), 3.25 (d, CH_3N) ^{13}C NMR ($CDCl_3$): 32.4, 36.6 CH_3N), 208.9 (C=Fe), 217.9 (CO)
*22	$Fe{=}C(NHCH_3)_2$ $P(C_6H_5)_3$	$B(C_6H_5)_4^-$ salt, IV (92%) ^1H NMR (CD_3COCD_3): 2.58, 3.28 (d's, CH_3N) I^- salt, IV ^1H NMR ($CDCl_3$): 2.62 (d, CH_3N), 3.25 (d, CH_3N), 7.0 to 8.5 (br, NH) ^{13}C NMR ($CDCl_3$): 34.1, 35.8 (CH_3N), 209.4 (C=Fe), 218.9 (CO)
*23	$Fe{=}C(NHCH_3)NHC_2H_5$ $P(CH_3)_3$	I^- salt, IV; isomer mixture A:B = 3:7 (see (Formula V), m.p. 214 °C (dec.) ^1H NMR (CD_3COCD_3, A/B): 3.26/2.96 (d's, CH_3NH), 3.45/3.80 (m's, CH_2N) ^{13}C NMR ($CDCl_3$; A/B): 13.7/16.7 (CH_3C), 36.8/32.0 (CH_3NH), 39.9/44.6 (CH_2N), 206.8 (C=Fe), 218.4 (CO)
*24	$Fe{=}C(NHCH_3)NHC_2H_5$ $P(C_6H_5)_3$	I^- salt, IV; isomer mixture A:B = 3:1 (see Formula V), m.p. 195 °C ^1H NMR (CD_3COCD_3; A/B): 0.58/1.26 (t, CH_3C), 3,30/2.69 (d's, CH_3NH), 3.17/3.82 (m's, CH_2N) ^{13}C NMR ($CDCl_3$, A/B): 12.7/16.4 (CH_3C), 36.0/33.8 (CH_3NH), 41.5/44.2 (CH_2N), 207.2 (C=Fe), 219.6 (CO)
*25	$Fe{=}C(NHCH_3)NHC_3H_7\text{-}i$ $P(C_6H_5)_3$	I^- salt, IV; isomer mixture A:B = 3:1 (see Formula V), m.p. 205 °C (dec.) ^1H NMR (CD_3COCD_3, A/B): 0.65, 0.85 (d's, CH_3C), 3.34/2.67 (d, CH_3NH), 4.01 (m, CHN) ^{13}C NMR ($CDCl_3$, A): 21.7, 22.7 (CH_3C), 36.3 (CH_3NH), 46.4 (CHN), 206.3 (C=Fe), 219.2 (CO)

Explanation for Table 4. Selected spectroscopic data are collected in the table. The appropriate group $Cp(CO)(^{2}D)Fe$ is abbreviated as Fe. The Cp ligand shows a signal in the ^{1}H NMR spectrum in the region 4.80 to 5.10 ppm in CD_3CN, 4.80 to 5.30 ppm in CD_2Cl_2, 4.95 to 5.05 ppm in $CDCl_3$, and 4.90 to 5.35 ppm in CD_3COCD_3 solution; J(P,H) values are of about 1.5 Hz. In the ^{13}C NMR spectrum shifts between 84.0 and 87.5 ppm are found. One strong ν(CO) band is found in the IR spectra between 1950 and 2000 cm^{-1}.

Further Information

$[Cp(CO)(P(C_6H_5)_3)Fe{=}CF_2]BF_4$ (Table **4**, No. **1**) is obtained in 11% yield by treatment of a solution of $Cp(CO)(P(C_6H_5)_3)FeCF_3$ in CH_2Cl_2 with two equivalents of BF_3. For further reactions, solutions of the extremely moisture-sensitive complex are used without isolation. The cation also forms at low temperature by treatment of the starting complex with BCl_3; see preparation of Nos. 2 and 3.

The ^{19}F NMR signal of the CF_2 ligand remains sharp from room temperature to $-80\,°C$, thus indicating the equivalence of the two fluorine atoms in solution by rapid rotation of the CF_2 ligand around the Fe–C bond throughout this temperature range.

The complex crystallizes in the space group $P2_1/c\text{-}C_5^{2h}$ (No. 14) with $a = 9.035(3)$, $b = 24.390(11)$, $c = 10.745(5)$ Å, $\beta = 102.20(3)°$; $Z = 4$, $d_c = 1.537$ g/cm^3. The plane of the CF_2 ligand is rotated from the vertical position (see the structure of the cation $[Fp{=}CCl_2]^+$ in Sect. 1.1.3.3, and General Remarks) toward CO, which is a better π-acceptor than $P(C_6H_5)_3$. It is nearly coplanar with CO but is tilted 18.0° toward the Fe–P bond. The very short Fe=C distance (1.724 Å) shows the ligand to be a good π-acceptor. The structure of the cation is depicted in Fig. 5.

The product of hydrolysis is $[FpP(C_6H_5)_3]^+$. For the reaction with BCl_3, see below.

$[Cp(CO)(P(C_6H_5)_3)Fe{=}CFCl]BF_{4-x}Cl_x$ and **$[Cp(CO)(P(C_6H_5)_3)Fe{=}CCl_2]BF_{4-x}Cl_x$** (Table **4**, Nos. **2** and **3**). These cations have only been detected by ^{19}F NMR spectroscopy in CD_2Cl_2 solution. Treatment of the neutral complex $Cp(CO)(P(C_6H_5)_3)FeCF_3$ with BCl_3 at low temperature gives first the cation of No. 1 by abstraction of F^-. On warming the solution a new signal appears at 207 ppm which is assigned to the CClF group of cation No. 2. At 40 °C the signal disappears, thus indicating the formation of the CCl_2 ligand of No. 3.

$[Cp(CO)(PCH_3(C_6H_5)_2)Fe{=}C_3H_4O_2]PF_6$ (Table **4**, No. **4**) is also obtained in 73% yield by the reaction of the corresponding dicarbonyl complex with $PCH_3(C_6H_5)_2$ in CH_2Cl_2 at ambient temperature (1 h) along with the by-products $[FpPCH_3(C_6H_5)_2]PF_6$ (12%) and $[Cp(CO)Fe(PCH_3(C_6H_5)_2)_2]PF_6$ (6%). A similar mixture of the complex (73%) with both these cations (21% and trace amounts, respectively) is formed when the reaction is carried out in THF solution and catalyzed by 0.1 equivalent of sodium naphthalenide.

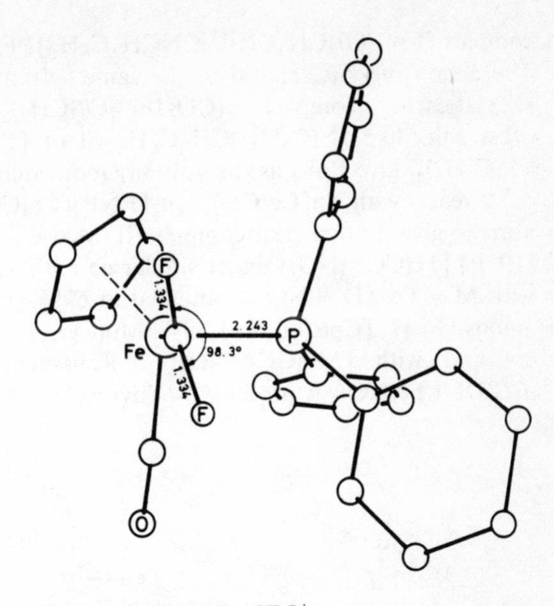

Fig. 5. Molecular structure of the cation $[Cp(CO)(P(C_6H_5)_3)Fe{=}CF_2]^+$

It reacts with excess $PCH_3(C_6H_5)_2$ in CH_2Cl_2 (20 h) to give the corresponding $[Cp(CO)Fe(^2D)_2]PF_6$ in 87% yield.

$[Cp(CO)(P(C_6H_5)_3)Fe{=}C_3H_4O_2]PF_6$ (Table **4**, No. **5**) is obtained in a procedure similar to that used for No. 4, along with $[FpP(C_6H_5)_3]PF_6$ from the corresponding dicarbonyl cation and $P(C_6H_5)_3$ (12 h).

$[Cp(CO)(S(CH_3)_2)Fe{=}C(SCH_3)_2]PF_6$ (Table **4**, No. **9**) is obtained by alkylation of II with $[O(CH_3)_3]BF_4$ in CH_2Cl_2 solution. After 45 min a 30-fold (approx.) excess of KPF_6 is added to the mixture; 74% yield.

The room-temperature ^1H NMR spectrum exhibits a broad signal for the coordinated $S(CH_3)_2$ ligand that sharpens at about 80 °C. On cooling to − 20 °C, the methyl groups being diastereotopic, exhibit two singlets. The activation energy, $\Delta G^{\ddagger}$, for the inversion at the sulfur is estimated to be 54.9 ± 0.6 kJ/mol.

It reacts with $P(OR)_3$ ($R = CH_3$, C_6H_5) at 40 °C in acetone (6 h) with replacement of $S(CH_3)_2$ to give No. 13 in quantitative yield and No. 12 in 74% yield.

$[Cp(CO)(CH_3CN)Fe{=}C(SCH_3)_2]X$ (Table **4**, No. **10**; $X = CF_3SO_3$, PF_6). Reduction of the PF_6 salt with sodium naphthalenide in the presence of CH_3SSCH_3 gives the neutral complex II in 70% yield. With $NaSCH_3$ complex II is obtained in only 30% yield. The CH_3CN ligand is a good leaving group and can be replaced by a variety of other bases; see Method II. Bases such as piperidine, $N(CH_2CH_2)_3N$, and $P(C_2H_5)_3$ cause slow decomposition. With a tenfold excess of $C_6H_5CH_2NH_2$ in CH_2Cl_2 solution, the cationic isonitrile complex $[Cp(CO)(CH_3CN)FeCNCH_2C_6H_5]PF_6$ is obtained in very low yield. The action of $[N(C_2H_5)_4]I$ in the same solvent produces a low yield of the neutral carbene complex $Cp(CO)IFe{=}C(SCH_3)_2$ (see No. 4 in Table 10), while other salts like $[N(C_2H_5)_3CH_2C_6H_5]Cl$ or [PNP]Cl (PNP = $(C_6H_5)_3P{=}N{=}P(C_6H_5)_3$) give FpCl as the only organometallic product.

It reacts with $Na[Co(CO)_4]$ or $[PNP][Co(CO)_4]$ in THF at room temperature to give the dinuclear complex III (M = Co, ^2D = CO) in 83% yield. With [PNP] $[Fe(CO)_3(NO)]$ under similar conditions the isoelectronic complex III with M = Fe, ^2D = NO is obtained in 69% yield. The complex oxidizes the anions $[Fp]^-$, $[Cp(CO)_2Ru]^-$, $[CpMo(CO)_3]^-$, and $[Mn(CO)_5]^-$ to the dimers and with $[Mn(CO)_2(NO)_2]^-$ Roussin's red methyl ester $(NO)_2Fe(\mu\text{-}SCH_3)_2Fe(NO)_2$ is produced in low yield.

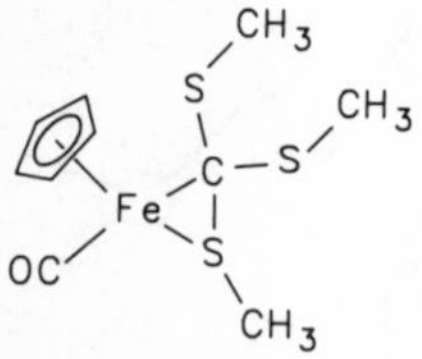

II 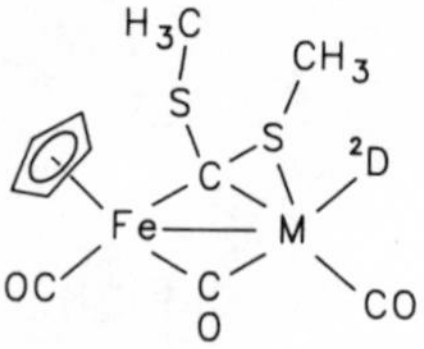 III 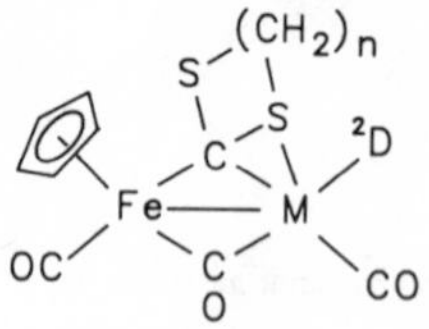 IV

$[Cp(CO)(P(OCH_3)_3)Fe{=}C(SCH_3)_2]PF_6$ (Table **4**, No. **12**) also forms in 72% yield along with some unidentified by-products, by reaction of No. 9 in acetone at 40 °C (2 h) with a 6 fold excess of $P(OCH_3)_3$ in a procedure similar to Method II.

$[Cp(CO)(P(OC_6H_5)_3)Fe{=}C(SCH_3)_2]PF_6$ (Table **4**, No. **13**) is obtained quantitatively from No. 9 and $P(OC_6H_5)_3$ as described above.

It reacts with $C_6H_5CH_2NH_2$ (1:2 mole ratio, 9 h) in CH_3CN solution to give the isocyanide complex $[Cp(CO)(P(OC_6H_5)_3)FeCNCH_2C_6H_5]PF_6$ in 87% yield.

$[Cp(CO)(PCH_3(C_6H_5)_2)Fe{=}C(SCH_3)_2]PF_6$ (Table **4**, No. **14**) was formed in 3.8% yield when $[Fp{=}C(SCH_3)_2]PF_6$ was allowed to react with an eightfold excess of the phosphine ligand in CH_2Cl_2 (2 h). The main products were $[Cp(CO)Fe(PCH_3(C_6H_5)_2)_2]PF_6$ (64%) and $[FpPCH_3(C_6H_5)_2]PF_6$ (13%); No. 14 was identified by ^1H NMR spectroscopy. If the reaction is carried out in the presence of catalytic amounts of sodium naphthalenide in THF solution, the yield of the title complex increases to 18%, whereas the other cationic compounds are formed in 32 and 48% yields, respectively.

$[Cp(CO)(As(C_6H_5)_3)Fe{=}C(SCH_3)_2]PF_6$ (Table **4**, No. **16**) reacts with $C_6H_5CH_2NH_2$ (1:1 mole ratio, 18 h) in CH_3CN solution to give the isocyanide complex $[Cp(CO)(As(C_6H_5)_3)FeCNCH_2C_6H_5]PF_6$ in 57% yield.

$[Cp(CO)(CH_3CN)Fe{=}C_3H_4S_2]PF_6$ (Table **4**, Nos. **18**) and **$[Cp(CO)(CH_3CN)Fe{=}C_4H_6S_2]PF_6$** (Table **4**, No. **20**). Reaction of these complexes with $Na[Co(CO)_4]$ or $[PNP][Fe(CO)_3NO]$ in THF produces the corresponding dinuclear complexes IV (M = Co, ^{2}D = CO; M = Fe, ^{2}D = NO) in 38 to 80% yields; see also reactions of No. 10. With M = Co similar Ru complexes have also been studied.

[Cp(CO)(PCH$_3$(C$_6$H$_5$)$_2$)Fe=C$_3$H$_4$S$_2$]PF$_6$ Table **4**, No. **19**) is obtained in 85% yield as a mixture with trace amounts of the corresponding compounds [Fp(^{2}D)]PF$_6$ and [Cp(CO) Fe(^{2}D)$_2$]PF$_6$ from the reaction of the corresponding dicarbonyl carbene complex and PCH$_3$(C$_6$H$_5$)$_2$ in CH$_2$Cl$_2$ solution (15 min).

[Cp(CO) P(CH$_3$)$_3$)Fe=C(NHCH$_3$)$_2$]I, [Cp(CO)(P(C$_6$H$_5$)$_3$)Fe=C(NH CH$_3$)$_2$]X (X = I, B(C$_6$H$_5$)$_4$), **[Cp(CO)(P(CH$_3$)$_3$)Fe=C(NHCH$_3$)NHC$_2$H$_5$]I, [Cp(CO)(P(C$_6$H$_5$)$_3$)Fe=C(NHCH$_3$)NHC$_2$H$_5$]I,** and **[Cp(CO)(P(C$_6$H$_5$)$_3$) Fe=C(NHCH$_3$)NHC$_3$H$_7$-*i*]I** (Table **4**, Nos. **21** to **25**). The compounds Nos. 21 and 22 exhibit two different signals in the ^{1}H NMR spectrum for the CH$_3$N protons, according to the orientation of the alkyl groups in the plane of the carbene ligand as depicted in Formula V (R = CH$_3$, A and B are identical). The rotational free energy barriers at the $C_{carbene}$–N bond, $\Delta G^{\ddagger}$ are 16.7 and 17.1 $\pm$ 0.3 kcal/mol, respectively, and the equilibrium is not influenced by the polarity of the solvent. With bulkier R groups and the presence of ^{2}D = P(C$_6$H$_5$)$_3$, the rotational isomers A are favored with the bulkier group *trans* to the iron atom, whereas with ^{2}D = P(CH$_3$)$_3$ or isocyanide, the population B is observed to be favored.

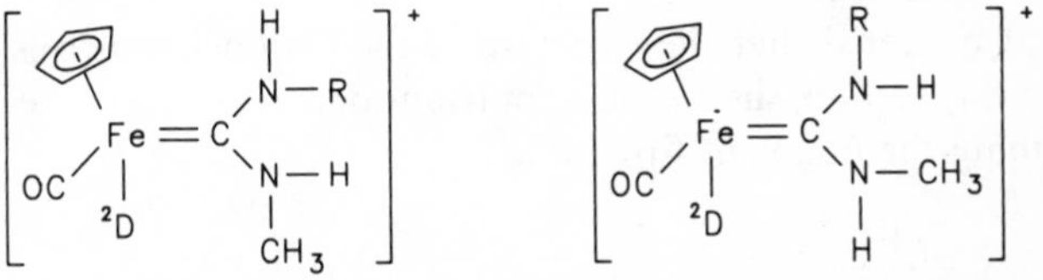

V

1.1.3 Cationic Complexes Containing the ^{5}L(CO)$_2$Fe=C Moiety

The compounds in this section mainly contain the fragment C$_5$H$_5$(CO)$_2$Fe, abbreviated as Fp. Some compounds with a ^{5}L ligand other than C$_5$H$_5$ (C$_5$(CH$_3$)$_5$, C$_5$H$_4$CH$_3$, or indenyl) are described at the end of each section.

The presence of only electron-withdrawing CO groups aside the ^{5}L ligand C$_5$H$_5$ makes the carbene carbon atom more electrophilic than the corresponding cations in the sections before. The compounds without heteroatoms in Sect. 1.1.3.1 have mainly carbenium character, indicated by ^{13}C NMR downfield shifts of the carbene carbon signal up to 420 ppm. With simple alkyl or H substituents at the carbene carbon atom, the cations are very unstable. The stability increases with introduction of a heteroatom at the carbene carbon atom in the series [Fp=CR$_2$]$^+$ < [Fp=C(R)ER]$^+$ < [Fp=C(ER)$_2$]$^+$.

The plane of the carbene moiety is arranged either within the plane of the Fp group or perpendicular to that plane; see also theoretical considerations in Sect. 1.1.

1.1.3.1 Cationic Complexes of the Type $[^5L(CO)_2Fe{=}CRR']^+$ without Heteroatom at the Carbene Carbon Atom

General Remarks. This section deals with cationic carbene complexes mainly of the general type $[Fp{=}CRR']^+$ with R and R' = H, alkyl, or aryl, in which the carbene carbon atoms are not stabilized by an electron donating heteroatom. Thus, especially compounds with H or alkyl substituents at the carbene carbon atom, were neither stable enough to be identified by spectroscopic methods nor could they be isolated. The formation of these compounds, however, is evidenced by chemical reaction, e.g. cyclopropanation reactions. We therefore have summarized all compounds in Table 6 which have been generated as intermediates according to the general methods given below.

The stability of the cations generally increases when the carbene carbon atom is part of an aromatic system (see General Remarks in Sect. 1.1) or conjugated to a π-bonding system. It apparently also increases if C_5H_5 is replaced by the bulkier ligand $C_5(CH_3)_5$ when the properties of the cations $[Fp{=}CH_2]^+$ and $[C_5(CH_3)_5(CO)_2Fe{=}CH_2]^+$ are compared; the latter is described at the end of this section.

It should be noted that the structure of the cation I, which is not further described in this section, shows some contribution of the resonance structure II, thus containing the fragment $Fp{=}CR_2$.

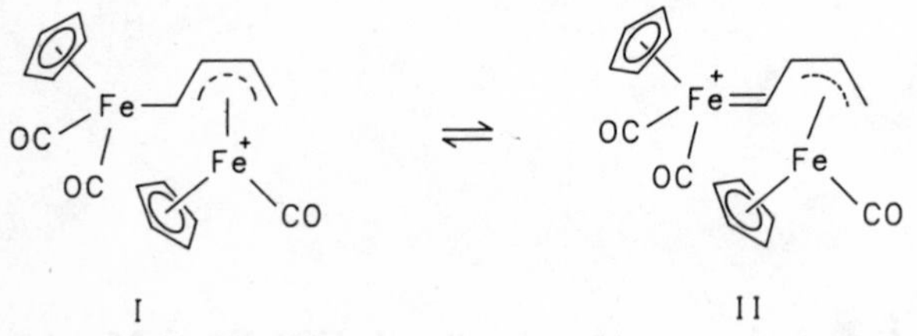

Fe OC OC Fe⁺ CO ⇌ Fe⁺ OC OC Fe CO

I II

Cyclopropanation reactions. Cations of the type $[Fp{=}CRR']^+$ (III; R = H, alkyl; R' = H, alkyl, aryl) are highly electrophilic and readily transfer the alkylidene ligand to unactivated alkenes to generate the corresponding cyclopropanes in high yield. Because of their instability, the cations are generated primarily at low temperature in the presence of the appropriate alkene. The reactions carried out with these compounds are summarized in Table 5. The cyclopropanation reactions are carried out according to the procedures A to H. In most cases with $[Fp{=}CHR]^+$ (R = alkyl, aryl), very high to moderate stereoselectivity is achieved to form *cis* or *syn* adducts with *cis*/*trans* ratios varying between 5/1 and >100/1, determined by the *cis*/*trans* approach ratios k_{cis}/k_{trans}, which are largely determined by steric effects.

Mechanistic studies have been carried out by the reaction of $[Fp=CHCH_3]^+$ with a series of styrenes (Formula IV, R = H, X = H, F, Cl, OCH_3, CH_3, CF_3) in which the nucleophilicity of the olefin was varied without altering the steric properties. Similar stereoselectivities (*cis*/*trans* ratios of about 6/1) for X = H, F, Cl, CH_3, and CF_3 show that the stereochemistry at the styrene double bond during cyclopropane formation is retained. The lowest stereoselectivity, found with X = OCH_3 (*cis*/*trans* ratio 0.9/1), is attributed to the formation of a fairly stable carbenium ion intermediate (Formula V for *cis* or VII for *trans*), stabilized by the strong electron donor ability of the OCH_3 group, which causes loss of stereochemistry by rotation about the alkene double bond.

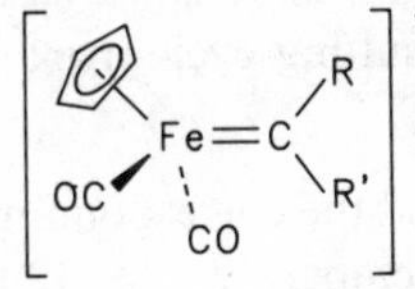

III

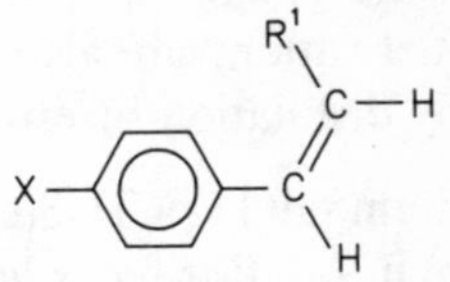

IV

A detailed mechanistic study has been undertaken by adding the carbene ligand to *cis*-β-deuterio-4-X-styrene (Formula IV, R = D; X = H, CH_3, OCH_3), resulting in the possible cyclopropane isomers IX to XII. With X = H and CH_3, only the isomers IX and XII are found with the *cis*/*trans* ratios given in Table 5, reactions Nos. 26 and 32, indicating retention of the stereochemistry at the

cis approach

trans approach

V VI VII VIII

IX X XI XII

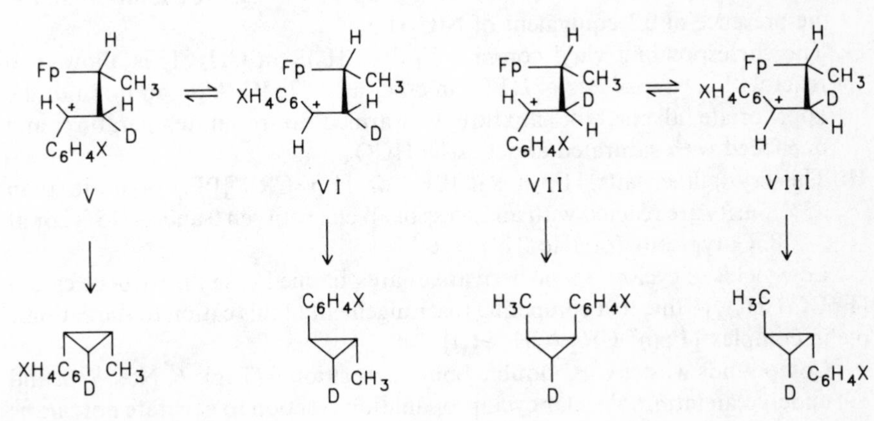

styrene. With X = OCH_3 (Table 5, reaction No. 29) the cyclopropane isomers IX to XII are produced in 41, 40, 6, and 13% yields, respectively, thus indicating reaction of the equilibrium systems V/VI and VII/VIII. The electron-donating OCH_3 group stabilizes the intermediate adduct in such a way that rotation about the former C=C bond of the styrene can occur before cyclopropanation takes place. From the sum of IX + X and the sum of XI + XII, a *cis/trans* approach ratio of about 4.3/1 can be calculated. The relative rates of bond rotation and cyclopropanation are competitive.

Explanation for Table 5. Cyclopropanation reactions are carried out with cations Nos. 1, 2, 6, 11, 22 to 28 from Table 6. Using the standard Methods A to H, the cations are generated in situ without isolation. In some cases (Nos. 6, 22, and 28) isolated cations are also used. The resulting cyclopropane derivatives are purified by distillation or estimated by GC.

A: The sulfonium salt $[FpCH_2S(CH_3)_2]BF_4$ and the corresponding alkene (2 : 1 mole ratio) in 1,4-dioxane is held at reflux temperature for 12 to 14 h under an atmosphere of N_2.

B: $FpCH(CH_3)SC_6H_5$ in CH_2Cl_2 is treated with the appropriate alkene (1 : 1 mole ratio) and CH_3SO_3F and the mixture stirred for 12 to 20 h at room temperature.

C: Solid $FpCH(CH_3)SC_6H_5$ and solid $[O(CH_3)_3]BF_4$ (1 : 1.25 mole ratio) under N_2 atmosphere are added to a mixture of dry CH_2Cl_2 and the appropriate alkene.

D: $FpCH_2OCH_3$ and HBF_4 at 0 °C are allowed to react in the presence of the appropriate alkene.

E: The carbene cation $[Fp{=}CHCH_3]^+$ is generated in situ from $FpCH(OCH_3)CH_3$ and $(CH_3)_3SiOSO_2CF_3$ in CH_2Cl_2 at − 78 °C with a fivefold excess of the appropriate styrene. Other carbene-to-alkene ratios are cited in Table 5.

F: The corresponding carbene precursor $FpCH(OCH_3)R$ is treated as described in E but at 0 °C (0.5 h) while bubbling propene through the solution and in the presence of 0.1 equivalent of $N(C_2H_5)_3$.

G: The corresponding vinyl complex $FpC({=}CH_2)R$ in CH_2Cl_2 is allowed to react with 2 equivalents of HBF_4 in ether at − 78 °C. After addition of the appropriate alkene, the mixture is warmed to room temperature and quenched with saturated aqueous $NaHCO_3$.

H: The crystalline salts $[Fp{=}CRR']BF_4$ or $[Fp{=}CRR']PF_6$ of more than 95% purity are reacted with an excess of alkene between 0 and − 15 °C or at − 78 °C, typically for 1 to 2 h.

Low yields of cyclopropane derivatives are obtained with the carbene cation $[Fp{=}C(CH_3)_2]^+$ due to a competing rearrangement of this cation to the cationic olefin complex $[Fp(\eta^2\text{-}CH_2{=}CHCH_3)]^+$.

Compounds with a C=C double bond in position 6 (Table 6, Nos. 7, 8, and 19) undergo an intramolecular cyclopropanation reaction to generate norcarane derivatives. Intramolecular stereoselective carbene C–H inertion in the C-5

Table 5. Cyclopropanation Reactions of the Cationic Complexes $[Fp{=}CRR']^+$, Nos. 1, 2, 6, 11, and 22 to 28 from Table 6, with Various Alkenes

No.	alkene	cyclopropane	reaction type (yield) conditions and remarks
alkylidene source is $[\mathbf{Fp{=}CH_2}]^+$ (Table 6, No. 1)			
1	 H_3C CH_3	 H_3C CH_3	D
2	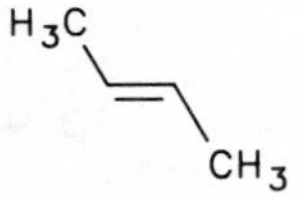 H_3C CH_3	 H_3C CH_3	D
3	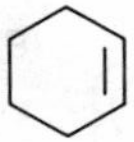	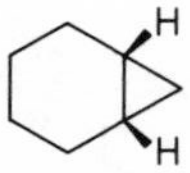 H H	D (46%)
4	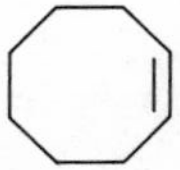	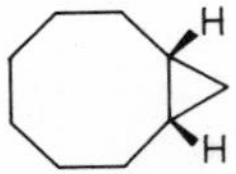 H H	A (92%), B (85%)
5	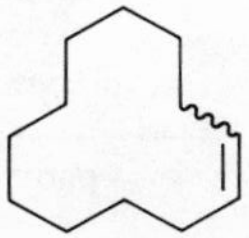	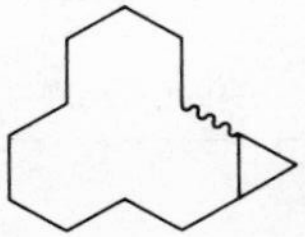	A (90%); alkene and product are 1:1 mixtures of *cis* and *trans* isomers

Table 5. Continued

No.	alkene	cyclopropane	reaction type (yield) conditions and remarks
6	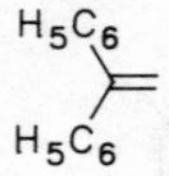 H_5C_6 H_5C_6	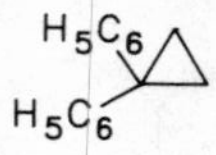 H_5C_6 H_5C_6	A (99%)
7	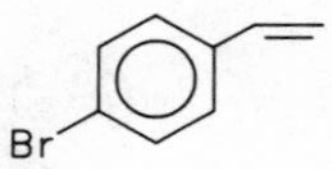 Br	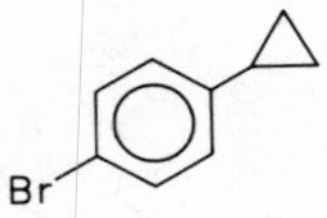 Br	A (> 90%) in CH_3NO_2 as solvent
8	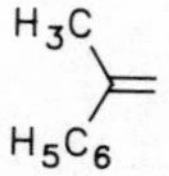 H_3C H_5C_6	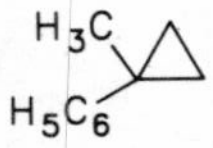 H_3C H_5C_6	A (64%)
9	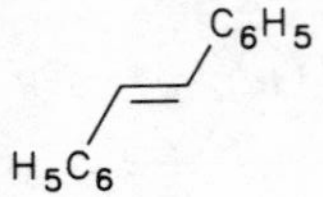 C_6H_5 H_5C_6	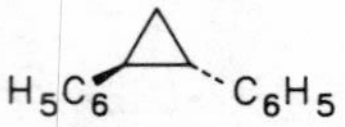 H_5C_6 C_6H_5	A (64%)
10	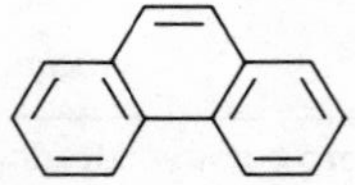	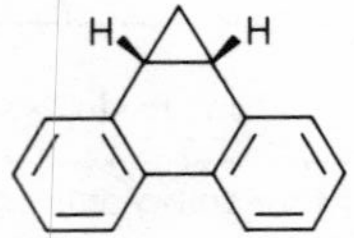 H H	A (82%); 38% alkene conversion

11	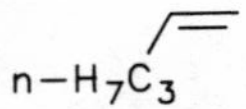 n−H_7C_3	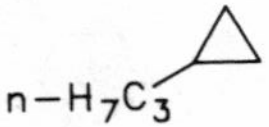 n−H_7C_3	A (70%)
12	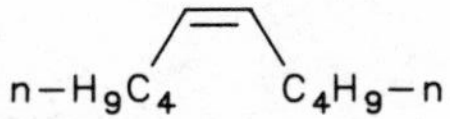 n−H_9C_4 C_4H_9−n	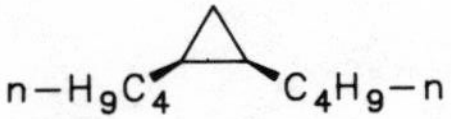 n−H_9C_4 C_4H_9−n	A (87%); 59% alkene conversion
13	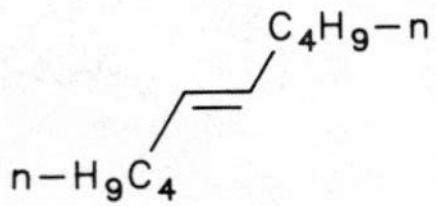 n−H_9C_4 C_4H_9−n	 n−H_9C_4 C_4H_9−n	A (67%); 49% alkene conversion
14	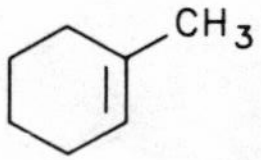 CH_3	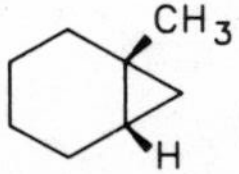 CH_3 H	A (26%)
15	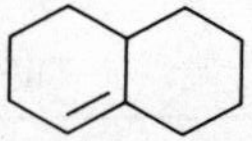	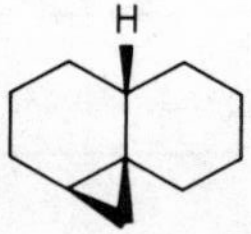 H	A
16	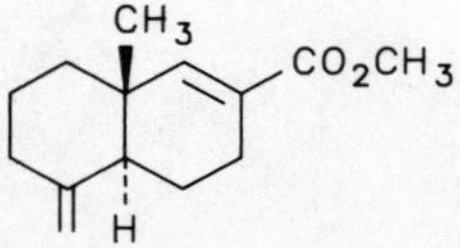 CH_3 CO_2CH_3 H	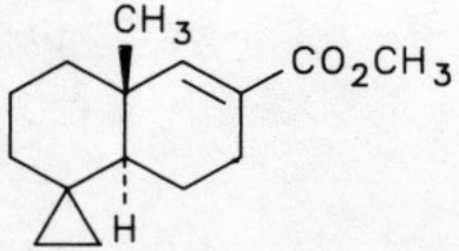 CH_3 CO_2CH_3 H	A (86%)

Table 5. Continued

No.	alkene	cyclopropane	reaction type (yield) conditions and remarks
17	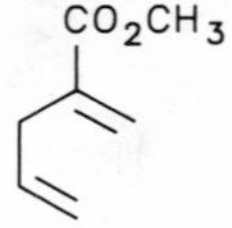 CO_2CH_3	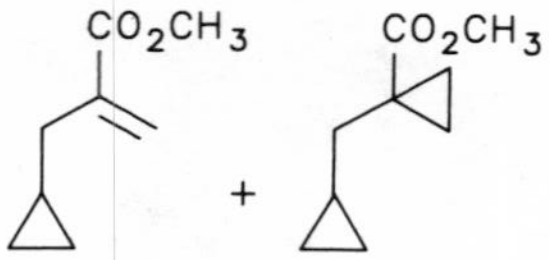 CO_2CH_3 + CO_2CH_3	A (67%)
18	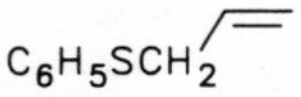 $C_6H_5SCH_2$	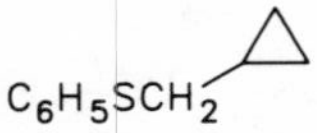 $C_6H_5SCH_2$	A (60%) along with 40% $C_6H_5SCH_3$
19	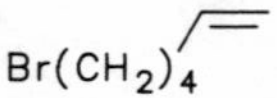 $Br(CH_2)_4$	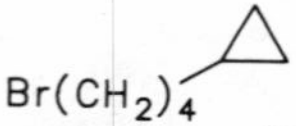 $Br(CH_2)_4$	A (55%)
20	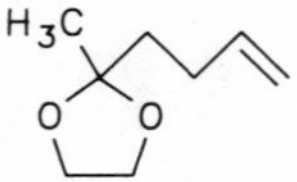 H_3C O O	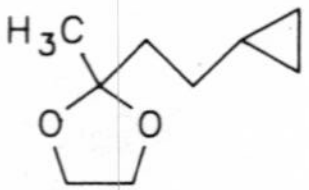 H_3C O O	A (22%)

alkylidene source: [**Fp=CHCH$_3$**]$^+$ (Table 6, No. 2)

No.	Alkene	Product	Remarks
21		H, CH_3, H	B (70%), C (73%) E (60, 25, 87%); alkene/carbene = 2/1, 1/1, 1/2; *cis*/*trans* > 25/1 G; *cis* isomer only
22		H, CH_3, H	B (70%); 49% alkene conversion
23	n−$H_{17}C_8$	n−$H_{17}C_8$, CH_3 *cis* : *trans* = 1 : 1	B (70%), C (87%); 55% alkene conversion
24	C_4H_9−n, n−H_9C_4	CH_3, C_4H_9−n, n−H_9C_4	C (48%); 29% alkene conversion

Table 5. Continued

No.	alkene	cyclopropane	reaction type (yield) conditions and remarks
25	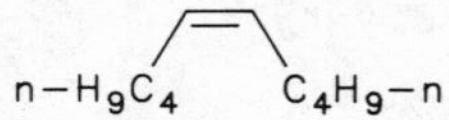 n–H_9C_4 C_4H_9–n	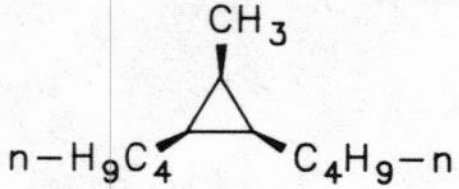 CH_3 n–H_9C_4 C_4H_9–n	B (44%), C (66%); 65% alkene conversion
26	H_5C_6	CH_3 H_5C_6	B, C (67%);
			E (75, 47, 91%); alkene/carbene = 2/1, 1/1, 1/2; *cis/trans* = 4.7/1 E (70%); *cis/trans* = 6.5/1;
27	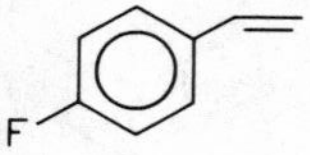 F	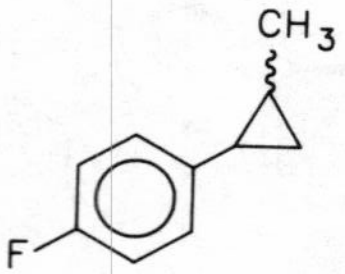 CH_3 F	E (83%); *cis/trans* = 7/1
28	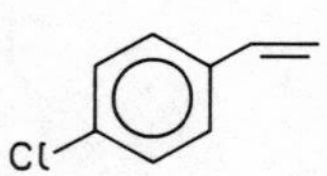 Cl	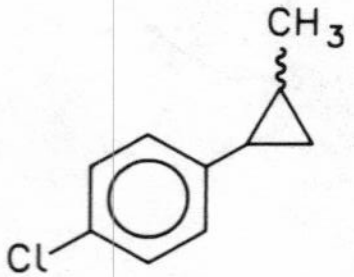 CH_3 Cl	E (35%); *cis/trans* = 6/1

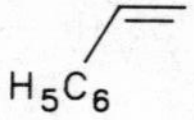

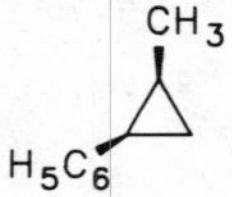

29	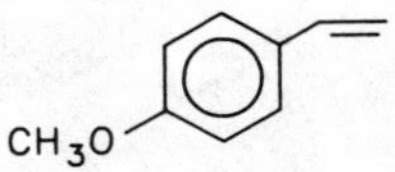 CH_3O	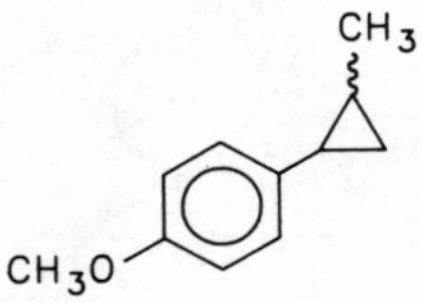 CH_3, CH_3O	E (50%); *cis/trans* = 0.9/1
30	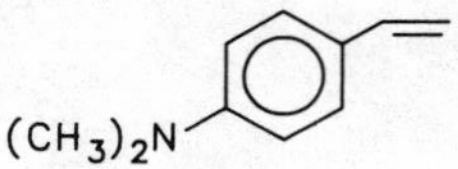 $(CH_3)_2N$	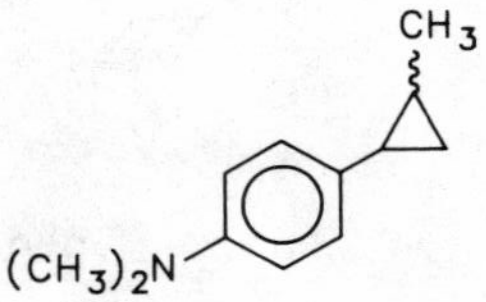 CH_3, $(CH_3)_2N$	E
31	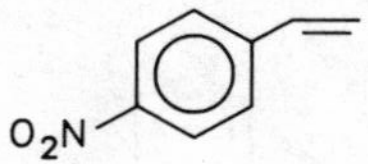 O_2N	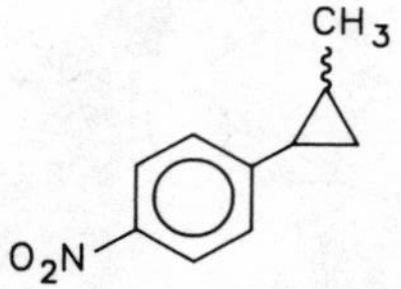 CH_3, O_2N	E
32	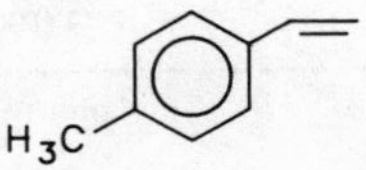 H_3C	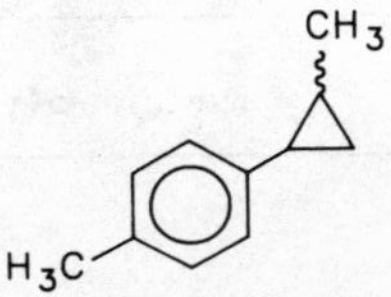 CH_3, H_3C	E (75%); *cis/trans* = 6/1

Table 5. Continued

No.	alkene	cyclopropane	reaction type (yield) conditions and remarks
33	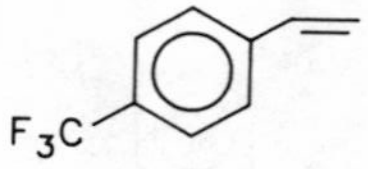 F_3C	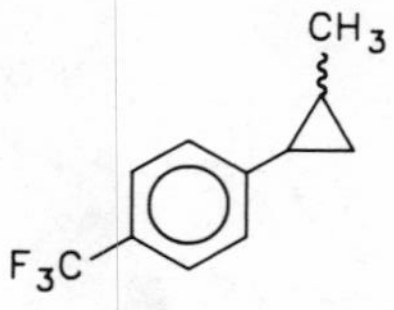 CH_3, F_3C	E (10%); *cis/trans* = 5/1
34	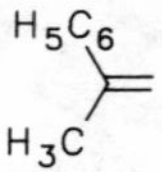 H_5C_6, H_3C	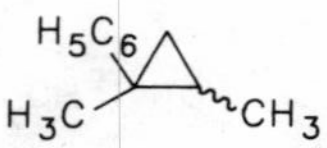 H_5C_6, H_3C, CH_3	B (58%), C (65%); Z/E = 41/17 G; mainly *Z* isomer
35	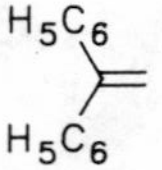 H_5C_6, H_5C_6	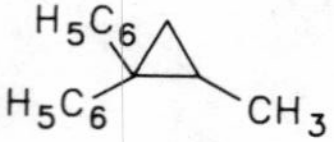 H_5C_6, H_5C_6, CH_3	B (81%); 59% alkene conversion
36	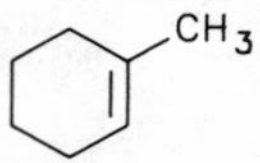 CH_3	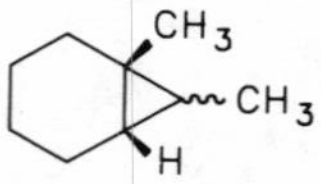 CH_3, CH_3, H	C (66%); 34% alkene conversion, *endo/exo* > 5.6/1

37	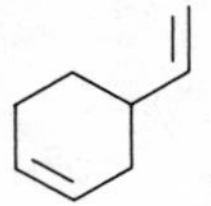	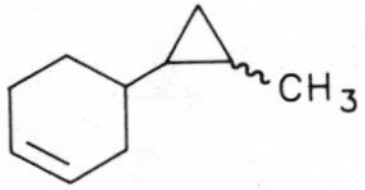 CH_3	C (80%); 50% alkene conversion *cis/trans* = 1/1
38	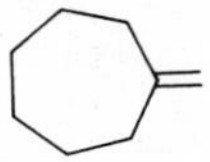	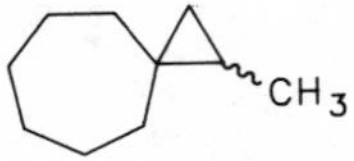 CH_3	C
39	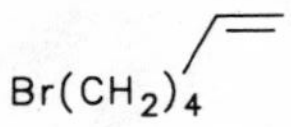 $Br(CH_2)_4$	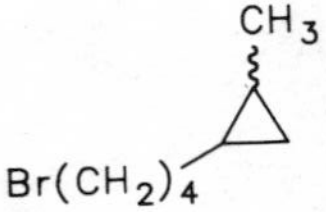 CH_3, $Br(CH_2)_4$	C (84%); 59% alkene conversion, *cis/trans* = 0.9/1
40	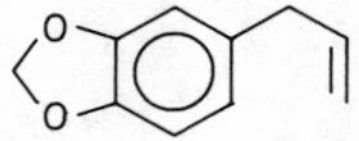 O, O	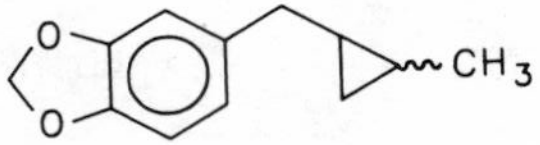 O, O, CH_3	C (46%); *cis/trans* = 1/1.3
41	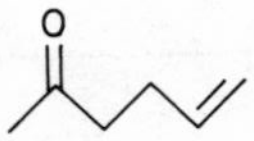 O	 O, CH_3	C (27%); 77% alkene conversion, *cis/trans* = 1/1.7
42	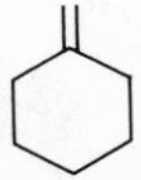	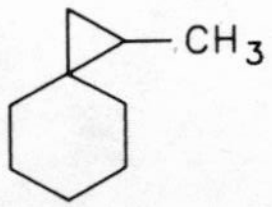 CH_3	E (86, 67, 99%); alkene/carbene = 2/1, 1/1, 1/2

Table 5. Continued

No.	alkene	cyclopropane	reaction type (yield) conditions and remarks
43	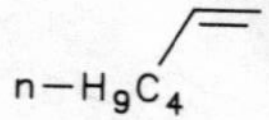 $n{-}H_9C_4$	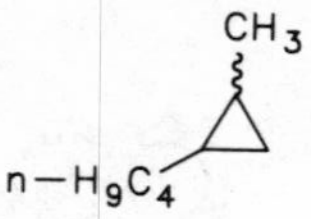 CH_3, $n{-}H_9C_4$	E (48, 28%); alkene/carbene = 4/1, 2/1; *cis*/*trans* = 1/1
44	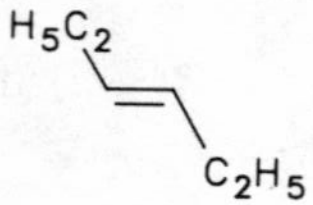 H_5C_2, C_2H_5	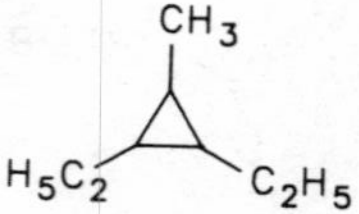 CH_3, H_5C_2, C_2H_5	E (3.4%); alkene/carbene = 2/1
45	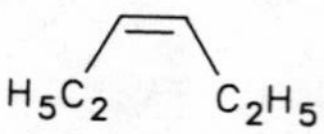 H_5C_2, C_2H_5	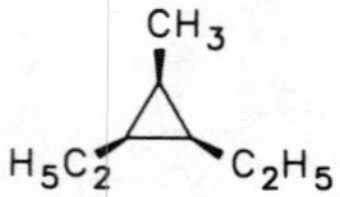 CH_3, H_5C_2, C_2H_5	E (58%); alkene/carbene = 2/1; all *cis* is main product (> 50/1)
46	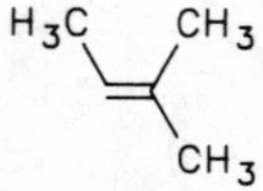 H_3C, CH_3, CH_3	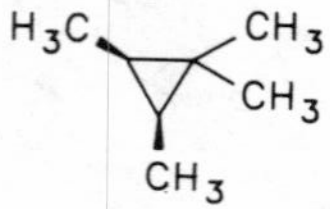 H_3C, CH_3, CH_3, CH_3	E (52%); alkene/carbene = 2/1; *cis* is main product (> 50/1)

alkylidene source: $[\mathbf{Fp{=}CHCH{=}C(CH_3)_2}]^+$ (Table 6, No. 6)

No.			
47	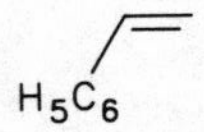	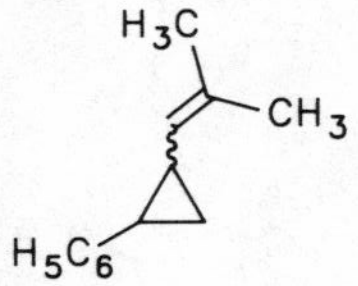	H (45%); alkene/isolated No. 6 = 2/1; *cis*/*trans* = 1/2
48	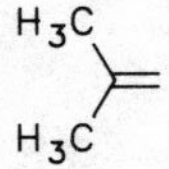	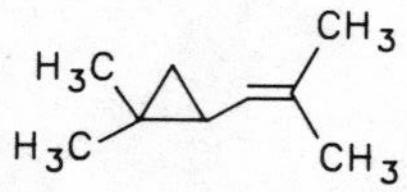	H (56%); alkene/No. 6 = 3/1
49	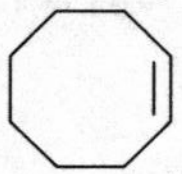	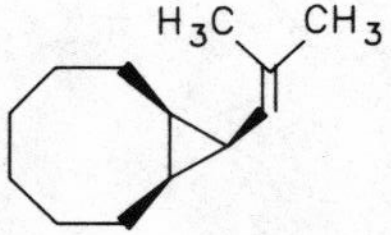	H (37%); alkene/No. 6 = 3/1; *syn* isomer

alkylidene source: $[\mathbf{Fp{=}CHC_3H_7\text{-}cyclo}]^+$ (Table 6, No. 11)

No.			
50	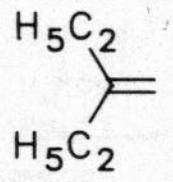	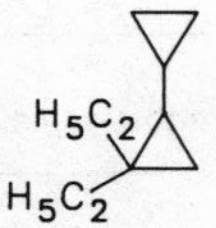	E (64%); alkene/carbene = 2/1

Table 5. Continued

No.	alkene	cyclopropane	reaction type (yield) conditions and remarks
51	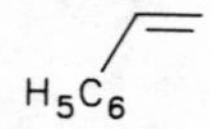 H_5C_6	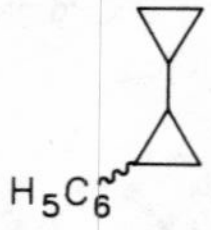 H_5C_6	E (66%); alkene/carbene = 2/1; *cis/trans* = 1/3.4
alkylidene source: **[Fp=CHC$_6$H$_5$]$^+$** (Table 6, No. 22)			
52	$H_2C = CH_2$	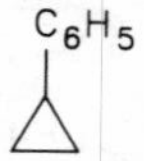 C_6H_5	H (75%)
53	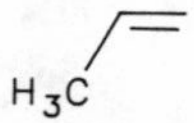 H_3C	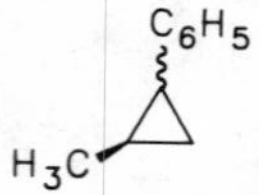 C_6H_5, H_3C	H (90%); *cis/trans* = 7.8/1
54	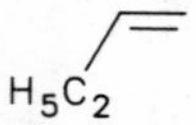 H_5C_2	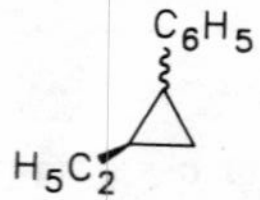 C_6H_5, H_5C_2	H (75%); *cis/trans* = 6.5/1

55	$i-H_7C_3$	C_6H_5, $i-H_7C_3$	H (76%); *cis/trans* = 4.6/1
56	H_5C_6	C_6H_5, H_5C_6	H (88%); *cis/trans* = 100/1
57	CH_3, H_3C	C_6H_5, CH_3, H_3C	H (89%)
58	H_3C, CH_3	C_6H_5, H_3C, CH_3	H (93%); *cis/trans* = >100/1
59		C_6H_5	H (78%); *cis/trans* = >200/1

Table 5. Continued

No.	alkene	cyclopropane	reaction type (yield) conditions and remarks
60	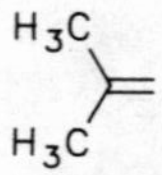 H_3C, H_3C	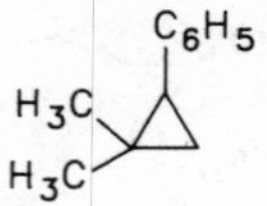 C_6H_5, H_3C, H_3C	H (82%)
61	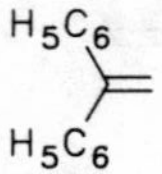 H_5C_6, H_5C_6	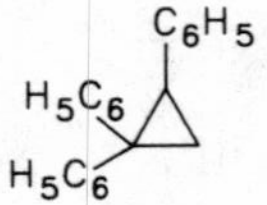 C_6H_5, H_5C_6, H_5C_6	H (75%)
62	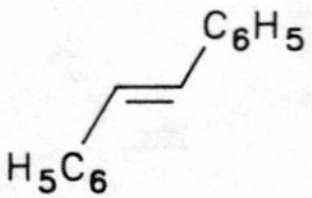 C_6H_5, H_5C_6	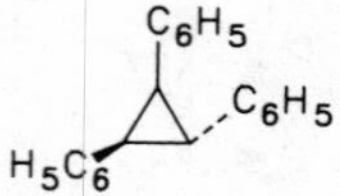 C_6H_5, C_6H_5, H_5C_6	H (96%)
63	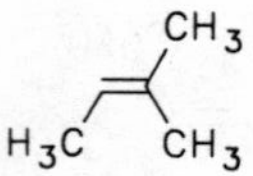 CH_3, H_3C, CH_3	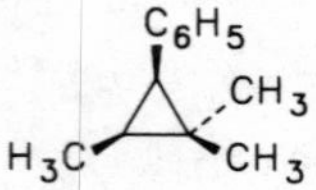 C_6H_5, CH_3, H_3C, CH_3	H (91%); *cis/trans* = >50/1

64	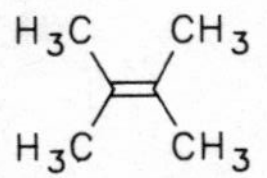 H_3C CH_3 H_3C CH_3	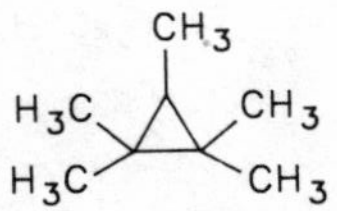 CH_3 H_3C CH_3 H_3C CH_3	**H** (59%)

alkylidene source is [**Fp=CHC$_6$H$_4$F**-*p*]$^+$ (Table 6, No. 23)

65	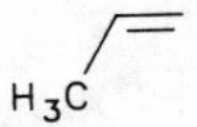 H_3C	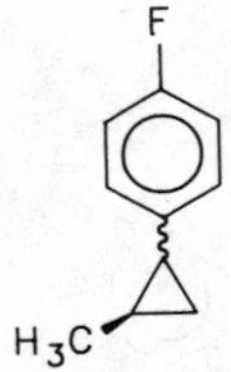 F H_3C	F (80%); *cis/trans* = 9.4/1

alkylidene source: [**Fp=CHC$_6$H$_4$OCH$_3$**-*p*]$^+$ (Table 6, No. 24)

66	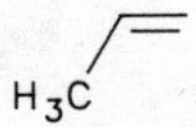 H_3C	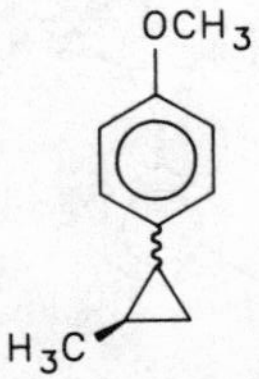 OCH_3 H_3C	F (85%); *cis/trans* = 2.0/1

Table 5. Continued

No.	alkene	cyclopropane	reaction type (yield) conditions and remarks
alkylidene source: $[\mathbf{Fp{=}CHC_6H_4OCH_3}\text{-}m]^+$ (Table 6, No. 25)			
67	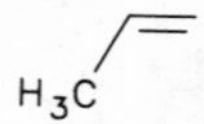 H_3C	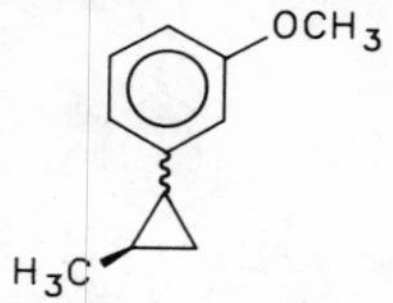 OCH_3, H_3C	F (81%); *cis/trans* = 9.6/1
alkylidene source: $[\mathbf{Fp{=}CHC_6H_4CH_3}\text{-}p]^+$ (Table 6, No. 26)			
68	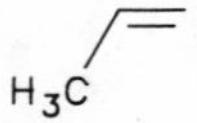 H_3C	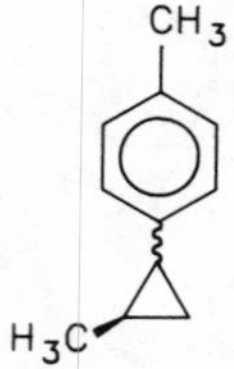 CH_3, H_3C	F (60%); *cis/trans* = 9.0/1
alkylidene source: $[\mathbf{Fp{=}CHC_6H_4CF_3}\text{-}p]^+$ (Table 6, No. 27)			
69	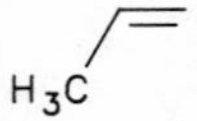 H_3C	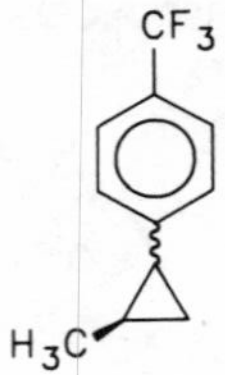 CF_3, H_3C	F (67%); *cis/trans* = 7.3/1

alkylidene source: $[\mathbf{Fp{=}C(CH_3)_2}]^+$ (Table 6, No. 28)

No.	Alkene	Product	Yield; conditions
70	$(H_3C)_2C{=}CH_2$	1,1-$(H_3C)_2$-2,2-$(CH_3)_2$-cyclopropane	H (20%); 6-fold excess at 0 °C G (40%); large excess alkene
71	$H_5C_6CH{=}CH_2$	2-H_5C_6-1,1-$(CH_3)_2$-cyclopropane	H (45%); as No. 70 G (40%); as No. 70
72	$n\text{-}H_{19}C_9CH{=}CH_2$	2-$n\text{-}H_{19}C_9$-1,1-$(H_3C)(CH_3)$-cyclopropane	G (40%)
73	$(H_3C)(H_5C_6)C{=}CH_2$	1-H_3C-1-H_5C_6-2,2-$(CH_3)_2$-cyclopropane	G (40%)
74	$(H_5C_6)_2C{=}CH_2$	1,1-$(H_5C_6)_2$-2,2-$(CH_3)_2$-cyclopropane	G (40%)

position (Table 6, Nos. 18 to 21) is also possible. The results of these reactions are described in further information for the individual compounds.

Preparation, Properties, and Other Reactions. The compounds described in Table 6 can be prepared or generated in situ according to the following methods:

Method I: Reaction of an appropriate ether complex of the general type $FpCR_2ER'$ (E = O, S) with electrophiles such as $[O(CH_3)_3]BF_4$, $(CH_3)_3SiOSO_2CF_3$, HBF_4, or $[C(C_6H_5)_3]PF_6$.
a. Starting material is $FpCR_2OR'$.
b. Starting material is $FpCR_2SR'$.

Method II: From the corresponding complex $FpCR_2Cl$ and $AgPF_6$ or $AgBF_4$ in CH_2Cl_2 at − 78 °C.

Method III: From Na[Fp] and cyclopropylium salt.

Method IV: Hydride abstraction from Fp-substituted derivatives of 1,3,5-cycloheptatriene with $[C(C_6H_5)_3]PF_6$.

Explanation for Table 6. Only those spectroscopic data are reported which are important for characterizing the carbene ligand. In the NMR spectra the Cp ligand appears as a singlet with δ values between 5.50 and 5.90 ppm (^1H) and 89 to 94 ppm (^{13}C). Two $\nu(CO)$ frequencies are observed in the IR spectra, separated by about 40 to 50 wave numbers. The band at the higher frequencies ranges between 2040 and 2080 cm^{-1} and the band at the lower frequencies is found between 1975 and 2035 cm^{-1}.

Further Information

$[Fp{=}CH_2]X$ (Table **6**, No. **1**; X = BF_4, PF_6) is too unstable to be isolated. It is considered to be an intermediate in the reaction of $FpCH_2OCH_3$ with HBF_4 to give $FpCH_3$ and $[Fp(\eta^2\text{-}CH_2{=}CH_2)]^+$ or of $FpCH_2Cl$ with $AgPF_6$ to give a mixture of $[C_5H_5Fe(CO)_3]^+$ and $[Fp(\eta^2\text{-}CH_2{=}CH_2)]^+$. Addition of $AgBF_4$ to $FpCH_2Cl$ at − 60 °C precipitates AgCl; addition of cyclohexene and warming the mixture to room temperature gives small amounts of norcarane. For CH_2 transfer from the stable carbene precursor $[FpCH_2S(CH_3)_2]BF_4$, see Table 5. In the electron-impact mass spectrum of $FpCH_2OCH_3$ the presence of protonating agents such as $[NH_4]^+$, $[C_2H_5]^+$, $[CH_5]^+$, or $[H_3]^+$ causes a dramatic increase in abundance of the ions $[Fp{=}CH_2]^+$ and $[C_5H_5(CO)Fe{=}CH_2]^+$. Addition of cyclohexene in the gase phase produces $[Fp]^+$ and norcarane. Addition of bases, B (B = NH_3, CH_3CN, CD_3CDO), generates the cations $[FpCH_2B]^+$ and no ligand displacement is observed. Hydride abstraction to yield $FpCH_3$ is also observed in the gas phase.

The cation generated either by Method Ia or II in CH_2Cl_2 solution at − 80 °C under 1 bar of CO, followed by immediate pressurization (5.5 bar CO) gives the ketene cation XIII with an unsymmetrically bonded (η^2-C,C) structure in essentially quantitative yield. Quenching this mixture with CH_3OH/

Table 6. Cationic Carbene Complexes of the Type $[Fp{=}CRR']^+$. Further information on numbers preceded by an asterisk is given at the end of the table.

No.	Fe=CRR′	anion, method of preparation (yield) properties and remarks
*1	$Fe{=}CH_2$	BF_4^- salt, Ia PF_6^- salt, II
*2	$Fe{=}CHCH_3$	FSO_3^- salt $CF_3SO_3^-$ salt, Ia PF_6^- salt
*3	$Fe{=}CHC_2H_5$	BF_4^- salt, Ia
4	$Fe{=}CHCH(CH_3)_2$	$CF_3SO_3^-$ salt, Ia isomerizes via intramolecular hydride migration to give $[Fp(\eta^2\text{-}CH_2{=}C(CH_3)_2]^+$
*5	$Fe{=}CHCH{=}CHC_2H_5$	$CF_3SO_3^-$ salt ^1H NMR (CD_2Cl_2 at $-100\,°C$): 17.16 (CH=Fe)
*6	$Fe{=}CHCH{=}C(CH_3)_2$	BF_4^- salt; red orange solid ^1H NMR (CD_2Cl_2 at $-55\,°C$): 2.22 (s, CH_3), 8.21 (d, CH=C; J = 14.7), 15.96 (d, CH=Fe) ^{13}C NMR (CD_2Cl_2 at $-60\,°C$): 154.0 (CH=), 178.8 (C=), 207.8 (CO), 316.7 (CH=Fe; J(H,C) = 148) $CF_3SO_3^-$ salt ^1H NMR (CD_2Cl_2 at $-60\,°C$): 8.33 (d, CH=C; J = 14.3), 15.88 (d, CH=Fe)
*7	$Fe{=}CH(CH_2)_4CH{=}CH_2$	BF_4^- salt, Ib; intermediate
*8	Fe=C(H)–CH2CH2CH2–(2-methylenecyclohexyl) [structure: Fe = C, H, CH2]	BF_4^- salt, Ib; intermediate
*9	$Fe{=}CHC(CH_3)_3$	BF_4^- salt, Ia ^1H NMR(CD_2Cl_2 at $-95\,°C$): 18.1 (CH)
10	$Fe{=}CHC(CH_3)_2C_6H_5$	BF_4^- salt, Ia rearranges to the cationic olefin complex $[Fp(C_6H_5CH{=}C(CH_3)_2)]BF_4$

Table 6. Continued

No.	Fe=CRR′	anion, method of preparation (yield) properties and remarks
*11	$Fe{=}CHC_3H_5$-*c*	$CF_3SO_3^-$ salt, Ia ^1H NMR (CD_2Cl_2 at − 78 °C): 2.63 (m, 2H of CH_2CH_2), 2.79 (m, 2H of CH_2CH_2), 4.09 (m, CH of C_3H_5), 15.47 (d, CH=Fe; J = 14) ^{13}C NMR (CD_2Cl_2 at − 78 °C): 37.2 (C_2H_4), 61.3 (CH of C_3H_5), 207.8 (CO), 365.4 (CH=Fe)
12	$Fe{=}CHC_4H_7$-*c*	$CF_3SO_3^-$ salt, Ia short-lived species; no alkylidene transfer observed
*13	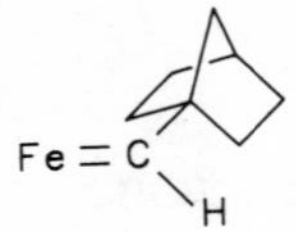 Fe=C H	BF_4^- salt, Ia ^{13}C NMR (CD_2Cl_2 at − 88 °C): 85.5 (C-1?), 205.7 (CO), 394.5 (CH=Fe)
*14	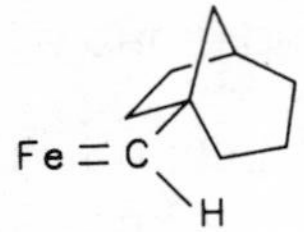 Fe=C H	BF_4^- salt, Ia ^1H NMR (CD_2Cl_2 at − 78 °C): 17.9 (CH=Fe) ^{13}C NMR (CD_2Cl_2 at − 78 °C): 397 (CH=Fe)
*15	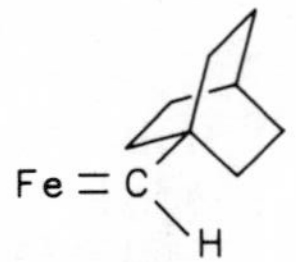 Fe=C H	BF_4^- or $CF_3SO_3^-$ salt, Ia ^1H NMR: 17.9 (CH=Fe) ^{13}C NMR (CD_2Cl_2 at − 78 °C?): 397 (CH=Fe)
*16	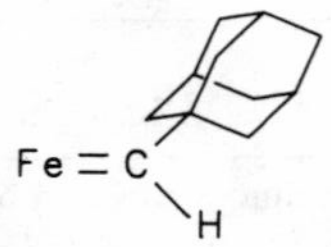 Fe=C H	BF_4^- or $CF_3SO_3^-$ salt, Ia ^1H NMR: 17.8 (CH=Fe) ^{13}C NMR (CD_2Cl_2 at − 78 °C): 397 (CH=Fe)

No.	Compound	Remarks
*17	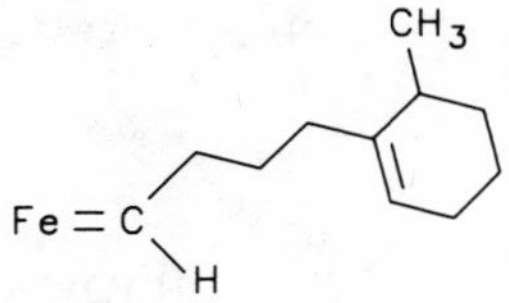	BF_4^- salt, Ib; intermediate
*18	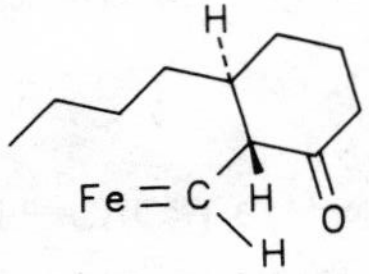	BF_4^- salt, Ib; intermediate
*19	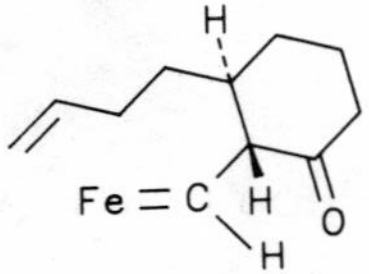	BF_4^- salt, Ib; intermediate
*20	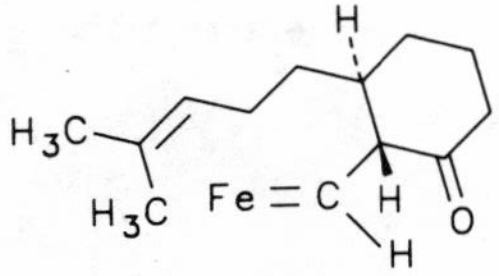	BF_4^- salt, Ib; intermediate
*21	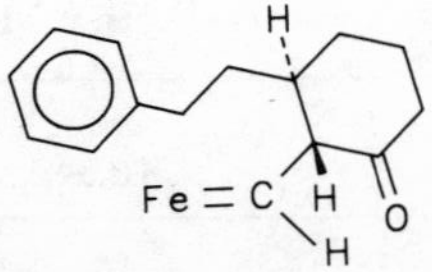	BF_4^- salt, Ib; intermediate

Table 6. Continued

No.	Fe=CRR′	anion, method of preparation (yield) properties and remarks
*22	$Fe{=}CHC_6H_5$	$CF_3SO_3^-$ salt, Ia ^1H NMR (CF_3SO_3H or CD_2Cl_2 at −25 °C): 16.86 (s, CH=Fe) ^{13}C NMR (CF_3SO_3H or CD_2Cl_2 at −40 °C): 151.8 (C-1), 206.5 (CO), 342.4 (CH=Fe) PF_6^- salt, Ia (77%); orange $CF_3CO_2^-$ salt, Ia FSO_3^- salt, Ia
*23	$Fe{=}CHC_6H_4F$-4	$CF_3SO_3^-$ salt, Ia ^1H NMR (CD_2Cl_2 at −40 °C): 16.78 (HC=Fe) ^{13}C NMR (CD_2Cl_2 at −40 °C): 150.8 (C-1), 170.7 (C-4), 207.9 (CO), 338.8 (C=Fe)
*24	$Fe{=}CHC_6H_4OCH_3$-4	$CF_3SO_3^-$ salt, Ia ^1H NMR (CD_2Cl_2 at −40 °C): 15.24 (s, HC=Fe); for variable-temperature spectrum, see "Further information" ^{13}C NMR (CD_2Cl_2 at −40 °C): 149.6 (C-1), 172.9 (C-4), 209.0 (CO), 310.0 (C=Fe) BF_4^- salt, Ia
*25	$Fe{=}CHC_6H_4OCH_3$-3	$CF_3SO_3^-$ salt, Ia
*26	$Fe{=}CHC_6H_4CH_3$-4	$CF_3SO_3^-$ salt, Ia ^1H NMR (CD_2Cl_2 at 0 °C): 16.61 (s, CH=Fe); for the variable-temperature spectrum, see "Further information" ^{13}C NMR (CD_2Cl_2 at −40 °C): 152.1 (C-1), 157.4 (C-4), 208.3 (CO), 335.5 (C=Fe)
*27	$Fe{=}CHC_6H_4CF_3$-4	$CF_3SO_3^-$ salt, Ia
*28	$Fe{=}C(CH_3)_2$	BF_4^- salt, Ia; yellow ^1H NMR (CD_2Cl_2 at −40 °C): 3.73 (s, CH_3) $CF_3SO_3^-$ salt, Ia; yellow ^1H NMR (CD_2Cl_2 at −40 °C): 3.69 (s, CH_3) ^{13}C NMR (CD_2Cl_2 at −60 °C): 61.4 (CH_3), 206.8 (CO), 419.0 (C=Fe)
*29	$Fe{=}C(CH_3)C_6H_5$	BF_4^- salt; not isolated

No.	Formula	Properties
30	Fe, C_4H_9–t, C_4H_9–t	BF_4^- salt, III (51%); yellow, m.p. 189 to 192 °C (dec.) ^{1}H NMR ($CDCl_3$): 1.49 (s, CH_3)
*31	Fe, C_6H_5, C_6H_5	BF_4^- salt, III; pale yellow, m.p. 185 °C (dec.) ClO_4^-, Br^-, and $SbCl_6^-$ salt; pale yellow, m.p.'s. 183, 175 to 185, and 188 °C (all dec.)
32	Fe, $N(C_2H_5)_2$, $N(C_2H_5)_2$	ClO_4^- salt, III (60%); pale yellow, m.p. 108 to 109 °C (dec.) ^{1}H NMR ($CDCl_3$): 1.33 (t, CH_3), 3.55 (m, CH_2)
*33	Fe, $N(C_3H_7$–i$)_2$, $N(C_3H_7$–i$)_2$	ClO_4^- salt, III (42%); pale yellow, m.p. 178 to 179 °C (dec.) ^{1}H NMR ($CDCl_3$): 1.41 (d, CH_3), 4.15 (m, CH)
*34	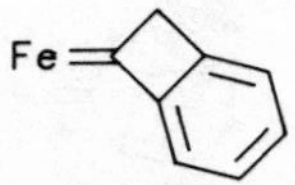 Fe	PF_6^- salt, IV (90%); deep orange red ^{1}H NMR (CD_3NO_2): 4.65 (s, CH_2) ^{13}C NMR: 352 (C=Fe)
*35	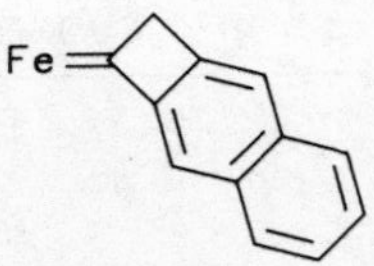 Fe	PF_6^- salt, IV (34%); m.p. 161 to 163 °C (dec.) ^{1}H NMR (CD_3NO_2): 4.86 (s, CH_2)

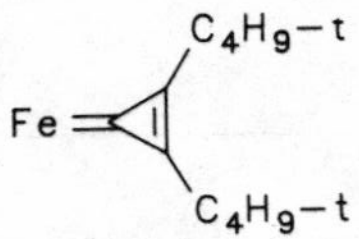

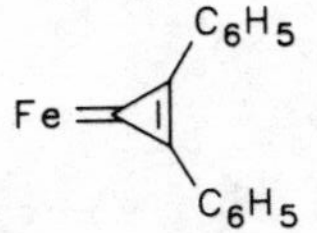

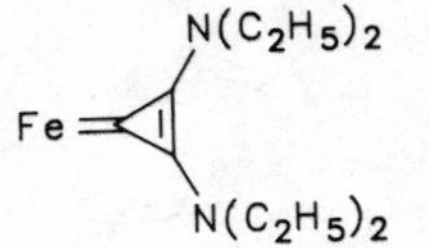

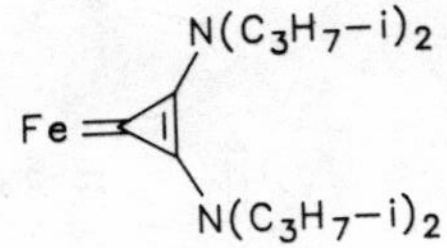

Table 6. Continued

No.	Fe=CRR′	anion, method of preparation (yield) properties and remarks
*36	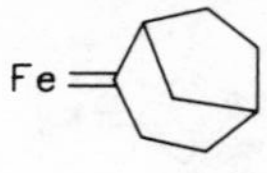	BF_4^- salt ^{13}C NMR (CD_2Cl_2 at −88 °C): 206.8, 207.3 (CO), 420.2 (C=Fe)
*37	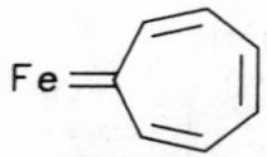	PF_6^- salt, IV (90%); yellow orange, m.p. 180 to 180.5 °C (dec.) ^1H NMR (CD_3COCD_3 at 0 °C): 7.94 to 8.3 (m, H-3, 6), 8.48 to 8.74 (m, H-4, 5), 10.01 (d, H-2, 7) ^{13}C NMR (CD_3COCD_3 at 0 °C): 138.2 (C-4, 5), 148.3 (C-3, 6), 170.0 (C-2, 7), 212.8 (CO), 242.3 (C=Fe)
*38	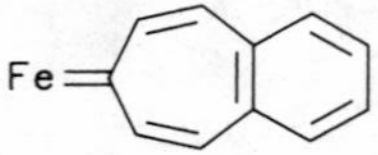	PF_6^- salt, IV (93%); red brown, m.p. 177 to 178 °C (dec.) ^1H NMR (CD_3CN at 0 °C): 8.26 (d, H-3, 10), 9.58 (d, H-2, 11) ^{13}C NMR (CD_3NO_2 at 0 °C): 212.7 (CO), 265.9 (C=Fe)
*39	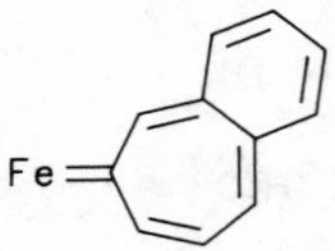	PF_6^- salt, IV (88%); red, m.p. 176 to 177 °C (dec.) ^{13}C NMR (CD_3NO_2 at 0 °C): 201 (C=Fe), 215 (CO)
*40	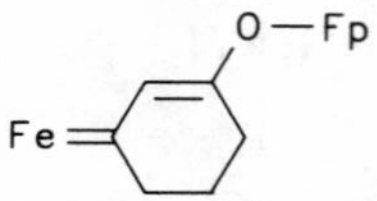	PF_6^- salt ^{57}Fe-γ (80 K): $\delta = 0.14$, $\Delta = 1.92$ (FeO), $\delta = -0.09$, $\Delta = 1.83$

$N(C_2H_5)_3$ produces $FpCH_2COOCH_3$ in 93% yield via the ketene semiacetal intermediate XIV along with 3% $[C_5H_5Fe(CO)_3]^+$ and 3% $[Fp(\eta^2\text{-}CH_2{=}CH_2)]^+$. An aqueous workup gives 88% of the acid $FpCH_2COOH$. With ^{13}CO-labeled $[Fp{=}CH_2]^+$, all labeled CO remains in the terminal carbonyl groups. Treatment of the freshly prepared PF_6^- salt at $-78\,°C$ with $S(CH_3)_2$ (3 min, excess) gives the ylide complex $[FpCH_2S(CH_3)_2]PF_6$ in 53% yield. With $FpCH{=}CH_2$ (excess) at $-78\,°C$ in CH_2Cl_2 solution, the cationic bimetallic complex XV (R = H, 57%) is produced along with $[Fp(\eta^2\text{-}CH_2{=}CH_2)]PF_6$ and $[C_5H_5Fe(CO)_3]PF_6$ (each 8%). A similar reaction with $FpCH_2CH{=}CH_2$ produces a mixture of $[Fp(\eta^2\text{-}CH_2{=}CH_2]PF_6$ (16%), $[C_5H_5Fe(CO)_3]PF_6$ (19%), $[Fp(\eta^2\text{-}CH_2{=}CHCH_3)]PF_6$ (18%), XVI (41%), and small amounts of Fp_2, $FpCH_3$, and the allyl complex $C_5H_5(CO)Fe(\eta^3\text{-}CH_2CHCH_2)$. Treatment of the title complex with the acyl complex $FpCOCH_3$ gives a mixture of the carbene complexes $[Fp{=}C(CH_3)OFp]PF_6$ (18%), probably $[Fp]PF_6$ (6%), $[Fp(\eta^2\text{-}CH_2{=}CH_2)]PF_6$ (32%), and $[C_5H_5Fe(CO)_3]PF_6$ (11%); similarly, $C_5H_5(CO)(P(C_6H_5)_3)FeCOCH_3$ generates the carbene complexes $[C_5H_5(CO)(P(C_6H_5)_3)Fe{=}C(CH_3)OCH_2Fp]PF_6$ (73%) and $[C_5H_5(CO)(P(C_6H_5)_3)Fe{=}C(CH_3)OFp]PF_6$ (15%) along with by-products as shown above.

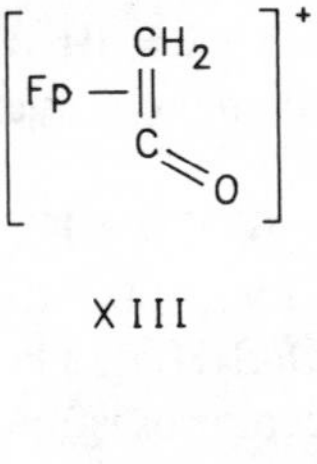

XIII

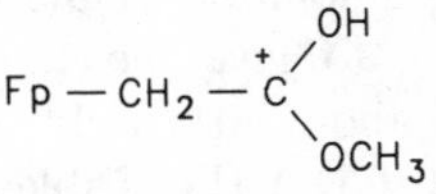

XIV

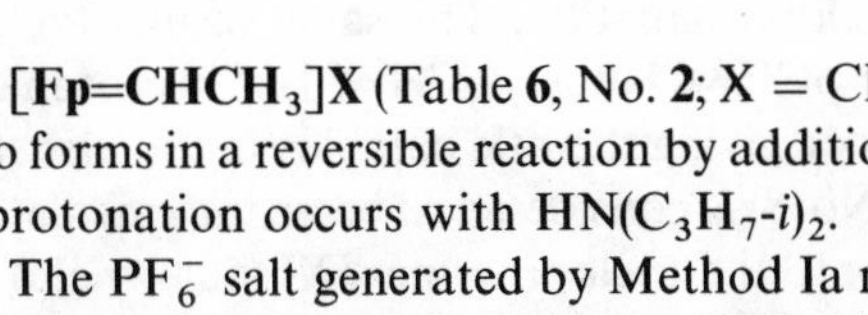

XV

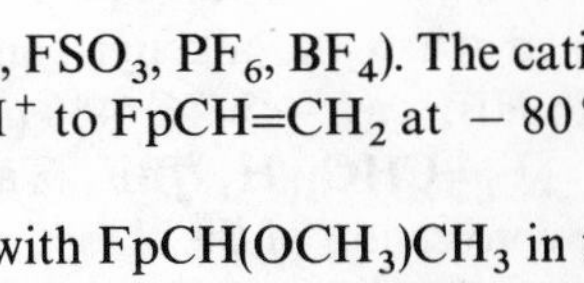

XVI

$[Fp{=}CHCH_3]X$ (Table **6**, No. **2**; X = CF_3SO_3, FSO_3, PF_6, BF_4). The cation also forms in a reversible reaction by addition of H^+ to $FpCH{=}CH_2$ at $-80\,°C$; deprotonation occurs with $HN(C_3H_7\text{-}i)_2$.

The PF_6^- salt generated by Method Ia reacts with $FpCH(OCH_3)CH_3$ in the presence of $HPF_6 \cdot O(C_2H_5)_2$ or $[C(C_6H_5)_3]PF_6$ in CH_2Cl_2 at $-25\,°C$ to produce the bimetallic cation XV (R = CH_3, 68 to 80%); this complex is also obtained from the title complex and $FpCH{=}CH_2$. On warming to room temperature the complex isomerizes into the olefin complex $[Fp(\eta^2\text{-}CH_2{=}CH_2)]^+$. Attempts to trap the cation with $P(C_6H_5)_3$ affords $[FpP(C_6H_5)_3]^+$ quantitatively. H^- addition gives FpC_2H_5. For cyclopropanation reactions, see Table 5.

$[Fp{=}CHC_2H_5]BF_4$ (Table **6**, No. **3**) has been proposed to be an intermediate in the cleavage of the C–C bond of XVII (R = H) with HBF_4 to give the olefin complex XVIII (R = H). Starting from XVII (R = D), via the intermediate $[Fp{=}C(D)C_2H_5]^+$, only the olefin complex XVIII (R = D) is formed and not XIX.

The complex generated according to Method Ia isomerizes via intramolecular hydride migration to give XVIII (R = H).

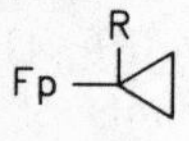

XVII 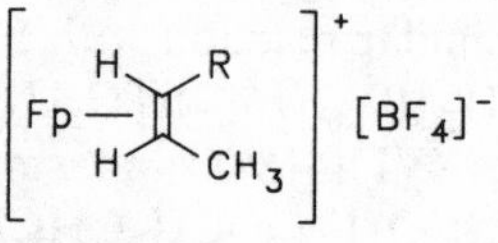 XVIII 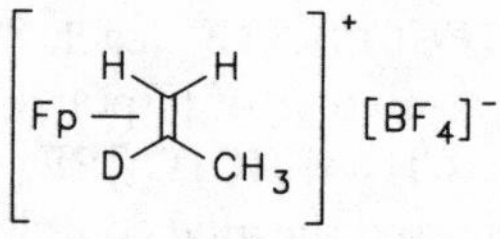 XIX

$[Fp{=}CHCH{=}CHC_2H_5]X$ (Table **6**, No. **5**, X = CF_3SO_3, BF_4). The BF_4^- salt is prepared by protonation of the neutral dienyl complex $FpCH{=}CHCH{=}CHCH_3$ with $HOSO_2CF_3$ in CH_2Cl_2 solution at − 20 to − 60 °C. An alternative structure of the cation, $[Fp{=}CHCH_2CH{=}CHCH_3]^+$, resulting from protonation of the β carbon atom of the starting complex, could not be excluded.

Upon warming a solution of the complex (X = BF_4) above − 60 °C, rearrangement occurs to give the 1,3-pentadiene complex $[Fp(CH_2{=}CHCH{=}CHCH_3)]BF_4$ in about 90% yield.

$[Fp{=}CHCH{=}C(CH_3)_2)]X$ (Table **6**, No. **6**, X = BF_4, CF_3SO_3). The BF_4^- salt is obtained as an orange precipitate by protonation of $FpCH{=}CHC(CH_3)_2OH$ in ether at − 15 °C with $HBF_4 \cdot O(C_2H_5)_2$. The complex is stable in solution at − 55 °C for several hours and decomposes upon warming to room temperature. The $CF_3SO_3^-$ salt is obtained by protonation of the dienyl complex $FpCH{=}CHC(CH_3){=}CH_2$ with $HOSO_2CF_3$ in CH_2Cl_2 at − 20 to − 60 °C.

For cyclopropanation reactions, see Table 5.

$[Fp{=}CH(CH_2)_4CH{=}CH_2]BF_4$ (Table **6**, No. **7**) probably forms as a nonisolable intermediate in the alkylation of the thioether $FpCH(SC_6H_5)(CH_2)_4CH{=}CH_2$ (Method Ib) with $[O(CH_3)_3]BF_4$ in CH_2Cl_2 solution to give norcarane in 25% yield via intramolecular cyclopropanation. The same results are obtained when the alkenyl complex $FpCH{=}CH(CH_2)_3CH{=}CH_2$ is protonated with HBF_4 at − 50 °C (4% norcarane), giving the title complex in the first step.

$[Fp{=}CHC_{10}H_{17}]BF_4$ (Table **6**, No. **8**) is considered to be an intermediate in the conversion of the thioether XX into the cyclopropane XXI (50% yield) via intramolecular cyclopropanation according to the procedure outlined for No. 7.

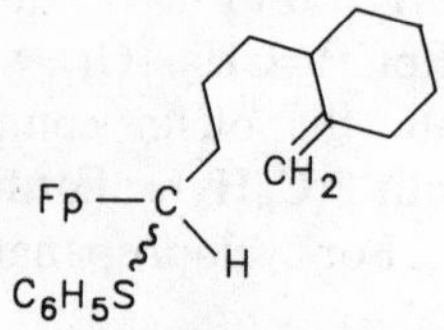

XX 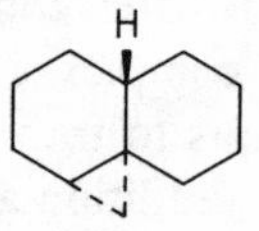 XXI

[Fp=CHC(CH$_3$)$_3$]BF$_4$ (Table **6**, No. **9**). The cation rearranges above $-95\,°C$ to the corresponding π complex [Fp(CH$_3$CH=C(CH$_3$)$_2$]BF$_4$ with an overall yield of 48%.

The rate of rearrangement was estimated by following the decrease of the proton signal at 18.1 ppm. The first order rate constant, k_1, is about $4 \times 10^{-3}\,s^{-1}$; activation parameter is $\Delta G^{\ddagger} = 12$ kcal/mol.

[Fp=CHC$_3$H$_5$-*c*)]CF$_3$SO$_3$ (Table **6**, No. **11**) is generated according to Method Ib and has been identified by low-temperature NMR spectroscopy. Variable ^1H NMR spectroscopy shows that decomposition begins at -30 to $-40\,°C$ and is complete at $0\,°C$. The nature of the decomposition products (and the decomposition products of the corresponding Ru complex) could not be identified.

The vicinal coupling between the Fe=CH proton and the cyclopropyl hydrogen was found to be 14 Hz. From this value a *trans* geometry of the cyclopropyl group and the organometallic fragment (s-*trans*-anticlinal, Formula XXII) was suggested; no evidence is found for the other possible isomers, s-*trans*-synclinal, s-*cis*-synclinal, and s-*cis*-anticlinal.

Cyclopropanation reactions are carried out with styrene and 2-ethylbutene-1; see Table 5.

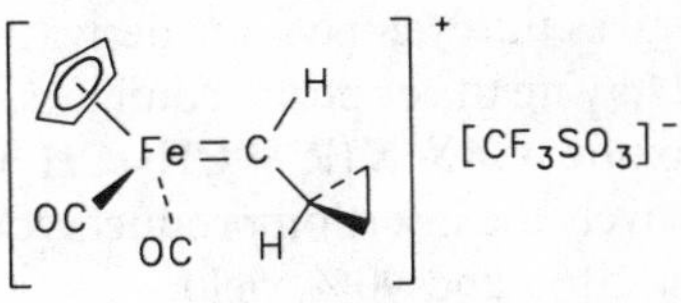

$[CF_3SO_3]^-$

XXII

[Fp=CHC$_7$H$_{11}$]BF$_4$ (Table **6**, No. **13**) has been identified by low temperature ^{13}C NMR measurements as the initial reaction product according to Method Ia from FpCH(OC$_2$H$_5$)C$_7$H$_{11}$. For rearrangement and final products, see No. 36.

Quenching with excess CH$_3$OH after formation at $-78\,°C$ and neutralization of the mixture with N(C$_2$H$_5$)$_3$ generates FpCH(OCH$_3$)C$_7$H$_{11}$ (90% yield).

[Fp=CHR]X (R = C$_8$H$_{13}$ (1-bicyclo-[3.2.1]octyl), C$_8$H$_{13}$ (1-bicyclo [2.2.2]octyl), C$_{10}$H$_{15}$ (1-adamantyl); X = BF$_4$ or CF$_3$SO$_3$; (Table **6**, Nos. **14** to **16**). These compounds are intermediates in the reaction of the corresponding neutral compounds FpCH(OCH$_3$) R with HBF$_4$ or CH$_3$OSO$_2$CF$_3$ to give the alkene complexes XXIII to XXV, respectively. Rates and activation parameters of β-to-α alkyl migration were determined for Nos. 14 to 16 to be k_1 $(-95\,°C) = 2.3 \times 10^{-6}$, 9.0×10^{-6}, and $9.4 \times 10^{-10}\,s^{-1}$ and $\Delta G^{\ddagger} = 15$, 14, and 18 kcal/mol; see also No. 9. The compounds could not be isolated but were observed at low temperatures by ^1H NMR and ^{13}C NMR spectroscopy.

Addition of Li[HB(C$_2$H$_5$)$_3$] in THF to a solution of No. 14 at $-100\,°C$ in CH$_2$Cl$_2$ generates the corresponding neutral complex FpCH$_2$R in 57% yield.

[Fp=CH(CH$_2$)$_3$C$_7$H$_{11}$]BF$_4$ (Table **6**, No. **17**). This intermediate, generated similarly to No. 7 from the thioether XXVI with [O(CH$_3$)$_3$]BF$_4$, decomposes to give the bicyclic compound XXVII.

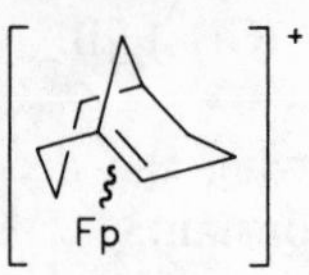

XXIII 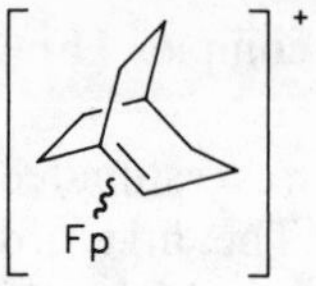 XXIV 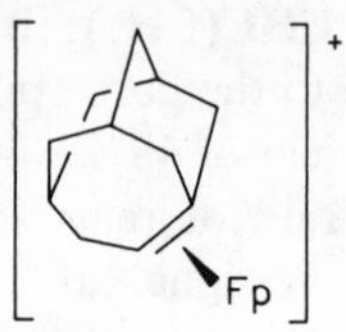 XXV

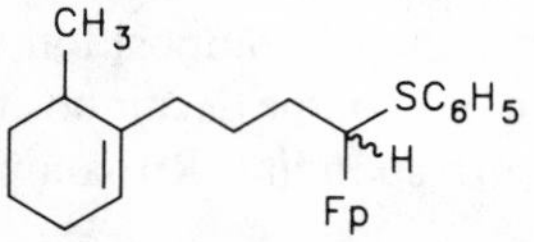

XXVI 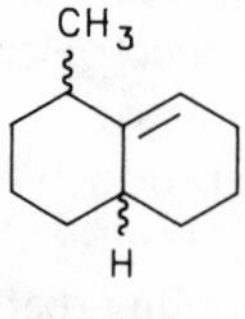 XXVII

[Fp=$C_{11}H_{18}O$]BF_4 (Table **6**, No. **18**), **[Fp=$C_{11}H_{16}O$]BF_4** (Table **6**, No. **19**), **[Fp=$C_{13}H_{20}O$]BF_4** (Table **6**, No. **20**), and **[Fp=$C_{15}H_{18}O$]BF_4** (Table **6**, No. **21**) are generated as intermediates according to Method Ib from the thioether derivatives XXVIII (R = C_2H_5, CH=CH_2, CH=C(CH_3)$_2$, or C_6H_5, respectively) to give products derived from intramolecular carbene insertion into a C–H bond. Thus, No. 18 gives the cyclopentane derivative XXIX (R = C_2H_5) in 39% yield (with one third having the opposite configuration of the C_2H_5 group); No. 19 produces a 2:3 mixture of XXIX (R = CH=CH_2) and XXX in 51% yield; Nos. 20 and 21 give exclusively the insertion products XXIX (R = CH=C(CH_3)$_2$ and C_6H_5, respectively) in 51% and 90% yield.

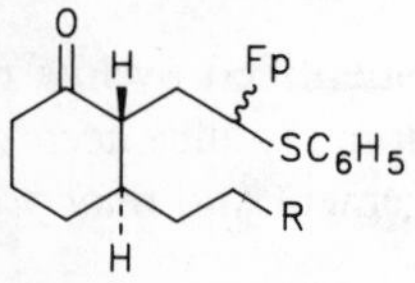

XXVIII 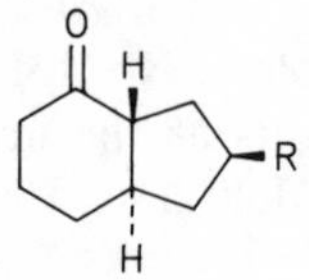 XXIX 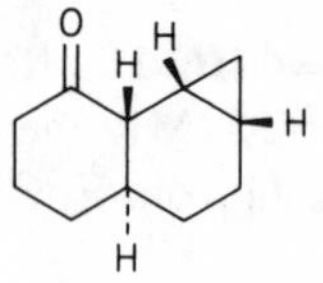 XXX

[Fp=CHC_6H_4-R)]X (Table **6**, No. **22**, R = H, X = CF_3SO_3, PF_6 BF_4; No. **23**, R = F-4, X = CF_3SO_3; No. **24**, R = OCH_3-4, X = CF_3SO_3, BF_4; No. **25**, R = OCH_3-3, X = CF_3SO_3; No. **26**, R = CH_3-4, X = CF_3SO_3; No. **27**, R = CF_3-4, X = CF_3SO_3) are obtained according to Method Ia at $-80\,°C$ as orange to red-orange solutions in CD_2Cl_2. The PF_6^- salt of No. 22 is also obtained by treatment of the ether precursor $FpCHC_6H_5OCH_3$ with [C(C_6H_5)$_3$]PF_6 in CH_2Cl_2 at $-20\,°C$, followed by precipitation with pentane or hexane (77% yield). The complexes decompose rapidly on warming with exception of No. 24, which survives for several hours at 25 °C. No. 22 is found to decompose at room temperature within 1 h in CF_3SO_3H solution to give *trans*-stilbene in about 50% yield.

Dynamic ^{1}H NMR studies were performed and barriers to aryl rotation were estimated as follows ($\Delta G^{\ddagger}_{rot}$ in kcal/mol): No. 22, 9.4; No. 23, 10.2; No. 24,

13.5; No. 26, 10.7. Variable-temperature ^{1}H NMR spectra in the aromatic region for Nos. 24 and 26 are depicted in Fig. 6. For No. 26 at $-76\,°C$ these resonances appear as two bands at 8.23 (d, 1 H; J = 8) and 7.55 (3 H) ppm. The two ortho protons are distinct, one overlapping with the meta resonances. At 56 °C the low-field *ortho* doublet averages with the *ortho* proton at 7.55 ppm. Above this temperature the signals sharpen to two doublets at 7.97 (H-2, 6) and 7.55 (H-3, 5; J = 8 Hz) ppm. Similarly, at $-100\,°C$ two distinct *ortho* resonances of No. 22 appear at 8.41 and 7.55 ppm (overlapped by H-3, 5, H-4) and coalesce at $-80\,°C$. The corresponding protons of No. 24 appear as a set of four distinct multiplets at 7 to 8.5 ppm, thus indicating the nonequivalence of the two *ortho* and the two *meta* protons at low temperature. Correlations of the ^{1}H chemical shifts of H-2 or the ^{13}C chemical shifts of $C_{carbene}$ with Hammet σ^{+} constants indicate that the barriers of rotation are largely electronic in nature. Low temperature ($-120\,°C$) ^{13}C NMR experiments with No. 24 show only one single CO resonance, thus indicating either a low Fe–$C_{carbene}$ rotational barrier or a vertical arrangement of the carbene ligand. From the highfield shift of the *para* carbon atom in the ^{13}C NMR spectrum of No. 22, a delocalization of the positive charge onto the aromatic ring with a perpendicular orientation of the phenyl ring to the *p* orbital of the carbene carbon atom is deduced.

No. 22 regenerates the starting $FpCH(OCH_3)C_6H_5$ with $CH_3OH/[CH_3O]^-$. Treatment of the PF_6^- salt with three equivalents of but-2-yne in CH_2Cl_2 at $-78\,°C$ results in the formation of $FpCH_2C_6H_5$, $[Fp(CH_3C{\equiv}CCH_3)]^+$, and the 1,2-dimethyl-3-phenyl-cyclopropenylium cation in about 75% yield. No. 24 reacts with nitrobenzene derivatives, O=N-Ar, ($Ar = C_6H_5$, $C_6H_4N(CH_3)_2$-*p*) at 0 °C in CH_2Cl_2 to form the oxygen-bonded nitrone compounds XXXI in about 50% yield. Similarly, with *cis*-azobenzene ($-30\,°C$) the insertion products XXXII ($X = BF_4$, CF_3SO_3) are obtained in about 70% yield.

$[Fp{-}O{-}N(Ar){=}C(H)Ar]^+ X^-$

XXXI

$[Fp{-}N(C_6H_5){-}N(C_6H_5){=}C(H)Ar]^+ X^-$

XXXII

$[Fp{=}C(CH_3)_2]X$ (Table **6**, No. **28**, $X = BF_4$, CF_3SO_3). Both salts have also been prepared by addition of $HBF_4 \cdot O(C_2H_5)_2$ at -23 or $-78\,°C$ or $HOSO_2CF_3$ at $-78\,°C$ to a solution of the vinyl complex $FpCCH_3({=}CH_2)$ in ether.

The observation of only one signal in both the ^{1}H NMR and ^{13}C NMR spectra for the two CH_3 groups is explained by a fast rotation about the Fe–$C_{carbene}$ bond due to a low barrier of rotation; the singlet persists down to $-110\,°C$.

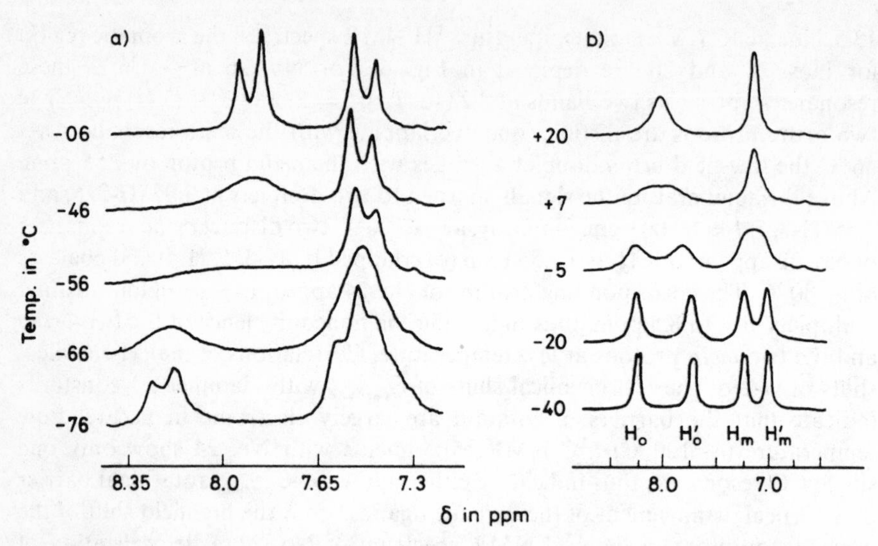

Fig. 6. Variable-temperature ^{1}H NMR spectra of $[Fp{=}CHC_6H_4R]^+$ (A: R = CH_3; B: R = OCH_3) in CD_2Cl_2

The complex is stable at 25 °C for at least 15 min. Solutions of the BF_4^- salt in CD_2Cl_2 decompose at − 11 °C to give the olefin complex $[Fp(\eta^2\text{-}CH_2{=}CHCH_3)]BF_4$ in nearly quantitative yields. Addition of $P(OCH_3)_3$ to the title complexes in ether or CH_2Cl_2 at − 78 °C produces the ylide complexes $[FpC(CH_3)_2P(OCH_3)_3]X$ in about 70 to 80% yield; a similar adduct is obtained with $P(C_6H_5)_3$ in 86% yield as the BF_4^- salt. For cyclopropanation reactions, see general remarks and Table 5.

$[Fp{=}C(CH_3)C_6H_5]BF_4$ (Table **6**, No. **29**) is probably a short-lived intermediate in the reaction of the vinyl complex $FpC({=}CH_2)C_6H_5$ in CH_2Cl_2 solution with an ether solution of HBF_4 at − 78 °C to give the olefin complex $[Fp(\eta^2\text{-}(CH_2{=}CHC_6H_5)]BF_4$ even in the presence of alkenes; no cyclopropanation product is obtained.

$[Fp{=}C_3R_2]X$ (Table **6**, No. **31**, R = C_6H_5, X = BF_4, ClO_4, Br, $SbCl_6$; No. **33**, R = $N(C_3H_7\text{-}i)_2$, X = ClO_4). The salts No. 31 are also nearly exclusively obtained by treatment of the neutral complexes XXXIII with various electrophiles, such as $[HO(C_2H_5)_2]BF_4$, HCl, Br_2, I_2, $SbCl_5$, $[C(OCH_3)_3]BF_4$. The BF_4^- salt forms in 40% yield from the cation $[(C_6H_5)_2ClC_3\text{-}c]^+$ and $FpSi(CH_3)_3$ in $CHCl_3$ under reflux temperature (1 h). The reaction of the carbenoid XXXIV with FpCl in THF at − 78 °C gives better yields of No. 33 (60%) than Method III. For an alternative formulation as Fp-substituted cyclopropenylium cations, see general remarks in Sect. 1.1.

$[Fp{=}CC_7H_6]PF_6$ and **$[Fp{=}CC_{11}H_8]PF_6$** (Table **6**, Nos. **34** (Formula XXXV) and **35**). No. 34 is also obtained from $FpCDC_7H_6$ (Formula XXXVI) according to Method IV without incoporation of the deuterium, thus indicating

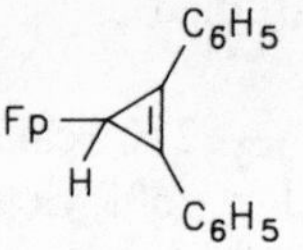

XXXIII 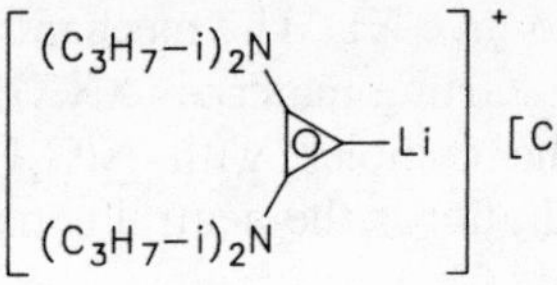 XXXIV

the preference of α-hydride rather than β-hydride abstraction. The compounds are indefinitely stable at room temperature when stored in the absence of moisture. No. 34 decomposes rapidly in nucleophilic solvents but is stable in CH_3NO_2, SO_2, and CH_2Cl_2. Rearrangement of the carbene cation to the isomeric cationic η^2-alkene complexes by 1,2-β-hydride shift is unfavorable because of the formation of an antiaromatic ligand.

No. 34 reacts with $LiAlH_4$ and $LiAlD_4$ to give 60% (H) and 57% (D) of the starting material. With CH_3OH, the CH_3O^- group is added at the carbene carbon atom to generate the corresponding ether $FpC(OCH_3)C_7H_6$ in 79% yield (similar reaction with No. 35). $P(C_6H_5)_3$ is added to give the ylide complex XXXVII (83%). $FpCH_2CH{=}CH_2$ produces the binuclear complex XXXVIII in 43% yield and H_2O eliminates the Fp group with formation of benzocyclobutenone (32% yield) and Fp_2. LiI in CH_3NO_2 gives 1-iodobenzocyclobutene, FpI, and the starting complex, $FpCHC_7H_6$. No. 34 is also a potent hydride abstractor, converting cycloheptatriene to the tropylium cation and $FpCH_2CH_3$ to the cationic ethylene complex.

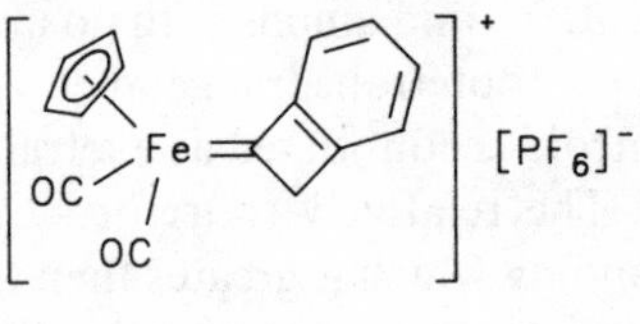

XXXV 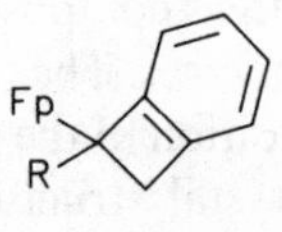 XXXVI

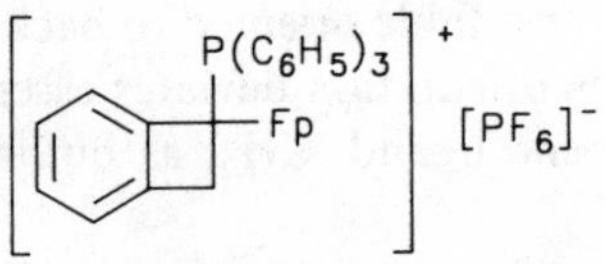

XXXVII 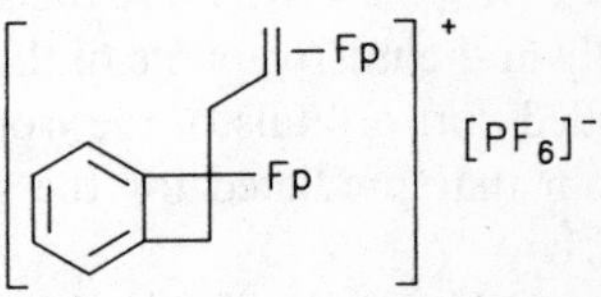 XXXVIII

$[Fp{=}C_8H_{12}]BF_4$ (Table **5**, No. **36**) was identified by variable-temperature ^{13}C NMR spectroscopy to be an intermediate in the reaction of $FpCH(OC_2H_5)C_7H_{11}$ (Formula XXXIX) with $HBF_4 \cdot O(C_2H_5)_2$ in CH_2Cl_2 resulting in the cationic olefin complex XL. In the first step, the carbene complex No. 13, prepared according to Method Ia, is formed and rearranges via ring enlargement to give No. 36, followed by migration of the Fp group and

β-hydrogen shift to give XL. The mechanism was studied using ^{13}C- and ^2H-labeled (at FeCH) starting materials XXXIX.

Reaction of the complex with $N(C_2H_5)_3$ at $-82\,°C$ causes reversible deprotonation and affords the neutral complex XLI in 75% yield.

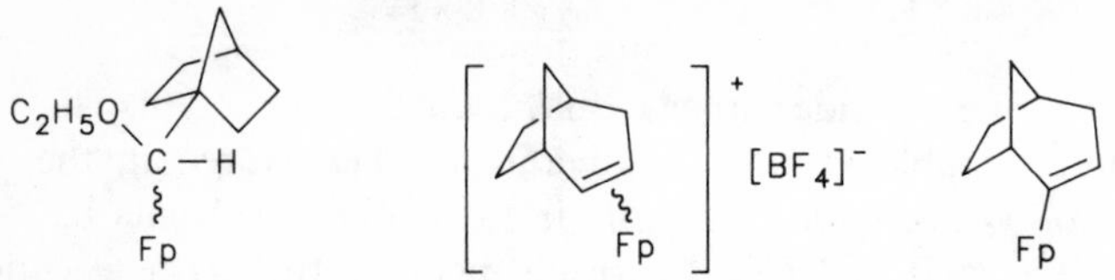

C_2H_5O C−H Fp [BF$_4$]$^-$ Fp Fp

XXXIX XL XLI

$[Fp{=}C_7H_6]PF_6$ (Table **6**, No. **37**) is prepared according to Method IV and is fairly air-stable in the solid if stored in a dark bottle, but decomposes slowly in solution. For an alternative formulation as Fp-substituted tropylium cation, see general remarks in Sect. 1.1.

The temperature-dependent NMR spectrum shows no kinetically induced broadening of signals down to $-105\,°C$; but see the $P(C_4H_9\text{-}n)_3$ substituted species in Sect. 1.1.2.1.

The complex crystallizes in the triclinic space group $P\bar{1}$-C_i^1 (No. 2) with a = 7.981(4), b = 14.378(3), c = 7.133(1) Å, α = 98.52(1)°, β = 100.75(1)°, γ = 93.33(1)°; Z = 2 and d_c = 1.728, while d_m = 1.70 g/cm^3. The data were collected at $-35\,°C$. The main bond angles and distances are given in Fig. 7. The crystal structure is composed of discrete cations and anions with no unusual close interionic distances. The cation shows a pseudotetrahedral geometry at the iron atom if the position of the C_5H_5 ring centroid is considered as a tetrahedral vertex; see also crystal structure of No. 39. The relatively short Fe—$C_{carbene}$ distance of 1.979(3) Å indicates $dp(\pi)$ back-bonding which is greater than that of No. 39 (1.996(2) Å). The C_7 rings of Nos. 37 and 39 are planar, and the dihedral angles between the mirror plane of the Fp fragment and the C_7 plane are 87.6 and 89.5°, respectively. Thus, the acceptor $p(\pi)$ orbitals of the carbene ligands lie virtually in the mirror plane of the Fp fragment, suitable oriented to back-bond with filled $d(\pi)$ orbitals of the iron atoms. This orientation deviates essentially 90° from that predicted for the simplest carbene ligand, CH_2, as outlined in Sect. 1.1.

$[Fp{=}C_8H_{11}]PF_6$ (Table **6**, Nos. **38** and **39**) are prepared according to Method IV and are fairly stable in air, similar to No. 37.

No. 39 crystallizes in the triclinic space group $P\bar{1}$-C_i^1 (No. 2) with a = 8.2986(6), b = 15.238(2), c = 7.4361(8) Å, α = 90.509(7)°, β = 104.396(5)°, γ = 94.676(6)°; Z = 2 and d_c = 1.692, while d_m = 1.68 g/cm^3. The data were collected at $-35\,°C$. Bond angles and distances are given in Fig. 8. The C_7 ring is planar and forms a dihedral angle of 1.1° with the fused C_6 ring. For further description of the structure, see No. 37.

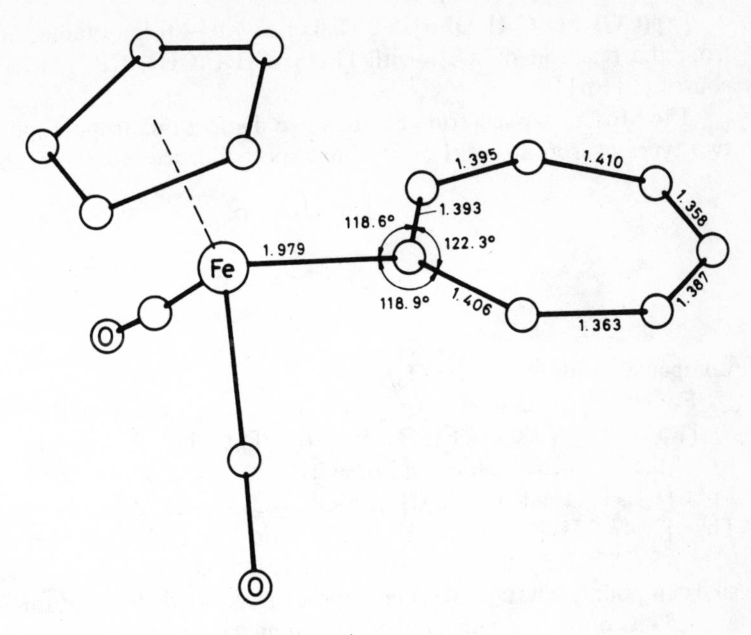

Fig. 7. Molecular structure of the cation $[Fp{=}C_7H_6]^+$

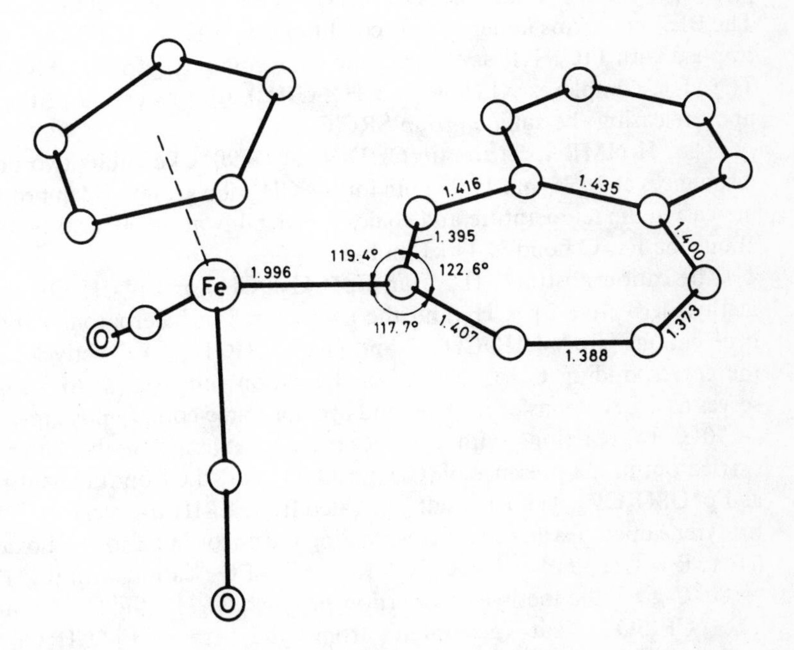

Fig. 8. Molecular structure of the cation $[Fp{=}C_8H_{11}]^+$

$[Fp(CO)_2Fe{=}C_6H_7OFp]PF_6$ (Table **6**, No. **40**) is obtained in 58% yield from the reaction of XLII with $[Fp(\eta^2\text{-}CH_2C(CH_3)_2)]PF_6$, which acts as a source of $[Fp]^+$.

The Mössbauer spectrum exhibits two distinct quadrupole splittings for the two types of iron atoms. For a similar spectrum, see No. 6 in Table 7.

Fp, O

XLII

Compounds with $^{5}L = C_5(CH_3)_5$

For clarity the $C_5(CH_3)_5(CO)_2Fe$ is abbreviated as Fp* in the following text.

$[Fp^*{=}CH_2]X$ ($X = CF_3SO_3$, BF_4, or PF_6). The $CF_3SO_3^-$ salt forms along with the carbene cation $[Fp^*{=}CHOH]^+$ when the neutral complex Fp^*CH_2OH is treated with $(CH_3)_3SiOSO_2CF_3$ in CH_2Cl_2 solution at $-90\,°C$. The $[Fp^*{=}CH_2]^+:[Fp^*{=}CHOH]^+$ ratio depends on the amount of $(CH_3)_3SiOSO_2CF_3$; no hydroxycarbene cation is found with an excess of the silyl compound. The complex decomposes at elevated temperatures and is stable at $-50\,°C$ only for a few minutes. It is similarly produced (as an intermediate) from $Fp^*CH_2OCH_3$ and has been formulated as a possible intermediate (source of a CH_2 fragment) in the reaction of $[Fp^*{=}CH(OCH_3)]^+$ with $(C_6H_5)_3SiH$ to give $(C_6H_5)_3SiOCH_3$ and $(C_6H_5)_3SiCH_3$; see also reactions of the cation, below. The BF_4^- salt forms under similar conditions with $HBF_4 \cdot O(C_2H_5)_2$ and can be trapped with $P(C_6H_5)_3$ as the cationic ylide complex $[Fp^*CH_2P(C_6H_5)_3]BF_4$. The ylide complexes XLIII ($R = CH_3$, C_6H_5) are precursors of the PF_6^- salt upon releasing the sulfide group $SRC(C_6H_5)_3$.

The ^1H NMR spectrum (in $CHDCl_2$ at $-90\,°C$) exhibits two one-protein resonances at 16.38 and 17.06 ppm for the $CH_2{=}Fe$ group, which broaden upon increasing the temperature and coalesce reversibly at $-60\,°C$; ΔG^* of rotation about the Fe=C bond is 44 kJ/mol.

The cation abstracts H^- from $Fp^*CH_2OCH_3$ or Fp^*CH_2OH to give the methyl derivative Fp^*CH_3 and the corresponding heteroatom-stabilized carbene cations $[Fp^*{=}CHOCH_3]^+$ and $[Fp^*{=}CHOH]^+$, respectively. Similar to the corresponding C_5H_5 derivative, the cation prepared at low temperature serves as a CH_2-transfer reagent, and cyclopropane compounds are obtained at $-90\,°C$ by reactions with cyclohexene and styrene; in the latter reaction, carried out in the presence of $(CH_3)_3SiOSO_2CF_3$, the iron fragment is trapped as $Fp^*OSO_2CF_3$. The PF_6^- salt generated from XLIII transfers the CH_2 group to styrene upon heating the corresponding precursor in refluxing dioxane for 2 h (80%, $R = CH_3$; 60%, $R = C_6H_5$). Reaction of the cation with $(C_6H_5)_3SiH$ at $-80\,°C$ gives the methylene insertion product, $(C_6H_5)_3SiCH_3$, quantitatively. The $CF_3SO_3^-$ salt generated from the ether $Fp^*CH_2OCH_3$ and $(CH_3)_3SiOSO_2CF_3$ reacts with the thioether Fp^*CH_2SR ($R = CH_3$, C_6H_5) to give the binuclear cation XLIV in about 70% yield.

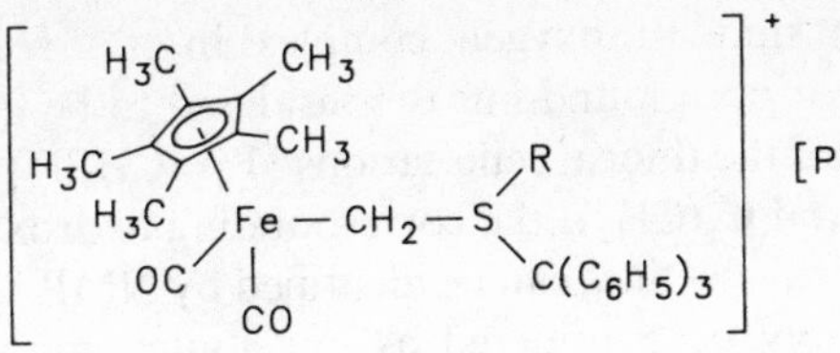

$$[C_5(CH_3)_5(CO)_2Fe{-}CH_2{-}S(R)C(C_6H_5)_3]^+[PF_6]^-$$

XLIII

$$[C_5(CH_3)_5(CO)_2Fe{-}CH_2{-}S(R){-}CH_2{-}Fe(CO)_2C_5(CH_3)_5]^+$$

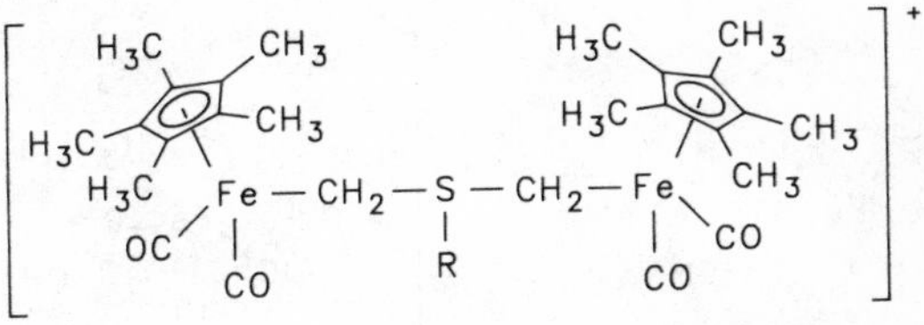

XLIV

1.1.3.2 Cationic Complexes of the Type $[{}^5L(CO)_2Fe{=}C(R)ER']^+$ with One Heteroatom at the Carbene Carbon Atom

This section comprises cationic carbene complexes in which the carbene carbon atom at the $C_5H_5(CO)_2Fe$ fragment (abbreviated as Fp) is additionally saturated by a partial double bond from one heteroatom. Only compounds with the heteroatoms E = O, S, and N have been described; R′ represents a set of ligands to achieve the rare gas shell. The complexes Nos. 6 to 9 and 12 can also be considered as adducts of the neutral acyl complexes FpCOR at the cationic 16 electron fragments $[Cp(CO)_3M]^+$ (M = Mo, W) or $[Cp(CO)_2M]^+$ (M = Fe); see also neutral carbene complexes in Sect. 1.2. For the description of these types of compounds by two resonance structures, see Sect. 1.1. For the contribution of an 'olefin' resonance form of Nos. 14 to 16, see further information.

Compounds with ^5L ligands other than Cp ($^5L = C_5H_4CH_3$, C_9H_7, $C_5(CH_3)_5$) are described at the end of this section.

The compounds listed in Table 7 can be prepared according to the following general methods:

Method I: H^- abstraction from FpCHR(ER′) by $[C(C_6H_5)_3]PF_6$.

Method II: Treatment of FpCOR compounds with cationic electrophiles.

a. The electrophile is H^+.
b. The electrophile is $R^+ = CH_3^+$, $C_2H_5^+$.
c. The electrophile is a cationic 16 electron transition metal fragment of the type $[Cp(CO)_3M]^+$ (M = Mo, W) or $[Fp]^+$.

Method III: Reaction of $[Fp{=}CHSCH_3]CF_3SO_3$ with R_2NH, RNH_2 and NH_3.

a. From $[Fp{=}CHSCH_3]CF_3SO_3$.
b. From its pyridine adduct $[FpCH(SCH_3)NC_5H_5]CF_3SO_3$.

General remarks. The simplest oxygen complex in this section, the cation $[Fp{=}CHOH]^+$, has not been found but is considered to be formed in the first step by the hydrolysis of the thiocarbene cations $[Fp{=}CHSR]^+$. However, with the more bulky ^{5}L ligand $C_5(CH_3)_5$ the corresponding hydroxy carbene cation, described at the end of this section, can be identified by NMR spectroscopy. The H^- salt II can probably be considered as one ionic canonical form of the hydroxymethyl complex $C_5(CH_3)_5(CO)_2FeCH_2OH$ (Formula I) which is a good H^- donor.

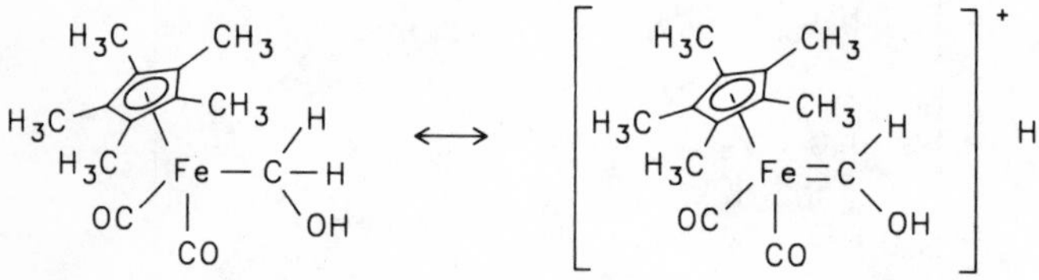

I II

The carbene plane can be oriented within the symmetry plane of the Fp fragment ("upright", see theoretical considerations in Sect. 1.1) or orthogonal to the Fp plane ("crosswise"). The true orientation probably depends on a combination of steric and electronic effects. Furthermore, in the upright conformation, *syn* and *anti* orientations of the heteroatoms are possible. All types of orientations are found with complexes containing the heteroatom E = S.

Explanation for Table 7. Only the most important spectroscopic data are presented. For the Cp ligand average NMR shifts (in ppm) are 5.30 in CD_3CN, CD_3NO_2, and CF_3COOH, 5.10 to 5.60 in CD_2Cl_2, and 5.50 in CD_3COCD_3 (^{1}H NMR spectra); 88.5 in CD_3CN, CD_2Cl_2, and CD_3COCD_3, and 90.5 in CD_3NO_2 (^{13}C NMR spectra). The two strong ν (CO) vibrations are centered at about 2020 and 2065 cm^{-1} for E = O, 2020 and 2055 cm^{-1} for E = S, and 2000 and 2035 cm^{-1} for E = N.

Further Information

[Fp=CH(OR)]PF$_6$ (Table 7, R = CH_3, C_2H_5; Nos. **1**, **2**). No. 1 can also be obtained similar to Method I by hydride abstraction from $FpCH_2OCH_3$ at $-20\,°C$ in CH_2Cl_2 solution with $[CpMo(CO)_3]PF_6$ in 58% yield along with the cation $[CpFe(CO)_3]^+$ (9%). No. 1 is prepared in 90% yield by protonation of $FpCH(OCH_3)_2$ with $HPF_6 \cdot O(C_2H_5)_2$; see reverse reaction below. The cation of No. 1 is also detected in the gas phase by ion cyclotron resonance spectroscopy by hydride abstraction from $FpCH_2OCH_3$ with the ion $[Fp{=}CH_2]^+$.

The compounds are slowly hydrolyzed in air and decompose rapidly in acetone solution and slowly in CH_3NO_2 to give mixtures of $[CpFe(CO)_3]^+$ and $FpCH_3$.

No. 1 reacts with LiR compounds (R = CH_3, C_4H_9-*n*, C_6H_5, $C_6H_4CH_3$-4) in ether to give α-methoxyalkyl iron complexes of the type $FpCHROCH_3$ in 70 to

Table 7. Cationic Carbene Complexes of the General Type $[Fp{=}C(R)ER']^+$; E is O, S, or N. Further information on numbers preceded by an asterisk is given at the end of the table.

No.	Fe=C(R)ER′	anion, method of preparation (yield) properties and remarks
heteroatom E is O		
*1	$Fe{=}CHOCH_3$	PF_6^- salt, I (88%); yellowish solid ^1H NMR (CD_3NO_2); 4.76 (s, CH_3), 13.32 (s, CH) ^{13}C NMR (CD_3NO_2); 79.4 (CH_3O), 207.3 (CO), 321.9 (C=Fe)
*2	$Fe{=}CHOC_2H_5$	PF_6^- salt, I (90%); yellow solid ^1H NMR (CD_3NO_2): 1.59 (t, CH_3), 4.97 (q, CH_2), 13.24 (s, CH)
*3	$Fe{=}C(CH_3)OH$	Cl^- salt, IIa; green powder $CF_3SO_3^-$ salt, not isolated BF_4^- salt, IIa (80%)
*4	$Fe{=}C(CH_3)OCH_3$	BF_4^- salt, IIb (94%); yellow, m.p. 141 to 243 °C ^1H NMR (CD_2Cl_2); 3.15 (s, CH_3), 4.60 (s, CH_3O) ^{13}C NMR (CD_3COCD_3): 68.7 (CH_3), 87.1 (CH_3O), 209.7 (CO), 336.0 (C=Fe) PF_6^- salt; pale yellow powder ^1H NMR (CF_3COOH): 3.18 (s, CH_3), 4.67 (s, CH_3O) $CF_3SO_3^-$ salt, IIb ^{13}C NMR (CD_2Cl_2): 46.81 (CH_3), 334.30 (C=Fe)
*5	$Fe{=}C(CH_3)OC_2H_5$	PF_6^- salt; yellow crystals ^1H NMR (CF_3COOH): 1.68 (t, CH_3 of C_2H_5), 3.11 (s, CH_3), 4.79 (q, CH_2) ^{13}C NMR (CH_3NO_2): 14.1 (CH_3 of C_2H_5), 46.2 (CH_3), 80.0 (CH_2), 210.2 (CO), 333.0 (C =Fe)
*6	$Fe{=}C(CH_3)OFp$	BF_4^- salt; red powder PF_6^- salt, IIc (50 to 85%); red solid ^1H NMR (CD_3NO_2): 2.61 (s, CH_3), 5.03 (s, Cp of Fe=C), 5.42 (s, Cp of FeO) ^{57}Fe-γ (80 K): $\delta = 0.07$, $\Delta = 2.09$ (Fe–O), $\delta = -0.01$ $\Delta = 1.64$ (Fe=C)
*7	$Fe{=}C(CH_3)ORe(NO)(CO)Cp$	no anion or properties reported

Table 7. Continued

No.	Fe=C(R)ER′	anion, method of preparation (yield) properties and remarks
*8	$Fe{=}C(CH_3)OMo(CO)_3Cp$	PF_6^- salt, IIc (50 to 85%); red powder ^1H NMR (CD_3NO_2): 2.61 (s, CH_3), 5.06 (s, CpFe), 6.08 (s, CpMo) SbF_6^- salt, IIc; red crystals ^1H NMR (CD_2Cl_2): 2.52 (s, CH_3), 5.03, 6.03 (s's, Cp)
*9	$Fe{=}C(CH_3)OW(CO)_3Cp$	PF_6^- salt, IIc (50 to 85%); red powder
*10	$Fe{=}C(C_2H_5)OCH_3$	$CF_3SO_3^-$ salt, IIb
*11	$Fe{=}C(C_3H_7\text{-}i)OCH_3$	$CF_3SO_3^-$ salt, IIb
12	$Fe{=}C(C_3H_7\text{-}i)OMo(CO)_3Cp$	PF_6^- salt, IIc ^1H NMR (CD_2Cl_2): 1.03 (d, CH_3)
*13	$Fe{=}C(CH_2OCH_3)OC_2H_5$	no anion reported, IIb
*14	$Fe{=}C(CH_2Fp)OH$	$CF_3SO_3^-$ salt, IIa (12%); reddish orange, m.p. 218 °C ^1H NMR (CD_2Cl_2): 3.03 (s, CH_2), 5.03, 5.14 (s's, Cp), 13.6 (br, HO) ^{13}C NMR (CD_2Cl_2): 32.24 (CH_2), 86.62, 87.95 (Cp), 211.90, 214.77 (CO), 302.74 (C=Fe)
*15	$Fe{=}C(CH_2Fp)OCH_3$	$CF_3SO_3^-$ salt, IIb (93%); orange, m.p. 111 °C ^1H NMR (CD_2Cl_2): 3.17 (s, CH_2), 4.03 (s, CH_3O), 5.19, 5.24 (s's, Cp) ^{13}C NMR (CD_2Cl_2): 30.14 (CH_2), 64.50 (CH_3O), 87.98, 89.02 (Cp), 212.01, 214.58 (CO), 299.07 (C=Fe)
*16	$Fe{=}C(CH_2Fp)OSi(CH_3)_3$	$CF_3SO_3^-$ salt, IIc (84%); orange crystals ^1H NMR (CD_2Cl_2): 3.27 (s, CH_2), 5.16, 5.17 (s's, Cp) ^{13}C NMR (CD_2Cl_2): 36.50 (CH_2), 86.95, 88.68 (Cp), 212.06, 214.89 (CO), 305.25 (C=Fe)
*17	$Fe{=}C(CH_2Fp)OFp$	BF_4^- salt, IIc (65%); deep purple red ^1H NMR (CD_2Cl_2): 2.44 (s, CH_2), 4.89, 5.07, 5.30 (s's, 15 H, Cp) ^{13}C NMR (CD_3NO_2): 33.33 (CH_2), 88.18 (10 C, Cp), 88.76 (Cp), 213.10, 215.44, 217.50 (CO), 297.46 (C=Fe)
*18	$Fe{=}C(CH_2Fp)OFe(CO)_2C_5H_4CH_3$	BF_4^- salt, IIc; 17:8 mixture with the carbene cation No. 20 ^1H NMR (CD_2Cl_2): 1.85 (s, CH_3), 2.45 (s, CH_2), 4.92, 5.05 (s's, 10 H, Cp) ^{13}C NMR (CD_3NO_2): 12.85 (CH_3), 34.50 (CH_2) 213.74, 215.98, 217.69 (CO), 297.90 (C=Fe)

No.	Compound	Data
*19	$Fe{=}C(CH_2Fp)OZr(Cp)_2CH_3$	$CF_3SO_3^-$ salt ^1H NMR (CD_3NO_2): 0.50 (s, CH_3Zr), 2.95 (s, CH_2), 5.15 (s, 10 H, CpFe), 6.38 (s, 10 H, CpZr) ^{13}C NMR (CD_3NO_2): 31.47 (CH_3), 33.39 (CH_2), 88.12, 89.04 (CpFe), 114.12 (CpZr), 214.69, 217.86 (CO), 312.99 (C=Fe)
*20	$Fe{=}C(CH_2Fe(CO)_2C_5H_4CH_3)OFp$	BF_4^- salt, IIc; 8:17 mixture with the carbene cation No. 18 ^1H NMR (CD_2Cl_2): 1.94 (s, CH_3), 2.34 (s, CH_2), 4.89, 5.31 (s's, 10 H, Cp) ^{13}C NMR (CD_3NO_2): 13.53 (CH_3), 36.91 (CH_2), 213.47, 215.82, 218.17 (CO), 296.63 (C=Fe)
21	$Fe{=}C(C_4H_9\text{-}t)OC_2H_5$	BF_4^- salt, IIb (73%) reduction with $Li[BH(C_2H_5)_3]$ in CH_2Cl_2 at − 78 °C (1 h) gives $FpCH(OC_2H_5)C_4H_9\text{-}t$
22	$Fe{=}C(C(CH_3)_2C_6H_5)OC_2H_5$	BF_4^- salt, IIb (68%) reduction with $Li[BH(C_2H_5)_3]$ in CH_2Cl_2 at − 78 °C (1 h) gives $FpCH(OC_2H_5)C(CH_3)_2C_6H_5$
23	$Fe{=}C(C_3H_5\text{-}c)OCH_3$	BF_4^- salt, IIb $CF_3SO_3^-$ salt, IIb ^1H NMR (CD_2Cl_2): 1.60 (m, 2H), 1.88 (m, 2H), 3.39 (s, CH_3O) ^{13}C NMR (CD_2Cl_2): 22.9 (CH_2), 44.3 (CH), 60.2 (CH_3O), 209.3 (CO), 315 (C=Fe) reduction with $NaBH_4$ in $NaOCH_3/CH_3OH$ at − 80 °C gives $FpCH(OCH_3)(C_3H_5\text{-}c)$ (50%)
24	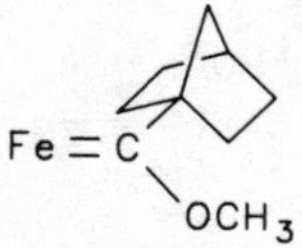 Fe=C, OCH_3 (structure)	$CF_3SO_3^-$ salt, IIb reduction with $NaBH_4$ in $NaOCH_3/CH_3OH$ (or $NaBD_4$ in $NaOCH_3/CH_3OD$) gives $FpCH(OCH_3)(C_7H_{11})$ (or $FpCD(OCH_3)(C_7H_{11})$)
*25	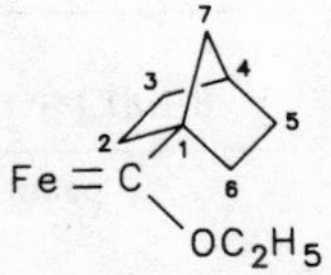 Fe=C, OC_2H_5 (structure; ring positions 1–7)	BF_4^- salt, IIb (80 to 90%); yellow ^1H NMR (CD_2Cl_2 at − 10 °C): 1.32 to 1.90 (m, CH_3 + 5 CH_2), 2.40 (s, CH of C_7H_{11}), 5.30 (q, CH_2O) ^{13}C NMR (CD_2Cl_2 at − 10 °C: 14.7 (CH_3), 78.7 (C-1), 83.9 (CH_2O), 209.4 (CO), 342.9 (C=Fe)

Table 7. Continued

No.	Fe=C(R)ER′	anion, method of preparation (yield) properties and remarks
26	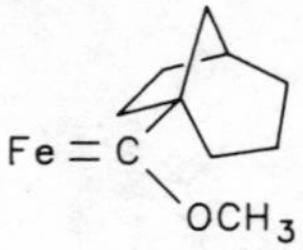 Fe=C, OCH_3	$CF_3SO_3^-$ salt, IIb reduction with $NaBH_4$ in basic CH_3OH gives a 1:1 diastereomeric mixture of $FpCH(OCH_3)C_8H_{13}$
27	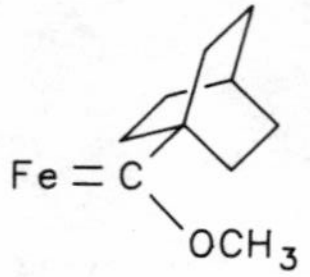 Fe=C, OCH_3	$CF_3SO_3^-$ salt, IIb reduction with $NaBH_4$ in basic CH_3OH gives a 1:1 diastereomeric mixture of $FpCH(OCH_3)C_8H_{13}$
28	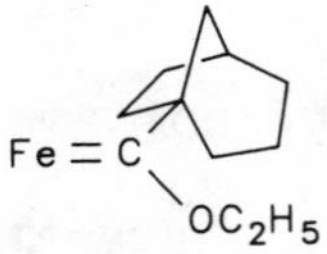 Fe=C, OC_2H_5	BF_4^- salt, IIb (94%); light yellow solid ^1H NMR (CD_2Cl_2): 1.2 to 2.1 (m, 12 H), 2.33 (s, br, CH), 5.2 (q, CH_2O) ^{13}C NMR (CD_2Cl_2 at − 10 °C): 35.2 (CH of C_8H_{13}), 75.3 (C of C_8H_{13}), 84.3 (CH_2O), 209.3, 209.5, (CO), 344.3 (C=Fe) reduction with $Li[HB(C_2H_5)_3]$ in CH_2Cl_2 at − 78 °C gives a 1:1 diastereomeric mixture of $FpCH(OC_2H_5)C_8H_{13}$
29	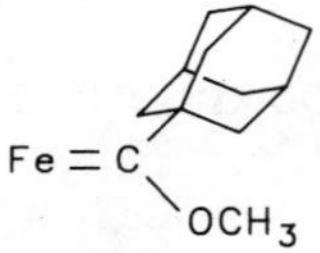 Fe=C, OCH_3	$CF_3SO_3^-$ salt, IIb reduction with $NaBH_4$ in basic CH_3OH gives a 1:1 diastereomeric mixture of $FpCH(OCH_3)C_{10}H_{15}$
*30	$Fe{=}C(C_6H_5)OCH_3$	PF_6^- salt, IIb; yellow crystals ^1H NMR (CF_3COOH): 4.58 (s, CH_3)

No.	compound	data
*31	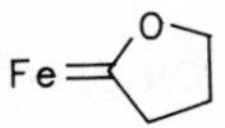 Fe=(2-oxacyclopentylidene)	BF_4^- salt; pale yellow PF_6^- salt; pale yellow, m.p. 115 to 117 °C ^{1}H NMR (CD_3COCD_3): 2.16 (q, CH_2), 3.99 (t, CH_2), 5.61 (t, CH_2O)
32	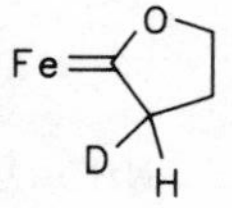 Fe=(2-oxacyclopentylidene, 3-D, 3-H)	BF_4^- salt, see No. 31

heteroatom E is S

No.	compound	data
*33	Fe=CHSCH$_3$	$CF_3SO_3^-$ salt; golden ^{1}H NMR (CD_2Cl_2): 3.00 (s, CH_3), 14.86 (br, CH); 15.79 (CH) in CF_3SO_3H BF_4^- salt PF_6^- salt; golden crystals ^{1}H NMR (CD_2Cl_2): 3.12 (s, CH_3), 15.24 (br, CH)
*34	Fe=CDSCH$_3$	$CF_3SO_3^-$ salt
*35	Fe=CHSC$_6$H$_5$	PF_6^- salt I (80%); golden yellow, m.p. 126 °C (dec.) ^{1}H NMR (CD_2Cl_2): 15.52 (s, CH=Fe) ^{13}C NMR (CD_3NO_2): 208.72 (CO), 317.35 (C=Fe)
*36	Fe=C(CH$_3$) SCH$_3$	$CF_3SO_3^-$ salt; yellow powder ^{1}H NMR (CD_3NO_2): 3.01 (s, CH_3), 3.56 (s, CH_3S) ^{13}C NMR (CD_3NO_2): 30.7 (CH_3S), 48.8 (CH_3), 210.5 (CO), 345.4 (C=Fe) PF_6^- salt; yellow, m.p. 136 °C (dec.) ^{1}H NMR (CD_2Cl_2): 3.06 (s, CH_3), 3.56 (s, CH_3S)
*37	Fe=C(CH$_3$)SC$_6$H$_5$	$CF_3SO_3^-$ salt PF_6^- salt; golden yellow, m.p. 92 to 96 °C (dec.) ^{1}H NMR (CD_3NO_2): 3.39 (s, CH_3) ^{13}C NMR (CD_3NO_2): 50.2 (CH_3), 210.0 (CO), 351.7 (C=Fe)

heteroatom E is N

No.	compound	data
*38	Fe=CHNH$_2$	$CF_3SO_3^-$ salt IIIb (20%); pale yellow ^{1}H NMR (CD_3CN): 11.30 (m, CH), 11.51 (m, NH)

Table 7. Continued

No.	Fe=C(R)ER′	anion, method of preparation (yield) properties and remarks
*39	Fe=CHNHCH$_3$	PF_6^- salt, IIIa (28%) ^{1}H NMR (CD_3CN): 3.33 (d, CH_3N), 10.69 (CH=Fe), 10.90 (NH) ^{13}C NMR (CD_3CN): 45.8, (CH_3N), 211.2 (CO), 238.6 (C=Fe)
*40	Fe=CHNHC$_3$H$_7$-*i*	PF_6^- salt, IIIa (30%) ^{1}H NMR ($CDCl_3$): 1.38 (d, CH_3), 3.89 (hept, CH) 10.64 (NH), 10.85 (CH=Fe) ^{13}C NMR (CD_3CN): 21.8 (CH_3), 63.3 (CH), 211.5 (CO), 234.5 (C=Fe)
*41	Fe=CHNHC$_4$H$_9$-*t*	$CF_3SO_3^-$ salt IIIa (38%), IIIb (47%) ^{1}H NMR ($CDCl_3$): 10.63 (NH), 10.83 (CH=Fe) ^{13}C NMR (CD_3CN): 212.5 (CO), 231.2 (CH=Fe)
*42	Fe=CHNHC$_6$H$_{11}$-*c*	$CF_3SO_3^-$ salt, IIIa (35%); yellow crystals ^{1}H NMR (CD_2Cl_2): 10.55 (NH), 10.76 (CH) ^{13}C NMR (CD_3CN): 211.3 (CO), 235.1 (C=Fe)
*43	Fe=C(D)NHC$_6$H$_{11}$-*c*	$CF_3SO_3^-$ salt, IIIa (32%) ^{1}H NMR (CD_2Cl_2): 10.63 (br, NH) ^{13}C NMR (CD_3CN): 211.3 (CO), 234.5 (CD=Fe)
44	Fe=CHN(CH$_3$)$_2$	PF_6^- salt, IIIa (30%) ^{1}H NMR (CD_3CN): 1.89 (d, CH_3N), 1.94 (d, CH_3N), 10.79 (CH=Fe) ^{13}C NMR (CD_3CN): 47.7, 55.9 (CH_3N), 211.8 (CO), 234.9 (C=Fe) reacts with $HN(CH_3)_2$ to give FpH, Fp_2 and $[HC(N(CH_3)_2)_2]^+$
*45	Fe=CHNCH$_3$C$_3$H$_7$-*i*	FSO_3^- salt; mixture of two rotamers A and B ^{1}H NMR (CD_3CN, A/B): 3.42/3.45 (d's, CH_3N), 3.90/4.20 (hept's, CH), 10.83/11.04 (CH=Fe)
46	Fe=CHN(C$_2$H$_5$)$_2$	$CF_3SO_3^-$ salt, IIIa (31%), IIIb (42%); pale yellow ^{1}H NMR (CD_3COCD_3): 3.96, 3.99 (q's, CH_2), 11.30 (CH=Fe) ^{13}C NMR (CD_3CN): 53.2, 58.8 (CH_2), 211.7 (CO), 232.4 (C=Fe)
*47	Fe=C(C$_3$H$_7$-*i*)NHC$_6$H$_5$	PF_6^- salt; orange yellow, mixture of the two rotamers A and B, m.p. 211 °C (dec.) ^{1}H NMR (CD_3CN, A/B): 3.75 (hept, CH), 11.6 (br, NH)
*48	Fe=C(CH(COOC$_2$H$_5$)$_2$)NHC$_6$H$_5$.	PF_6^- salt; bright yellow, mixture of the two rotamers A and B, m.p. 118 °C (dec.) ^{1}H NMR (CD_3CN, A/B): 3.83 (s, CH), 5.11/5.56 (s's, Cp), 12.3 (br, NH)

No.	Compound	Properties
*49	$Fe{=}C(CH(C_6H_5)_2)NHCH_3$	PF_6^- salt; yellow, m.p. 246 °C (dec.) ^1H NMR (CD_3CN): 3.53 (d, CH_3), 6.16 (s, CH)
*50	$Fe{=}C(CH(C_6H_5)_2)NHC_6H_5$	PF_6^- salt; bright yellow, m.p. 188 °C (dec.) ^1H NMR (CD_3CN): 6.26 (s, CH), 10.8 (br, NH)
*51	$Fe{=}C(C_6H_5)NHCH_3$	PF_6^- salt; yellow, m.p. 155 °C (dec.) ^1H NMR (CD_3COCD_3): 3.88 (d, CH_3N)
*52	$Fe{=}C(C_6H_5)NHC_3H_7\text{-}i$	PF_6^- salt; yellow, *Z*/*E* isomer mixture, m.p. 203 °C (dec.) ^1H NMR (CD_3COCD_3, Z/E): 1.53/131 (d, CH_3), 4.70 (m, CH)
*53	$Fe{=}C(C_6H_5)\ NHCH_2C_6H_5$	PF_6^- salt; yellow, m.p. 148 °C (dec.) ^1H NMR (CD_3COCD_3): 5.38 (d, CH_2)
*54	$Fe{=}C(C_6H_5)NHC_6H_5$	PF_6^- salt; yellow, *Z*/*E* isomer mixture m.p. 195 °C (dec.) ^1H NMR (CD_3COCD_3, *Z*/*E*): 5.23/5.64 (s's, Cp)
*55	$Fe{=}C(C_6H_5)N(CH_3)C(=NCH_3)C_6H_5$	PF_6^- salt; yellow needles, m.p. 137 °C (dec.) ^1H NMR (CD_3COCD_3): 3.38 (s, CH_3N), 4.23 (s, CH_3N) ^{13}C NMR (CD_3COCD_3): 39.7, 48.3 (CH_3), 210.9 (CO)
*56	$Fe{=}C(C_6H_5)N(C_3H_7\text{-}i)C(=N{-}C_3H_7\text{-}i)C_6H_5$	Cl^- salt
*57	$Fe{=}C{-}CH{=}C(OC_2H_5){-}N(CH_3){-}C(=O){-}N(CH_3)$ (ring)	BF_4^- salt; yellow crystals; m.p. >192 °C (dec.)
*58	$Fe{=}C{-}CH{=}C(N(CH_3)_2){-}N(CH_3){-}C(=O){-}N(CH_3)$ (ring)	BF_4^- salt; yellow crystals; m.p. >200 °C (dec.) from No. 57 and $(CH_3)_3SiN(CH_3)_2$

86% yields. Vinylmagnesium bromide or vinyllithium gives the rearrangement product $FpCH_2CH{=}CHOCH_3$, but with 2-methylprop-1-enyllithium only Fp_2 is formed. Addition of CH_3O^- to No. 1 generates the acetal complex $[FpCH(OCH_3)_2$ in 77% yield. Treatment of the complexes with $[CH_3P(C_6H_5)_3]$ I yields $[CpFe(CO)_3]^+$, $FpCH_2OR$ and an alkyl iodide. Rapid hydride transfer from an intermediate η^1-formyl complex FpCHO may be responsible for the products observed. Upon reaction $[CH_3P(C_6H_5)_3]BH_4$ in CH_2Cl_2 solution, No. 2 produces a mixture of $FpCH_3$ and $FpCH_2OC_2H_5$. With $Cp(CO)(P(C_6H_5)_3)FeH$ and No. 1, hydride transfer occurs quantitatively to give the starting material $FpCH_2OCH_3$, which is also formed by treatment of the complex with $[BH_4]^-$.

$[Fp{=}C(CH_3)OH]X$ (Table **7**, No. **3**, X = Cl, CF_3SO_3, BF_4). The Cl^- salt is obtained as a cream-colored powder by the action of dry HCl on $FpCOCH_3$, but it dissociates rapidly outside an HCl atmosphere. Similarly, the action of $HBF_4 \cdot O(CH_3)_2$ gives the BF_4^- salt (80% conversion). Treatment of No. 14 with CF_3SO_3H in CH_2Cl_2 (12 h) generates the $CF_3SO_3^-$ salt along with $FpOSO_2CF_3$. Both compounds regenerate the acyl complex $FpCOCH_3$ by treatment with $N(C_2H_5)_3$.

$[Fp{=}C(CH_3)OCH_3]X$ (Table **7**, No. **4**, X = CF_3SO_3, PF_6, BF_4). The PF_6^- salt is obtained in 81% yield similar to Method IIb with $[CH(OCH_3)_2]PF_6$ as the alkylation agent; see also preparation of No. 5. The $CF_3SO_3^-$ complex forms by acidification of the carbene complex $[Fp{=}C(CH_2Fp)OCH_3]CF_3SO_3$ (No. 15) with CF_3SO_3H.

The reaction of the BF_4^- salt with $NaOCH_3$ in CH_2Cl_2 at $-78\,°C$ produces a mixture of the vinyl ether $FpC({=}CH_2)OCH_3$, $FpCOCH_3$, and the ketal $FpCCH_3(OCH_3)_2$ in a ratio of 55:6:39, respectively, whereas the more sterically hindered base KOC_4H_9-*t* gives the vinyl ether in 87% yield. Grignard or Grignard analogue reagents produce various products. Thus, with $LiCH_3$ in CH_2Cl_2 at $-78\,°C$ a 1:1 mixture of the addition product $FpC(CH_3)_2OCH_3$ and the vinyl ether $FpC({=}CH_2)OCH_3$ is formed, whereas CH_3MgI predominantly gives the demethylation product $FpCOCH_3$. $LiCu(CH_3)_2$ in CH_2Cl_2 at $-78\,°C$ generates the addition product in 45 to 50% yield along with small amounts of the vinyl ether complex. H^- donors such as $Li[HB(C_2H_5)_3]$ (in THF at $-80\,°C$), $NaBH_4$, or $Cp(CO)(P(C_6H_5)_3)FeH$ produce high yields of the complex $FpCH(CH_3)OCH_3$.

$[Fp{=}C(CH_3)OC_2H_5]PF_6$ (Table **7**, No. **5**). A freshly prepared suspension of $[HC(OC_2H_5)_2]PF_6$ (from $HC(OC_2H_5)_3$ and $[C(C_6H_5)_3]PF_6$) in CH_2Cl_2 is allowed to react with $FpCOCH_3$ for about 18 h; 84% yield.

$[Fp{=}C(CH_3)OFe(CO)_2Cp)]X$ (Table **7**, No. **6**, X = PF_6, BF_4). The BF_4^- salt is prepared by allowing $FpCOCH_3$ to react with the olefin complex $[Cp(CO)(O(C_2H_5)_2)Fe(\eta^2\text{-}CH_2{=}CHCH_3)]BF_4$ in CH_2Cl_2 solution at $-78\,°C$. After warming the mixture to room temperature the title complex is precipitated from the resulting solution with ether (29% yield). The PF_6^- salt also forms in 18% yield by the reaction of the carbene complex $[Fp{=}CH_2]PF_6$ (generated at low temperature, see Sect. 1.1.2.1) with $FpCOCH_3$ under similar conditions. The deep burgundy solution becomes orange on warming to room temperature. Filtration from a solid (a mixture of $[Fp(\eta^2\text{-}CH_2{=}CH_2)]PF_6$ (32%) and $[CpFe(CO)_3]PF_6$ (11%)) and removal of the solvent gives a gummy orange residue; the starting material (79%) is extracted with ether. The remaining ether-insoluble solid consists of a mixture of the title complex and about 6% of an uncharacterized material. The initially expected 'alkylation product', the cation $[Fp{=}C(CH_3)OCH_2Fp]^+$ (but see Sect. 1.1.2.2, synthesis of the analog $[Cp(CO)(P(C_6H_5)_3)Fe{=}C(CH_3)OCH_2Fp]^+$), cannot be detected. When $FpCH_3$ is allowed to react with HBF_4^- a complex mixture is obtained also containing complex No. 6. The BH_4^- salt was presumed to be one of the products obtained by the reaction of $FpCH_3$ with $[C(C_6H_5)_3]BF_4$.

For a similar Mössbauer spectrum of an analogous complex, described in Sect. 1.1.3.1, see Table 6, No. 40.

The complex readily incorporates $C_5H_4CH_3(CO)_2FeCOCH_3$ to give the substituted carbene complex $[C_5H_4CH_3(CO)_2Fe{=}C(CH_3)OFp]BF_4$. From this reaction an equilibration according to $[Fp{=}C(CH_3)OFp]BF_4 \rightleftharpoons FpCOCH_3 + FpFBF_3$ is deduced. Transformation of the O-bonded Fp group to the acyl complex $Cp(CO)(P(C_6H_5)_3)FeCOCH_3$ occurs in CH_2Cl_2 (18 h at 20 °C, 80% yield) to give the new carbene complex $[Cp(CO)(P(C_6H_5)_3)Fe{=}C(CH_3)OFp]^+$ (Sect. 1.1.2.2); see also similar reactions with the cations Nos. 7 and 9. The complex does not catalyze the carbonylation of $FpCH_3$.

[Fp=C(CH₃)ORe(NO)(CO)₃Cp]X (Table 7, No. 7; X not reported) reacts with the acyl complex $Cp(CO)(P(C_6H_5)_3)FeCOCH_3$ in CH_2Cl_2 (18 h, 20 °C) with exchange of the Re fragment to give the cation $[Cp(CO)(P(C_6H_5)_3)Fe{=}C(CH_3)ORe(NO)(CO)Cp]^+$ in 100% yield (see Sect. 1.1.2.2).

$[Fp{=}C(CH_3)OMo(CO)_3Cp]X$ (Table 7, No. **8**; X = PF_6, SbF_6). The SbF_6^- salt can also be prepared similar to Method IIc, but with $FpCH_3$ as starting material. Thus, $Cp(CO)_3MoFSbF_5$ is allowed to react with $FpCH_3$ in CH_2Cl_2 solution at − 30 °C. The mixture is then brought to room temperature within 24 h and stirred for 2 days. The resulting solid, containing $[Fp(CO)]SbF_6$ and $[(CpMo(CO)_3)_2H]SbF_6$, is removed and the title complex precipitated from the solution with hexane at − 30 °C. The solution contains $Cp(CO)_3MoCH_3$ and $FpCOCH_3$. A 4:5 mixture of unidentified $[CpMo(CO)_3(solvate)]^+$ and No. 8 is obtained by treatment of $[Cp(CO)_3Mo{=}CH_2]PF_6$ (from $Cp(CO)_3MoCH_2Cl$ and $AgPF_6$) with $FpCOCH_3$ at − 78 °C.

The SbF_6^- salt crystallizes in the orthorombic space goup Pbca (No. 61) with a = 12.413(4), b = 15.674(3), c = 23.462(8) Å; (Z) = 8, d_c = 2.04 g/cm³, while d_m = 2.06(2) g/cm³. Disorder for both Cp rings is found. The structure of the cation is depicted in Fig. 9.

With $[N(C_4H_9\text{-}t)_4]I$, conversion to a mixture of $Cp(CO)_3MoI$ and $FpCOCH_3$ occurs. $FpCOCH_3$ also forms upon decomposition of the SbF_6^- salt.

$[Fp{=}C(CH_3)OW(CO)_3Cp]X$ (Table 7, No. **9**; X not reported) reacts with the acyl complex $Cp(CO)(P(C_6H_5)_3)FeCOCH_3$ in CH_2Cl_2 (18 h, 20 °C) with exchange of the W fragment to give the cation $[Cp(CO)(P(C_6H_5)_3Fe{=}C(CH_3)OW(CO)_3Cp]^+$ in 30% yield.

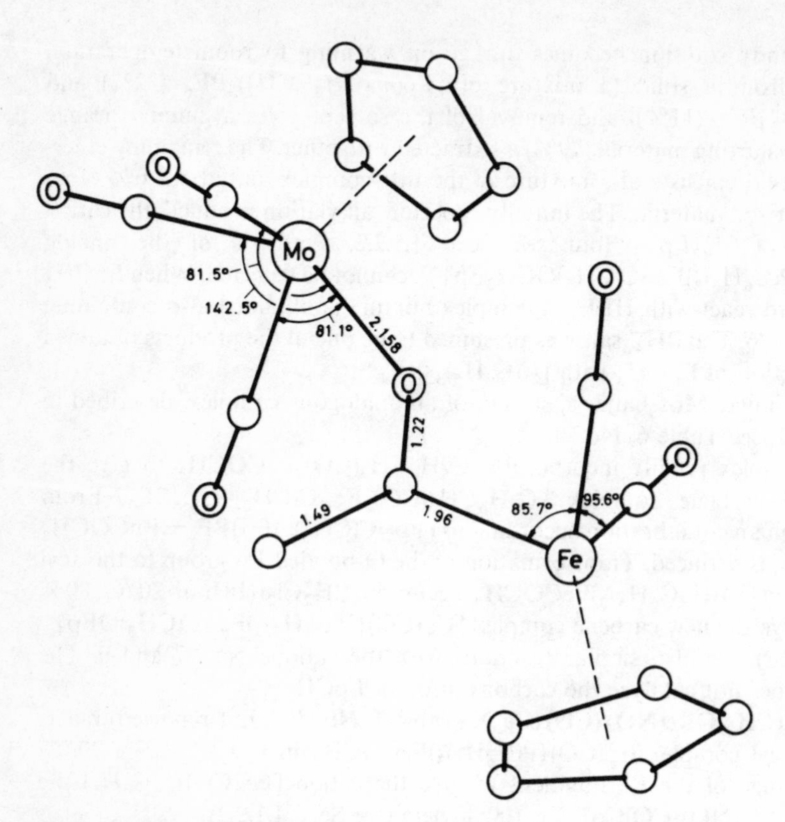

Fig. 9. Molecular structure of the cation of [Fp=C(CH_3)OMo(CO)$_3$Cp]SbF_6

[Fp=C(R)OCH$_3$] CF$_3$SO$_3$ (Table 7, R = C_2H_5, C_3H_7-*i*, Nos. **10**, **11**) can be reduced to the corresponding complexes FpCH(R)OCH$_3$ by quenching a CH_2Cl_2 solution into CH_3OH/[CH_3O]$^-$/BH_4^- at − 78 °C; yield 80 to 90%.

[Fp=C(CH$_2$OCH$_3$)OC$_2$H$_5$]X (Table 7, No. **13**; X not reported) undergoes a monohydride reduction with Li[BH(C_2H_5)$_3$] in THF at − 80 °C to give FpCH(OCH_3)CH_2OCH_3 in about 40% yield. The cation rearranges to the η^2-*cis*-α,β-dialkoxyethylene compound, [Fp(η^2-CH(OCH_3)=CH(OC_2H_5))]$^+$.

[Fp=C(CH$_2$Fp)OR]CF$_3$SO$_3$ (Table 7, R = H, CH_3, Si(CH_3)$_3$; Nos. **14** to **16**). The compounds are prepared according to Methods IIa, IIb, and IIc, respectively, and are thermally stable. The complexes can be regarded as hybrids of the two resonance forms III and IV. No. 16 hydrolyzes in wet ether to give No. 14 (72%) which can be transferred into No. 15 with CH_2N_2 (68%).

The reactivity reflects some contribution of the resonance form IV to the carbenoid structure III of these compounds. Whereas soft nucleophiles react at the alkyl-side Fp, hard nucleophiles such as H^- react at the electrophilic carbene center. Thus, with P(C_6H_5)$_3$ in CH_2Cl_2 and No. 15 the compounds [Fp(P(C_6H_5)$_3$)]$^+$ and FpC(=CH_2)OCH_3 are quantitatively obtained, probably via resonance form IV. With $[HBR_3]^-$ hydride attack occurs at the carbene carbon atom of No. 15, resulting in Fp_2 and CH_2=$CHOCH_3$, while with $NaBH_4$ ($NaBD_4$) in basic media (CH_3O^-/CH_3OH), a mixture of $FpCH(CH_3)OCH_3$ ($FpCD(CH_3)OCH_3$; 27%), Fp_2, and CH_2=$CHOCH_3$ is formed. With $NaOCH_3$ a mixture of Fp_2 and $FpC(=CH_2)OCH_3$ is obtained. The hydroxycarbene complex No. 14 can be quantitatively transferred into the starting complex with pyridine. Protonation of Nos. 14 and 15 with CF_3SO_3H produces the corresponding cations $[Fp{=}C(CH_3)OR]^+$. $Al(CH_3)_3$ and No. 14 generate a mixture of $FpCH_3$ and $FpCOCH_3$. With $(Cp)_2Zr(CH_3)_2$, methane elimination occurs to give complex No. 19.

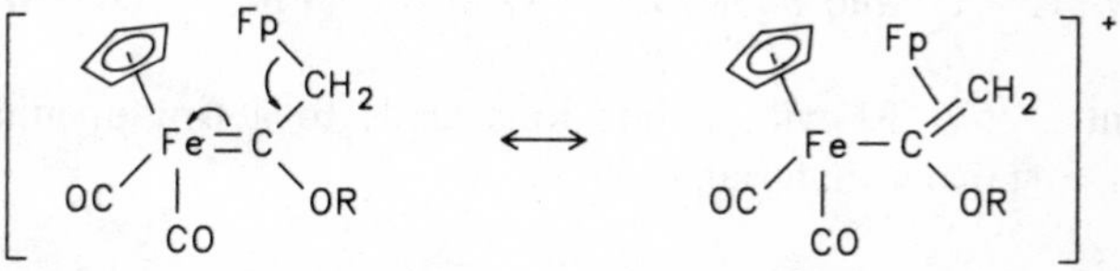

III ⟷ IV

$[Fp{=}C(CH_2Fp)OFp]BF_4$ (Table **7**, No. **17**) reacts with slight excess $P(C_6H_5)_3$ in CH_2Cl_2 to generate the starting acyl complex $FpCOCH_2Fp$ and the cationic complex $[Fp(P(C_6H_5)_3)]BF_4$ in quantitative yield. Incorporation of the cation $[C_5H_4CH_3(CO)_2Fe]^+$ (abbreviated as Fp*) at the positions OFp or CH_2Fp (see Nos. 18 and 20) by treatment of the complex with $[Fp^*(THF)]^+$ in CH_2Cl_2 (12 h) occurs only to a small extent.

$[Fp{=}C(CH_2Fp)OFe(CO)_2C_5H_4CH_3]BF_4$ and **$[Fp{=}C(CH_2Fe(CO)_2C_5H_4CH_3)OFp]BF_4$** (Table **7**, Nos. **18** and **20**). A 68:32 mixture of both compounds is generated according to Method IIc if the labeled μ-ketene complex $FpCOCH_2Fp^*$ is treated with $[Fp(THF)]BF_4$ or if the complex $FpCOCH_2Fp$ is treated with $[Fp^*(THF)]BF_4$.

Addition of $P(C_6H_5)_3$ to the isomer mixture produced a 68:32 mixture of $FpCOCH_2Fp$ and $FpCOCH_2Fp^*$, respectively. For exchange reactions with the cations $[Fp]^+$ or $[Fp^*]$, see No. 17.

$[Fp{=}C(CH_2Fp)OZr(Cp)_2CH_3]CF_3SO_3$ (Table **7**, No. **19**) is obtained by the reaction of $[Fp{=}C(CH_2Fp)OH]CF_3SO_3$ (No. 14) with $(Cp)_2Zr(CH_3)_2$ in CH_2Cl_2 solution in 58% yield; CH_4 is eliminated.

$[Fp{=}C(C_7H_{11})OC_2H_5]BF_4$ (Table **7**, No. **25**) is reduced with $NaBH_4$ in $CH_3OH/NaOCH_3$ (or $NaBD_4$ in $CH_3OD/NaOCH_3$) to give the neutral complex $FpCH(OC_2H_5)C_7H_{11}$ (or $FpCD(OCH_3)C_7H_{11}$). Similar results are obtained when $Li[HB(C_2H_5)_3]$ or $Li[DB(C_2H_5)_3]$ in THF solution are added to a solution of the complex in CH_2Cl_2 at $-78\,°C$ (65%).

$[Fp{=}C(C_6H_5)OCH_3]PF_6$ (Table **7**, No. **30**) is also obtained by reaction of $[HC(OCH_3)_2]PF_6$ (from $HC(OCH_3)_3$ and $[C(C_6H_5)_3]PF_6$) with the acyl complex $FpCOC_6H_5$ in CH_2Cl_2 solution at room temperature (18 h, 91%); see also Nos. 4 and 5.

The reaction with $Cp(CO)(P(C_6H_5)_3)FeH$ produces a mixture of $FpCH(OCH_3)C_6H_5$ and $[Cp(CO)(P(C_6H_5)_3)Fe(solvate)]PF_6$.

[Fp=C$_4$H$_6$O]X (Table 7, No. **31**; X = BF_4, PF_6). Both compounds are prepared by treatment of the acyl complex $FpCOCH_2CH_2CH_2Cl$ with $AgPF_6$ or $AgBF_4$ in acetone. The PF_6^- salt can be precipitated with ether (59%). A mixture of equal amounts of the cationic olefin complex V (R = H) and the BF_4^- salt No. 31 is obtained when the olefin complex $[Fp(\eta^2\text{-}CH_2{=}C(CH_3)_2)]BF_4$ is allowed to react with $RC{\equiv}CCH_2CH_2OH$ in 1,2-dichlorethane at 70 °C for 15 min (90% yield). The formation of the intermediate cation $[Fp(\eta^2\text{-}RC{\equiv}CCH_2CH_2OH)]^+$ is proposed. Similar experiments with the acetylene $DC{\equiv}CCH_2CH_2OH$ in the presence of excess $HOC_4H_9\text{-}t$ show complete deuteration at V (R = D) and 50% loss of D at VI (**[FpC$_4$H$_5$DO]BF$_4$**, Table **7**, No. **32**).

Treatment of No. 31 with halide ions leads to a ring-opening reaction, resulting in the starting material.

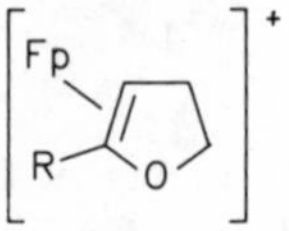

V

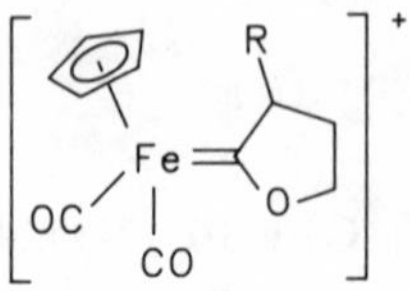

VI

[Fp=CHSCH$_3$]X (Table **7**, No. **33**; X = PF_6, CF_3SO_3, BF_4). The $CF_3SO_3^-$ and BF_4^- salts can be obtained in about 75% yield by dropwise treatment of a stirred solution of $FpCH(SCH_3)_2$ in ether with CF_3SO_3H or $HBF_4 \cdot O(C_2H_5)_2$, respectively. The PF_6^- salt is obtained similarly by removal of a SCH_3^- group from the starting material with $[C(C_6H_5)_3]PF_6$. Thus, a solution of $FpCH(SCH_3)_2$ in ether is added to a rapidly stirred solution of the salt in CH_2Cl_2 at − 40 °C, 70% yield. Small and varying amounts of the cation are obtained using Method I with $FpCH_2SCH_3$ as starting material.

The complexes are stable to O_2 but sensitive to moisture. The PF_6^- and the $CF_3SO_3^-$ salts are sparingly soluble in CH_2Cl_2 but dissolve readily in THF, CH_3CN and CF_3SO_3H.

Addition of pyridine gives the ylide complex $[FpCH(SCH_3)NC_5H_5]CF_3SO_3$ in 38% yield. Secondary and primary amines replace the SCH_3 group with NR_2 (see Method IV); $FpCH(SCH_3)_2$ and $[H_2NR_2]^+$ are also formed with similar yields, probably by the reaction of the resulting $HSCH_3$ with the title complex in the presence of HNR_2. A fivefold excess of $HN(CH_3)_2$ produces mainly Fp_2 and the cation $[HC(N(CH_3)_2]^+$. With phosphorus donor ligands PR_3 (PR_3 = $PH_2(C_6H_{11}\text{-}c)$, $PH_2(C_6H_5)$, $PH(C_6H_{11}\text{-}c)_2$, $PH(C_6H_5)_2$, $PCH_3(C_6H_5)_2$, $P(C_6H_5)_2Cl$, PCl_3, $P(OCH_2)_3CCH_3$, $P(OC_6H_5)_3$) the cation forms ylide complexes of the type $[FpCH(SCH_3)PR_3]^+$; no adducts were obtained with CH_3CN, THF, $S(CH_3)_2$, $As(C_6H_5)_3)_3$, or 1,4-diazabicyclo[2.2.2.]octane. The title complex (X = CF_3SO_3) reacts with H_2O to yield $[CpFe(CO)_3]^+$ and

$FpCH_2SCH_3$ in a 1:1 mole ratio; for the mechanism, see No. 34. Addition of an ethereal solution of CH_2N_2 to the complex in CH_2Cl_2 results in the formation of the sulfur-bonded complex $[FpSCH_3(CH{=}CH_2)]CF_3SO_3$ (76% yield).

$[Fp{=}C(D)SCH_3]CF_3SO_3$ (Table **7**, No. **34**) is prepared following the above workup procedure starting with $FpCD(SCH_3)_2$ (80% yield).

Hydrolysis of this complex (see above) gives $FpCD_2SCH_3$ and $[CpFe(CO)_3]^+$. From these results a mechanism for the hydrolysis of Nos. 33 and 34 involving H^- (or D^-) transfer via the intermediate formyl complex FpCHO (probably formed from deprotonation of $[Fp{=}CH(OH)]^+$, the initial product of hydrolysis) to the starting material is proposed. With cyclohexylamine the carbene complex No. 43 is obtained.

$[Fp{=}CHSC_6H_5]PF_6$ (Table **7**, No. **35**) is slightly air- and moisture-sensitive but is stable under N_2 at 25 °C.

It crystallizes in the monoclinic space group $P2_{1/c}C^5_{2h}$ (No. 14) with a = 7.267(2), b = 14.245(4), c = 16.668(3) Å, β = 96.84(2)°; (Z) = 4 and d_c = 1.722 g/cm³. The cation forms two molecules with different carbene fragments, with random distributions of 80 and 20% for molecule A and B, respectively. As depicted in Fig. 10 the structures differ in the relationship of the phenylthio group to the Cp ring (*syn* and *anti*, see also General Remarks). The angle of the plane of the phenyl rings and the carbene plane (Fe–$C_{carbene}$–S) is 23° in A and 33° in B. The deviation from planarity is probably caused by interaction of the *ortho* H atoms with the carbene hydrogen atom. The dihedral angle between the carbene plane and the Fp plane (center of Cp, Fe, and $C_{carbene}$) is 4°.

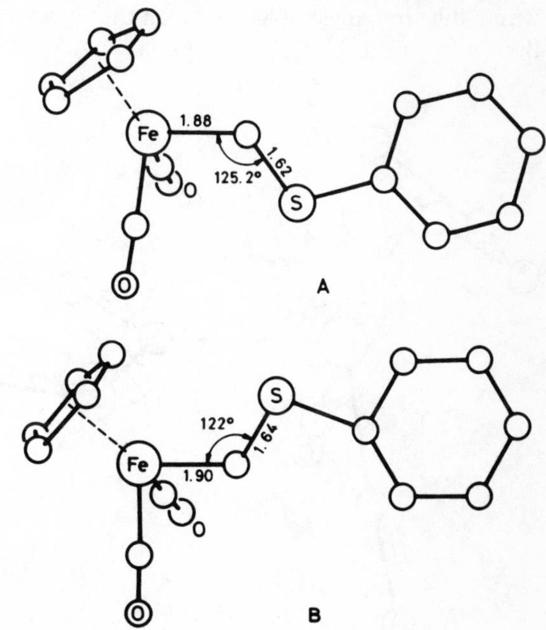

Fig. 10. Molecular structures of the molecules A (80%) and B (20%) of the cation of $[Fp{=}CHSC_6H_5]PF_6$

The compound reacts with nucleophilic reagents of the type RM (RM $= CH_3Li$, CH_3MgBr, $n\text{-}C_5H_{11}MgBr$, C_6H_5MgBr) in THF at $-78\,°C$ to give the corresponding neutral alkyl complexes $FpCHR(SC_6H_5)$ in about 70 to 93% yield. Addition of the enolate VII produces a mixture of the diastereomers VIII in 91% yield. It slowly abstracts hydride from ether at ambient temperature.

OLi — VII

C_6H_5S, H, O, Fp — VIII

MgBr, CH_3 — IX

$[Fp{=}C(CH_3)SCH_3]X$ (Table 7, No. **36**; X = PF_6, CF_3SO_3). Dropwise addition of an ethereal solution of $O(CF_3SO_2)_2$ to the acetyl complex $FpCOCH_3$ in ether at $-78\,°C$ generates a yellow precipitate (probably the vinylidine complex $[Fp{=}C{=}CH_2]CF_3SO_3$). Onto this suspension is condensed CH_3SH and the mixture warmed to room temperature (83% yield). The complex decomposes at 25 °C within 3 h. The PF_6^- salt is obtained by metathesis of this complex with NH_4PF_6 in CH_3NO_2 (2 h at 25 °C, 46% yield).

A crystal structure determination was performed on the PF_6^- salt. It crystallizes in the orthorombic space group $P2_12_12_1\text{-}D_2^4$ (No. 19) with the unit cell dimensions a = 7.818(2), b = 11.696(2), c = 16.228(4) Å; (Z) = 4 and d_c = 1.773 g/cm^3. As shown in Fig. 11 an 'orthogonal' conformation is present with a dihedral angle of 84° between the carbene plane and the mirror plane of the Fp fragment.

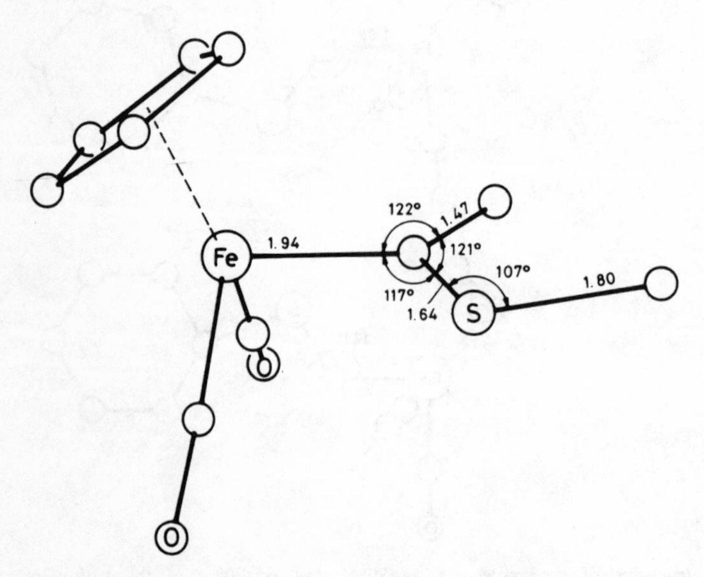

Fig. 11. Molecular structure of the cation $[Fp{=}C(CH_3)SCH_3]^+$

The action of $LiCH_3$ produces mainly $FpC(=CH_2)SCH_3$ and the acyl complex $FpCOCH_3$ and only small amounts of the adduct $FpC(CH_3)_2SCH_3$. The less basic cuprates, $LiCuR_2$ (R = CH_3, C_4H_9-n), at − 78 °C in CH_2Cl_2 produce the corresponding adducts $FpC(CH_3)(SCH_3)R$ in 63 and 51% yield, respectively.

[Fp=C(CH$_3$)SC$_6$H$_5$]X (Table **7**, No. **37**, X = PF_6, CF_3SO_3). Both complexes are obtained and purified as outlined for No. 36, but with addition of C_6H_5SH to the vinylidene intermediate (43% yield).

The cations react with $LiCH_3$ similar to No. 36 and with the cuprate $LiCu(CH_3)_2$ in CH_2Cl_2 at − 78°C, the complex $FpC(CH_3)_2SC_6H_5$ is formed in 60% yield.

[Fp=CHNH$_2$]CF$_3$SO$_3$ (Table **7**, No. **38**) is quite stable to air and only soluble in very polar solvents, such as CH_2Cl_2 and CH_3CN.

The ^{1}H NMR spectrum shows complex unresolvable patterns in the NH and CH region, presumably owing to coupling among the three nonequivalent protons resulting from restricted rotation around the C–N bond.

[Fp=CHNHR]PF$_6$ (Table **7**, R = CH_3, C_3H_7-*i*; Nos. **39**, **40**) and **[Fp =CHNHR]CF$_3$SO$_3$** (Table **7**, R = C_4H_9-*t*, C_6H_{11}-*c*, Nos. **41**, **42**) can be deprotonated by $NaOH/C_2H_5OH$ (in the case of No. 42 also by NaH) to give the corresponding neutral compounds FpCH=NR as mixtures of the (*Z*) (*syn*) and (*E*) (*anti*) isomers (Formulas X and XI, respectively) in about 78% yield. Regeneration of the starting complex proceeds with CF_3SO_3H. With 5 equivalents of the corresponding primary amine H_2NR (not reported for No. 40) in CH_2Cl_2 at room temperature, the cations are converted into FpH and the formamidinium ion $[HC(NHR)_2]^+$; the mixed formamidinium cation is obtained with H_2NCH_3 and No. 42. The rate of reaction decreases in the order $H_2NCH_3 > H_2NC_6H_{11}$-*c* $> H_2NC_4H_9$-*t*. Rapid NH exchange occurs in the title cations by addition of D_2O. Irradiation of No. 42 in CH_3CN solution or reaction of the complex with $(CH_3)_3NO$ in the same solvent produces the substituted carbene complex $[Cp(CO)(CH_3CN)Fe{=}CHNHC_6H_{11}$-*c*$]CF_3SO_3$ (Sect. 1.1.2.2) in 93 and 85% yield, respectively.

X XI XII XIII

[Fp=CDNHC$_6$H$_{11}$-*c*]CF$_3$SO$_3$ (Table **7**, No. **43**) is obtained according to Method IV from $[Fp{=}CDSCH_3]CF_3SO_3$ (32% yield).

With $H_2NC_6H_{11}$ (5 equivalents), both FpH and the ion $[CD(NHC_6H_{11}$-*c*$)_2]^+$ are formed, thus indicating a transfer of a NH proton to the iron atom.

[Fp=CHN(CH$_3$)C$_3$H$_7$-*i*]FSO$_3$ (Table **7**, No. **45**) forms in 77% yield as a mixture of two isomers (presumably (*Z*) and (*E*) isomers, see also Formula XII and XIII) by the addition of CH_3OSO_2F to an ethereal solution of the isomer mixture X and XI (R = C_3H_7-*i*). The isomer distribution varies with the amount of alkylation agent added. Thus, with 6 equivalents or 3 equivalent of CH_3OSO_2F, the relative amounts of 62:32 or 95:5, respectively, are formed. The different isomer distributions probably arise from different rates of methylation of X and XI and of *syn* and *anti* isomerization of the starting complex during alkylation.

The NMR data of the two isomers (A and B in Table 7) are not assigned to a specific (*Z*) or (*E*) configuration.

[Fp=C(CHR$_2$) NHR′]PF$_6$ (Table **7**, No. **47**, R = CH_3, R′ = C_6H_5; No. **48**, R = $COOC_2H_5$, R′ = C_6H_5; No. **49**, R = C_6H_5, R′=CH_3; No. **50**, R = R′=C_6H_5). The cations are prepared by treatment of the appropriate ferraazetidine complex XIV (adducts of the ketimines R_2C=C=NR′ with $[Fp]^-$) with 2 equivalents HCl in ether at − 78 °C (50 to 60% yield).

Nos. 47 and 48 are 3:1 and 4:1 mixtures of isomers differing in the position of the group R’ at the N atom (Formula XII and XIII). The NMR data of the two isomers A and B are not assigned to a specific (*Z*) or (*E*) configuration. From the coalescence temperature, T_c = 78 °C, of the Cp signals of No. 47, a rotational barrier around the C—N axis of 7712 kJ/mol has been determined. The other two compounds bearing the sterically crowding benzhydryl substituent (R = C_6H_5) form only one isomer (presumably the (*E*) isomer, Formula XIII).

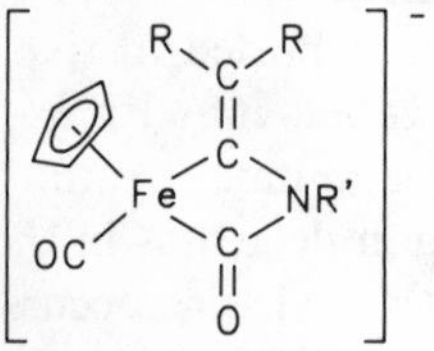

XIV 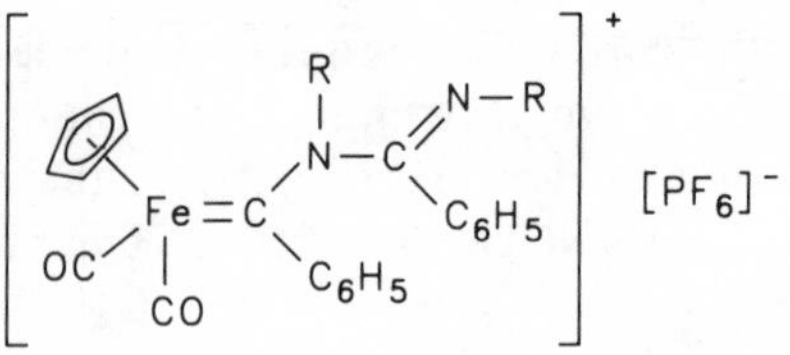 XV

[Fp=C(C_6H_5)NHR]PF$_6$ (Table **7**, No. **51**, R = CH_3; No. **52**, R = C_3H_7-*i*; No. **53**, R = $CH_2C_6H_5$; No. **54**, R = C_6H_5). The compounds are prepared from Na[Fp] in THF and the appropriate benzimide chloride $C_6H_5C(Cl)$=NR followed by addition of HCl. No. 52 forms also by hydrolysis of No. 56.

All compounds are present as (*Z*)/(*E*) isomer mixtures (Formulas XII and XIII, respectively; CHR_2 replaced by C_6H_5); Nos. 52 (*Z*) and 54 (*E*) can be obtained pure by repeated recrystallization.

[Fp=C(C_6H_5)N(CH_3)C(=NCH_3)C_6H_5)]PF$_6$ (Table **7**; No. **55**). The complex (Formula XV, R = CH_3) is prepared by addition of 2 equivalents $C_6H_5C(Cl)$ =NCH_3 to 1 equivalent Na[Fp] in THF and warming the mixture to 40 °C (1 h). Addition of water to an ethanol solution precipitates a mixture of mainly Fp_2 and some XVI. Addition of NH_4PF_6 to the filtrate give the title complex (29% yield). If the above reaction is carried out at 65 °C, the cationic complex XVII can be isolated by the same procedure.

Under the influence of moisture, hydrolysis occurs to give No. 51.

XVI XVII

[Fp=C(C_6H_5)N(C_3H_7-*i*)C(=NC_3H_7-*i*)C_6H_5]Cl (Table 7, No. **56**; Formula XV, R = C_3H_7-i) is prepared with C_6H_5C(Cl)=NC_3H_7-*i* as described above for No. 55. Attempts to create the PF_6^- salt result in hydrolysis at the C–N bond to give the carbene complex No. 52.

[Fp =$C_8H_{12}N_2O_2$]BF_4 (Table **7**, No. **57**) is obtained according to preparation Method IIb from the neutral complex XVIII and [O(C_2H_5)$_3$] BF_4 in CH_2Cl_2 at 20 °C.

The compound reacts with $(CH_3)_3SiN(CH_3)_2$ in $CHCl_3$ to produce No. 58.

XVIII

Compounds with [5]L other than Cp

Only compounds with E = O (5 compounds) and S (1 compound) are known. They are arranged in the order of substitution at the Cp ligand. For better readability the specific [5]L(CO)$_2$Fe groups are abbreviated as Fp*.

[$C_5H_4CH_3$(CO)$_2$Fe=C(CH_3)OFp]BF_4 is obtained by the reaction of the carbene complex [Fp=C(CH_3)OFp]BF_4 (see Table 7, No. 6) with Fp*$COCH_3$.

[$C_5H_4CH_3$(CO)$_2$Fe=C(CH_2Fp)OCH_3]CF_3SO_3 is obtained by alkylation of the acyl complex Fp*$COCH_2$Fp with $CF_3SO_3CH_3$ in CH_2Cl_2 solution at ambient temperature (2 h); orange powder, m.p. 75 °C (49%). For the contribution of other tautomers, see Nos. 14 to 16 (Formula III and IV; the Cp ligand is replaced by $C_5H_4CH_3$).

^{1}H NMR ($CDCl_3$) in ppm: 1.99 (s, CH_3 at C_5H_4), 3.19 (s, CH_2), 4.03 (s, CH_3O), 4.85 to 5.03, 5.03 to 5.14 (m, C_5H_4), 5.23 (s, Cp).

When the complex is treated with various borohydride reagents [HBR_3^-] in THF solution, a 1:2:1 mixture of iron dimers Fp_2, FpFp*, and Fp_2^*, respectively, is produced. Reduction of the complex in basic media (Na-[OCH_3]/CH_3OH) with $NaBH_4$ gives the corresponding Fp*CH(OCH_3)CH_3. Al$(CH_3)_3$ in toluene at − 78 °C produces a mixture of Fp*$COCH_3$ (54%) (via Fp*C(=CH_2)OCH_3, followed by hydrolysis under chromatographic conditions) and FpCH_3 (46%).

[C_9H_7(CO)$_2$Fe=C(CH_3)OC_2H_5)]PF_6 (C_9H_7 = indenyl) is prepared according to Method IIb by addition of [O(C_2H_5)$_3$]PF_6 to a CH_2Cl_2 solution of the corresponding acyl complex; bright yellow, moisture-sensitive powder (87% yield).

^{1}H NMR spectrum in CD_3NO_2 (in ppm): 3.12 (s, CH_3CFe), 5.54 (t, H-2; J = 2.8), 6.09 (d, H-1, 3; J = 2.8). ^{13}C NMR in the same solvent: 46.8 (CH_3), 77.5 (C-1, 3), 98.0 (C-2).

The reaction with iodide in CH_2Cl_2 quantitatively regenerates the starting complex. Treatment with Li[HB(C_2H_5)$_3$] in CH_2Cl_2 or with $NaBH_4$ in methanolic $NaOCH_3$, both at $-80\,°C$, reduces the complex to the neutral complex Fp*CH(CH_3)OC_2H_5 in 60 and 90% yield, respectively.

[$C_5(CH_3)_5(CO)_2Fe{=}CHOH$]$CF_3SO_3$. This complex was detected by ^{1}H NMR spectroscopy to be formed along with the carbene cation [Fp*=CH_2]$^+$ when the hydroxymethylene complex Fp*CH_2OH was treated with $CF_3SO_3Si(CH_3)_3$ in CD_2Cl_2 at $-90\,°C$. As the first step, HO^- abstraction with the formation of the very reactive [Fp*=CH_2]$^+$ cation is suggested; this abstracts H^- from the starting complex to give the title complex and the corresponding Fp*CH_3. The ratio of the two carbene species depends on the ratio of starting complex to $CF_3SO_3Si(CH_3)_3$, and no title complex is observed if excess silyl compound is used. Attempts to directly synthesize the title complex by hydride abstraction with [C(C_6H_5)$_3$]$^+$ give radical products which decompose at $-20\,°C$ to give the decomposition products of [Fp*=CHOH]$^+$ probably via the radical cation [Fp*CH_2OH]$^+$.

The ^{1}H NMR spectrum at $-90\,°C$ shows a singlet at 1.86 ppm (CH_3) and two signals in the ratio 80 to 20 at 16.65 and 16.40 ppm, respectively, for the CH proton. The two signals were attributed to *syn* (Formula XIX) and *anti* (Formula XX, R=H) isomers. On raising the temperature, slow irreversible isomerization of the major kinetic isomer into the thermodynamically more

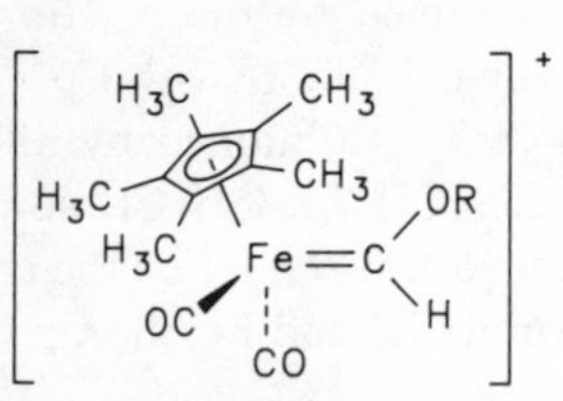

[$(C_5(CH_3)_5)(OC)(CO)Fe{=}C(OR)H$]$^+$

XIX

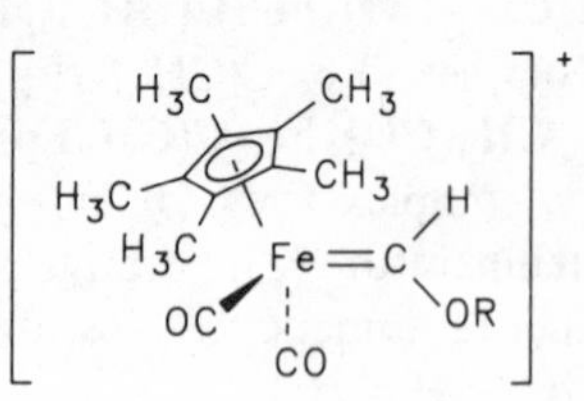

[$(C_5(CH_3)_5)(OC)(CO)Fe{=}C(H)OR$]$^+$

XX

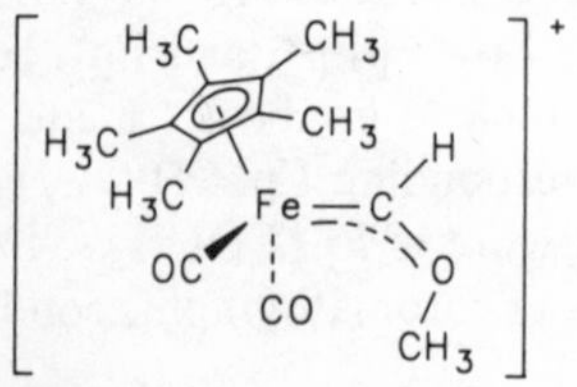

[$(C_5(CH_3)_5)(OC)(CO)Fe{=}C(H){=}O{-}CH_3$]$^+$

XXI

[$(C_5(CH_3)_5)(OC)(CO)Fe{=}C(H){=}O{-}CH_3$]$^+$

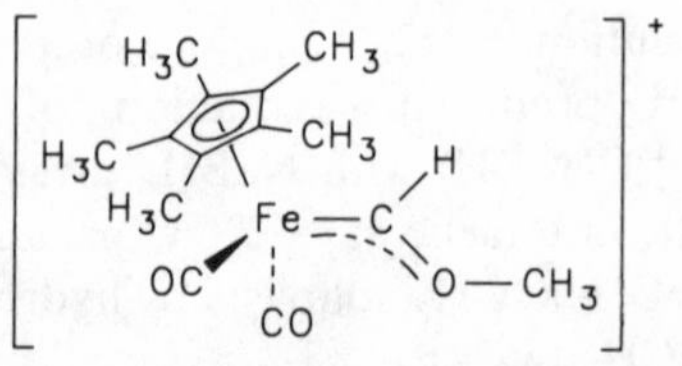

XXII

stable isomer (16.4 ppm) occurs; it is complete at − 40 °C; see also general remarks.

$[C_5(CH_3)_5(CO)_2Fe{=}CHOCH_3]X$ ($X{=}PF_6$, CF_3SO_3, BF_4). The PF_6^- salt is obtained according to Method I from the corresponding ether $Fp^*CH_2OCH_3$ and $[C(C_6H_5)_3]PF_6$ at − 80 °C in CH_2Cl_2 solution as a pale yellow air-stable solid, but is contaminated with a small amount (5 to 10%) of $[C_5(CH_3)_5Fe(CO)_3]PF_6$ (overall yield 71%). The $CF_3SO_3^-$ and BF_4^- salts are generated in an NMR tube in CD_2Cl_2 solution at − 80 °C from the starting complex and $(CH_3)_3SiOSO_2CF_3$ or $HBF_4 \cdot O(CH_3)_2$, respectively. At first, $[CH_3O]^-$ abstraction occurs to give the carbene cation $[Fp^*{=}CH_2]^+$ which then abstracts H^- from the starting material to generate the title complex and Fp^*CH_3. The compounds obtained at − 80 °C are formed as a mixture of *cis* and *trans* isomers XXI and XXII (90:10 ratio), as shown by variable-temperature NMR spectroscopy (see below), owing to restricted rotation about the carbene oxygen bond. On warming to room temperature, isomerization occurs to the thermodynamically more stable *trans* isomer XXII. Similar results are reported from the corresponding Ru compound.

The ^1H NMR spectrum in CD_2Cl_2 of the PF_6^- salt at 26 °C shows the following signals (in ppm): 1.94 (s, CH_3C), 4.67 (s, CH_3O), 12.27 (s, CH). The ^1H NMR spectrum of a sample prepared at − 80 °C exhibits at this temperature two broad signals at 12.23 and 12.68 ppm corresponding to the isomers XXII and XXI, respectively. At − 60 °C the signal at 12.68 irreversibly decreases, while the other one increases and at − 50 °C the isomerization is complete. ^{13}C NMR (CD_2Cl_2): 9.8 (CH_3C), 78.5 (CH_3O), 103.4 (C_5), 209.6 (CO), 325.3 (C=Fe).

The complex slowly decomposes in solution to give the cation $[C_5(CH_3)_5Fe(CO)_3]^+$. The complex adds $P(C_6H_5)_3$ at − 80 °C to give quantitatively the unstable ylide complex $[Fp^*CH(OCH_3)P(C_6H_5)_3]PF_6$. With the silanes $(CH_3)_2C_6H_5SiH$ and $(C_2H_5)_3SiH$, insertion of the methoxymethylene into the SiH bond occurs with formation of the corresponding $R_3SiCH_2OCH_3$ compounds in about 60% yield; in the presence of alkenes $RCH{=}CH_2$ (ethylene or styrene) the iron is isolated as the corresponding alkene complexes $[Fp^*(\eta^2\text{-}CH_2{=}CHR)]^+$. $(C_6H_5)_3SiH$ gives a mixture of $(C_6H_5)_3SiOCH_3$ and $(C_6H_5)_3SiCH_3$. The PF_6^- salt reacts with $(C_5(CH_3)_5)Fe(\eta^5\text{-}C_6(CH_3)_5{=}CH_2)$ in THF at − 80 °C to yield the bimetallic complex XXV in 90% yield.

$[C_5(CH_3)_5(CO)_2Fe{=}CHSC_6H_5]PF_6$ is generated from the corresponding thioether $Fp^*CH_2SC_6H_5$ and $[C(C_6H_5)_3]PF_6$ (see Method I) as a mixture with the ylide complex XXIV in a 40:60 molar ratio. Monitoring the reaction by variable-temperature NMR spectroscopy shows that both compounds are produced simultaneously. The spectroscopic data exhibit the presence of *cis* (XXIIIa, 90%) and *trans* (XXIIIb, 10%) isomers.

^1H NMR in CD_2Cl_2 (in ppm), XXIIIa/XXIIIb: 1.98/1.87 (s, CH_3), 14.76/14.30 (s, CH=Fe). ^{13}C NMR (same conditions): 9.9/9.7 (CH_3), 104.3/104.5 (C of C_5), 211.2 (CO), 319.1/317.0 (C=Fe).

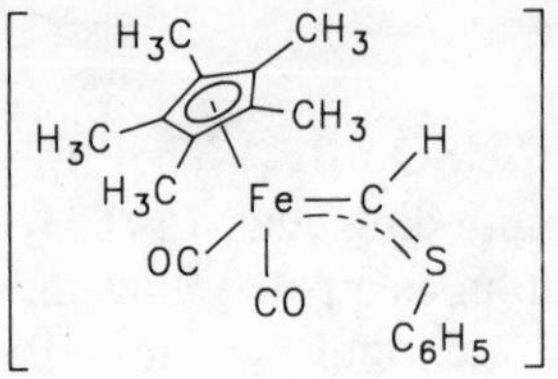

XXIIIa

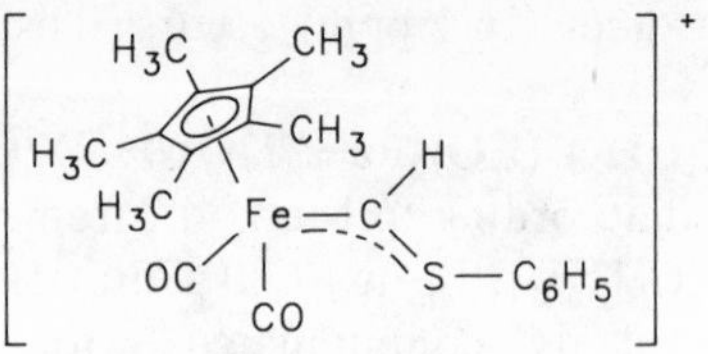

XXIIIb

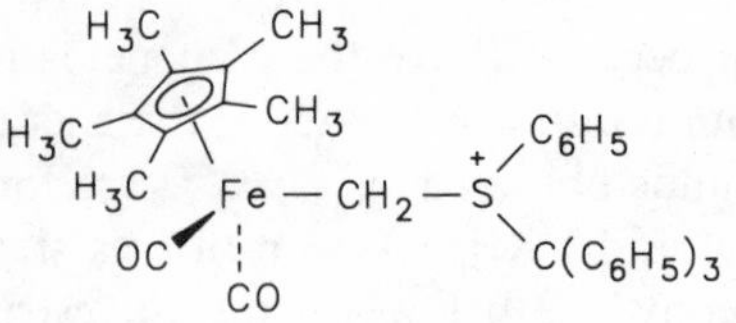

XXIV

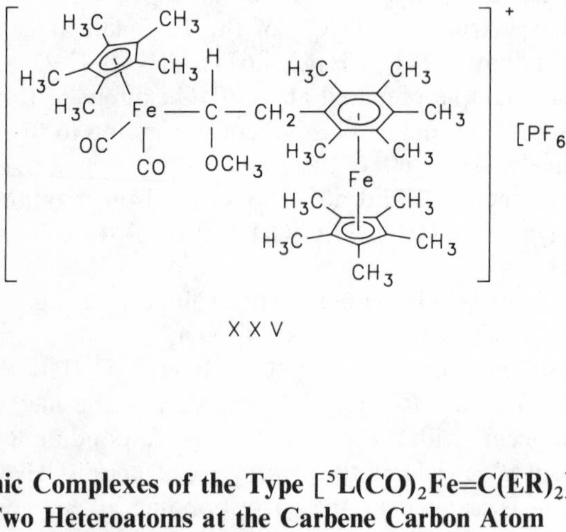

XXV

1.1.3.3 Cationic Complexes of the Type $[^5L(CO)_2Fe{=}C(ER)_2]^+$ with Two Heteroatoms at the Carbene Carbon Atom

In this section compounds of the general type $[^5L(CO)_2Fe{=}C(ER)_2]X$ are described with ER = F, Cl, Br, OR, SR, SeR, and NR_2. The two heteroatoms, E, at the carbene carbon atom can be equal or different.

R means alkyl, aryl, as well as various neutral 17-electron transition metal fragments ML_n, or a five-membered ring, formed with the =CEE unit, as depicted in Formula I, for which the resonance Formulas Ia to Ic can be formulated. As shown by the resonance Formula Ib the endocyclic iron atom can also be viewed as a cationic carbene complex of the general type $[^5L(CO)(^2D)Fe{=}C(ER)_2]^+$ described in Sect. 1.1.3.2.

Also included are cationic compounds in which the Fp=CEE fragment coordinates at a transition metal fragment with the heteroatoms E as a chelating ligand, forming a four-membered ring system (see compounds Nos. 8 and 40).

With exception of one complex containing the $C_5(CH_3)_5$ ligand at the end of this section, all other complexes possess the C_5H_5 (Cp) ring as ^{5}L ligand and the $C_5H_5(CO)_2Fe$ group is abbreviated as Fp.

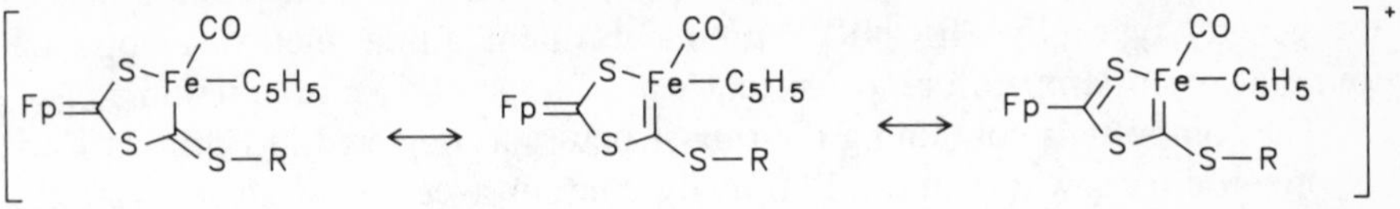

Ia Ib Ic

The compounds listed in Table 8 can be prepared according to the following general methods:

Method I: Alkylation of the neutral complex FpC(S)ER (E=O, S, or Se) with $ROSO_2CF_3$, $ROSO_2F$, or RX compounds.

Method II: Reaction of the complexes $[FpCO]CF_3SO_3$ or $[FpCS]CF_3SO_3$ with the 3-membered heterocycles II (Z = O, S, NH).
a. The cation is $[Fp(CO)]^+$
b. The cation is $[Fp(CS)]^+$

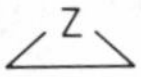

II

Method III: ER_n substitution by NR_2 in appropriate carbene compound.
a. Replacement of the SCH_3 group from $[Fp{=}C(OCH_3)SCH_3]^+$
b. Replacement of one SCH_3 group from $[Fp{=}C(SCH_3)_2]^+$
c. Replacement of both SCH_3 groups from $[Fp{=}C(SCH_3)_2]^+$

Method IV: Reaction of the carbene complex I (R = CH_3; Table 8, No. 42) with HNR_2.

General remarks. Both groups R are *syn* and *anti* oriented as shown in Formula III according to the ability of the heteroatom E to donate some π-electron density to the carbene carbon atom. This also decreases the barrier to rotation about the C=Fe bond compared to similar cations without heteroatoms. However, if both R's are CH_3's the room-temperature NMR spectrum exhibits only one signal for the CH_3 groups because of rapid rotation about the carbene-E bond on the NMR time scale. If E = S (No. 24) at low temperature the CH_3 groups become nonequivalent, resulting in two signals which broaden

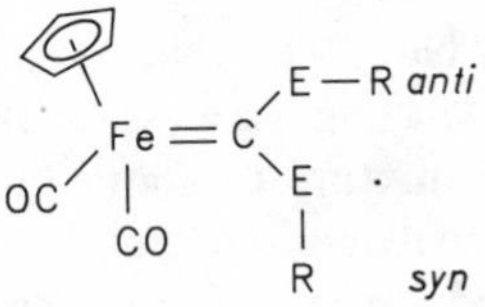

III

on warming and finally coalesce at $-2.5\,°C$. A similar temperature dependence is observed for other groups R (higher rotation barrier with bulkier phenyl groups) and also if E = Se, but not observed for E = O with weaker π donation ability. If one ER group is $N(CH_3)_2$ (see complex No. 51 or other compounds of the general type $[Fp{=}(ER)NR_2]^+$), nonequivalent amine methyl groups are found at room temperature, consistent with a large N–C π interaction.

The compounds containing the ring structure as depicted in Formula I are asymmetric as a whole and additionally contain a center of chirality at the endocyclic iron atom. Based on the two favored orientations of the carbene plane (dihedral angle 90 or 0° with respect to the Fp symmetry plane), four idealized conformations can be obtained by rotation about the Fp=C bond; see also general remarks in Sect. 1.1. The dihedral angles of the compounds in this section are 0° (No. 2), 41° (No. 49), 73° (No. 43), and 90° (No. 24) as shown by X-ray structure determinations.

The compounds generally do not form stable adducts with neutral bases, resulting in cationic ylide complexes. However, these ylides are considered to be intermediates formed by nucleophilic attack of primary or secondary amines at the carbene carbon atom during various substitution reactions described in preparation method III. The reactivity of such an intermediate may be controlled by the relative leaving-group ability of the carbene substituents ER. A correlation between the increasing pK_a values of the corresponding HER compounds and the leaving-group ability is found to follow the order $C_6H_5Se^-$ $=C_6H_5S^-$, $>C_6H_5O^-$ $>CH_3S^-$ $>CH_3O^-$. If both ER's are good leaving groups, a competitive reaction with primary amines can occur to give also, or even exclusively, the isocyanide compounds, $[Fp(CNR)]^+$; see also a report on the formation of product mixtures with both ER groups of similar leaving-group abilities at the carbene carbon atom.

A stepwise replacement of CH_3S^- with diamines, amino alcohols, and amino thiols (preparation method IIIc) via a carbene intermediate of the type $[Fp{=}C(SCH_3)NH{-}R{-}EH]^+$ (E = NH, O, S) is discussed.

Explanation for Table 8. The Cp rings appear as singlets in the NMR spectra (in ppm). ^1H NMR: 5.50 to 5.65 in acetone-d_6, 5.15 to 5.30 in CD_3CN, 6.00 in CD_2Cl_2, and 5.10 in $CDCl_3$; ^{13}C NMR: 88 to 89.5 in acetone-d_6, 88 to 89.5 in CD_3CN, and 95.5 in CD_2Cl_2. Due to the presence of the Fp group all compounds exhibit two strong ν(CO) stretching frequencies in the IR spectrum in the range of 1900 to 2100 cm^{-1}.

Further Information

$[Fp{=}CF_2]BF_4$ (Table **8**, No. **1**) is obtained as an extremely hygroscopic white solid by condensation of a slight excess of BF_3 into a benzene solution of $FpCF_3$ and warming of the mixture to room temperature; (81% yield). The cation was also detected by the reaction of $FpCF_3$ with $SnCl_4$ and it is generally suggested to be an intermediate in the halogene-exchange reaction of this complex with BCl_3 or $AlCl_3$ to give $FpCCl_3$. The $B(C_6H_5)_4^-$ salt forms in about

Table 8. Cationic Carbene Complexes of the Type $[Fp{=}C(ER)_2]^+$. Further information on numbers preceded by an asterisk is given at the end of the table.

No.	$Fe{=}C(ER)_2$	anion, method of preparation (yield) properties and remarks
heteroatoms E = F		
*1	$Fe{=}CF_2$	BF_4^- salt, white solid IR (Nujol): 2071, 2115; ν(CF) 1200, 1233 $B(C_6H_5)_4^-$ salt, white solid
heteroatoms E = Cl		
*2	$Fe{=}CCl_2$	BCl_4^- salt, orange crystals ^{1}H NMR (CD_2Cl_2): 5.99 (Cp) ^{13}C NMR (CD_2Cl_2 at − 40 °C): 95.9 (Cp), 202.2 (CO), 319.5 (C=Fe)
heteroatoms E = Br		
3	$Fe{=}CBr_2$	detected by the preparation of $FpCBr_3$
heteroatoms E = O		
*4	$Fe{=}C(OCH_3)_2$	PF_6^- salt; pale yellow, m.p. 200 to 205 °C (dec.) ^{1}H NMR (CD_3COCD_3): 4.42 (s, CH_3) ^{13}C NMR (CD_3COCD_3): 63.6 (CH_3O), 210.3 (CO), 251.9 (C=Fe) $CF_3SO_3^-$ salt, pale yellow, m.p. 128 to 131 °C ^{1}H NMR (CD_3COCD_3): 4.41 (s, CH_3) ^{13}C NMR (CD_3COCD_3/CD_3CN): 63.8/63.9 (CH_3), 210.3 (CO), 251.8/251.9 (C=Fe)
*5	$Fe{=}C(OCH_3)OC_2H_5$	$CF_3SO_3^-$ salt; yellow needles, m.p. 66 to 68 °C ^{1}H NMR (CD_3COCD_3): 1.50 (t, CH_3), 4.37 (s, CH_3O), 4.82 (q, CH_2) ^{13}C NMR (CD_3CN): 14.3 (CH_3), 74.9 (CH_2), 63.4 (CH_3O), 210.2 (CO), 251.9 (C=Fe)
*6	$Fe{=}C(OCH_3)OC_6H_{11}$-*c*	BF_4^- salt ^{1}H NMR (CD_3NO_2): 4.35 (s, CH_3O)
*7	$Fe{=}C(OCH_2CH_2O)$ (cyclic)	PF_6^- salt, IIa (81%); pale yellow, m.p. 123 to 126 °C ^{1}H NMR (CD_3CN): 4.74 (s, CH_2) ^{13}C NMR (CD_3CN): 73.93 (CH_2), 209.66 (CO), 242.51 (C=Fe)

Table 8. Continued

No.	$Fe{=}C(ER)_2$	anion, method of preparation (yield) properties and remarks
*8	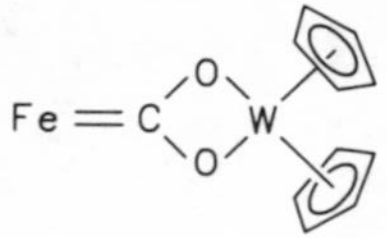	BF_4^- salt ^{1}H NMR (CD_3COCD_3): 5.21 (CpFe), 6.13 (CpW) ^{13}C NMR (CD_3COCD_3): 87.7 (Cp), 98.4 (Cp), 211.9 (CO), 259.5 (C=Fe) ^{17}O NMR: 168 (CO_2), 369 (CO)
heteroatoms E = O, S		
*9	$Fe{=}C(OCH_3)SCH_3$	PF_6^- salt, I (41%); golden yellow, m.p. 105 °C ^{1}H NMR (CD_3COCD_3): 2.67 (s, CH_3S), 4.91 (s, CH_3O) $CF_3SO_3^-$ salt, I (92%); yellow, m.p. 92 to 95 °C ^{13}C NMR (CD_3COCD_3): 21.5 (CH_3S), 71.8 (CH_3O), 210.0 (CO), 297.9 (C=Fe)
*10	$Fe{=}C(OC_6H_5)SCH_3$	PF_6^- salt, I (43%); yellow, m.p. 153 to 154 °C ^{1}H NMR (CD_3CN): 2.78 (s, CH_3) ^{13}C NMR (CD_3COCD_3, CD_3CN): 22.5 (CH_3), 209.0 (CO), 301.1 (C=Fe) FSO_3^- salt, I
heteroatoms E = O, N		
11	$Fe{=}C(OCH_3)NH_2$	$CF_3SO_3^-$ salt IIIa (81%); yellow, m.p. 119 to 121 °C ^{1}H NMR (CD_3COCD_3): 4.08 (s, CH_3), 9.18, 9.90 (br, NH_2) ^{13}C NMR (CD_3COCD_3): 59.5 (CH_3), 211.9 (CO), 227.2 (C=Fe)
*12	$Fe{=}C(OCH_3)NHCH_3$	$CF_3SO_3^-$ salt, IIIa (50%); yellow crystals, mixture with No. 11 (20%), m.p. 128 to 134 °C ^{1}H NMR (CD_3COCD_3): 2.97 (d, CH_3N), 4.26 (s, CH_3O), 9.70 (br, NH) ^{13}C NMR (CD_3COCD_3): 31.8 (CH_3N), 63.7 (CH_3O), 211.5 (CO), 224.6 (C=Fe)
13	$Fe{=}C(OCH_3)NHCH_2CH_2OH$	$CF_3SO_3^-$ salt, IIIa (68%); yellow, m.p. 124 to 128 °C ^{1}H NMR (CD_3COCD_3): 3.62 (m, CH_2), 4.03 (t, HO), 4.25 (s, CH_3O), 9.63 (br, NH) ^{13}C NMR (CD_3COCD_3): 48.5 (CH_2N), 60.1 (CH_2O), 63.8 (CH_3O), 211.6 (CO), 225.9 (C=Fe) reflux in CH_3CN gives No. 18

No.	Compound	Data
*14	$Fe{=}C(OCH_3)NHCH_2C_6H_5$	$CF_3SO_3^-$ salt, IIIa (53%); yellow, m.p. 118 to 122 °C ^1H NMR (CD_3COCD_3): 4.27 (s, CH_3O), 4.63 (d, CH_2N), 10.23 (NH) ^{13}C NMR (CD_3COCD_3): 31.8 (CH_2N), 63.9 (CH_3O), 211.5 (CO), 226.1 (C=Fe)
15	$Fe{=}C(OCH_3)N(CH_3)_2$	$CF_3SO_3^-$ salt, IIIa (68%); yellow, m.p. 174 to 178 °C (dec.) ^1H NMR (CD_3COCD_3): 3.27, 3.62 (s's, CH_3N) 4.25 (s, CH_3O) ^{13}C NMR (CD_3COCD_3): 40.9, 46.6 (CH_3N), 63.4 (CH_3O), 211.7 (CO), 223.6 (C=Fe)
16	Fe=C(N)(OCH$_3$) (N in piperidine ring)	$CF_3SO_3^-$ salt, IIIa (60%); yellow orange, m.p. 141 to 143 °C ^1H NMR (CD_3COCD_3): 3.94, 4.13 (t's, CH_2N) 4.25 (s, CH_3O) ^{13}C NMR (CD_3CN): 50.1, 56.6 (CH_2N), 63.4 (CH_3O), 211.6 (CO), 221.4 (C=Fe)
*17	Fe=C(N)(OC$_6$H$_5$) (N in piperidine ring)	PF_6^- salt; yellow ^1H NMR (CD_3COCD_3): 4.23 (m, CH_2N)
*18	Fe=C(N–H)O (oxazolidine ring)	PF_6^- salt, IIa (99%), IIIc (73%); yellow, m.p. 134 °C ^1H NMR (CD_3CN): 3.70 (t, CH_2N), 4.60 (t, CH_2O), 9.50 (s, HN) ^{13}C NMR (CD_3CN): 46.11 (CH_2N), 73.11 (CH_2O), 211.31 (CO), 220.24 (C=Fe) $CF_3SO_3^-$ salt, IIa (90%); pale yellow, m.p. 118 °C ^1H NMR: 3.89, 4.80 (AA'BB', CH_2) in CD_3COCD_3; 3.71 (t, CH_2N), 4.62 (t, CH_2O) in CD_3CN ^{13}C NMR (CD_3COCD_3): 46.20 (CH_2N), 73.24 (CH_2O), 211.24 (CO), 220.08 (C=Fe)
*19	Fe=C(N–CH$_3$)O (oxazolidine ring)	PF_6^- salt; light yellow ^1H NMR (CD_3CN): 3.27 (s, CH_3), 3.77 (t, CH_2N), 4.60 (t, CH_2O) ^{13}C NMR (CD_3CN): 37.2 (CH_3), 53.0 (CH_2N), 72.8 (CH_2O), 211.4 (CO), 218.8 (C=Fe)
*20	Fe=C(N–CH$_2$–CH=CH$_2$)O (oxazolidine ring)	PF_6^- salt; light yellow ^1H NMR (CD_3CN): 3.76 (t, CH_2N), 4.32 (d, CH_2 of C_3H_5), 4.66 (t, CH_2O) ^{13}C NMR (CD_3CN): 50.4 (CH_2N of C_3H_5), 53.2 (CH_2N), 73.2 (CH_2O), 220.3 (C=Fe), 221.2 (CO)

Table 8. Continued

No.	$Fe{=}C(ER)_2$	anion, method of preparation (yield) properties and remarks
*21	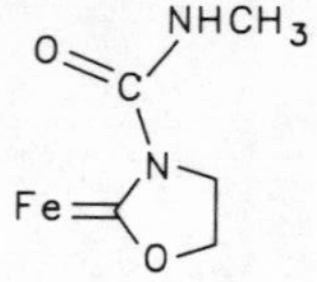	PF_6^- salt; orange ^1H NMR (CD_3CN): 4.00 (t, CH_2N), 4.78 (t, CH_2O) ^{13}C NMR (CD_3CN): 50.4 (CH_2N), 74.9 (CH_2O), 147.5 (C=O), 210.9 (CO)
*22	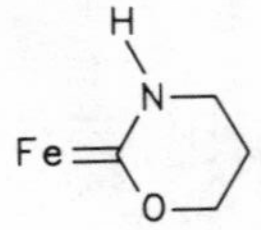	PF_6^- salt; yellow-brown, m.p. 116 °C ^1H NMR (CD_3CN): 2.13 (m, CH_2), 3.30 (m, CH_2N), 4.34 (t, CH_2O), 9.04 (s, HN) ^{13}C NMR (CD_3CN): 20.71 (CH_2), 42.05 (CH_2N), 70.98 (CH_2O), 212.09 (CO), 218.76 (C=Fe)
heteroatoms E = S		
*23	Fe=C(SH)SFp	$CF_3SO_3^-$ salt ^1H NMR (CD_3COCD_3): 5.55, 5.63 (Cp) BF_4^- salt; yellow brown, m.p. 115 to 118 °C (dec.) ^1H NMR (CD_3COCD_3): 5.34, 5.50 (Cp), 9.2 (br, HS) ^{13}C NMR (CD_3CN): 85.0, 86.6 (Cp), 209.9, 211.2 (CO)
*24	$Fe{=}C(SCH_3)_2$	PF_6^- salt, I (69%); yellow, m.p. 163 to 165 °C ^1H NMR (CD_3COCD_3): 3.27 (s, CH_3); at − 55 °C: 3.13, 3.47 (s's, CH_3, coalescence at − 2.5 °C) ^{13}C NMR (CD_3COCD_3): 29.6 (CH_3), 209.5 (CO), 303.1 (C=Fe) $CF_3SO_3^-$ salt, I (87%); dark yellow, m.p. 76 to 78 °C ^1H NMR (CD_3COCD_3): 3.31 (s, CH_3) ^{13}C NMR (CD_3CN): 30.3 (s, CH_3), 210.5 (CO), 304.2 (C=Fe)
25	$Fe{=}C(SCH_3)SC_2H_5$	PF_6^- salt, I (60%); yellow, m.p. 117 to 119 °C ^1H NMR (CD_3COCD_3): 3.31 (s, CH_3S), 3.84 (q, CH_2)
26	$Fe{=}C(SCH_3)SCH_2C_6H_5$	PF_6^- salt, I (64%); bright yellow, m.p. 150 °C (dec.) ^1H NMR (CD_3COCD_3): 3.37 (s, CH_3), 5.01 (s, CH_2)

No.	compound	properties
*27	$Fe{=}C(SCH_3)SC_6H_5$	PF_6^- salt, I (75%); bright yellow, m.p. 155 to 157 °C ^1H NMR (CD_3CN): 3.12 (s, CH_3) ^{13}C NMR (CD_3COCD_3, 20 °C/CD_3CN, 73 °C): 30.4/31.3 (CH_3), 209.6/209.7 (CO), 308.9/309.4 (C=Fe) $CF_3SO_3^-$ salt, I ^1H NMR (CD_3COCD_3): 3.28 (s, CH_3)
*28	$Fe{=}C(SCH_3)SFp$	$CF_3SO_3^-$ salt, I; yellow, m.p. 140 °C (dec.) ^1H NMR (CD_3COCD_3): 3.35 (s, CH_3), 5.51, 5.43 (Cp) ^{13}C NMR (CD_3COCD_3): 29.3 (CH_3), 88.3, 89.0 (Cp), 211.5, 2.11.8 (CO), 315.4 (C=Fe) PF_6^- salt, I (60%) ^1H NMR (CD_3COCD_3): 3.35 (s, CH_3), 5.43, 5.51 (Cp) ^{13}C NMR (CD_3COCD_3): 29.3 (CH_3), 88.3, 89.0 (Cp), 206.8 (CO), 315.4 (C=Fe) I^- salt, I (84%); m.p. 118 °C (dec.) ^1H NMR (CD_3COCD_3): 3.38 (s, CH_3), 5.49, 5.58 (Cp)
29	$Fe{=}C(SCH_3)SRu(CO)_2Cp$	$CF_3SO_3^-$ salt, I; yellow, m.p. 134 °C (dec.) ^1H NMR (CD_3COCD_3): 3.30 (s, CH_3), 5.28, 5.67 (Cp)
*30	$Fe{=}C(SCH_3)SRe(CO)_5$	$CF_3SO_3^-$ salt, I; yellow, m.p. 122 °C (dec.) ^1H NMR (CD_3COCD_3): 3.49 (s, CH_3) ^{13}C NMR (CD_3COCD_3): 187.1 (CORe-*cis*), 206.8 (CORe- *trans*), 211.8, (COFe), 321.4 (C=Fe) PF_6^- salt, I (> 95%); yellow ^1H NMR ($CDCl_3$ and CD_3COCD_3): 3.49 (s, CH_3) BF_4^- salt; light yellow ^1H NMR (CD_3COCD_3): 3.5 (s, CH_3)
31	$Fe{=}C(SC_2H_5)SFp$	$CF_3SO_3^-$ salt, I (92%); yellow, m.p. 137 °C (dec.) ^1H NMR (CD_3COCD_3): 1.58 (t, CH_3), 4.00 (q, CH_2), 5.51, 5.58 (Cp) ^{13}C NMR (CD_3COCD_3): 13.3 (CH_3), 41.4 (CH_2), 88.4, 89.1 (Cp), 211.6, 2.11.8 (CO), 315.5 (C=Fe) reacts with I^- to give $FpCSSC_2H_5$ (see No. 28)
32	$Fe{=}C(SC_2H_5)SRe(CO)_5$	$CF_3SO_3^-$ salt, I; yellow, m.p. 116 °C (dec.) ^1H NMR (CD_3COCD_3): 4.02 (q, CH_2) ^{13}C NMR (CD_3COCD_3): 18.3 (CH_3), 35.3 (CH_2), 187.6 (CORe-*cis*), 206.1 (CORe-*trans*), 211.8, (COFe), 323.5 (C=Fe)
33	$Fe{=}C(SCH_2C_6H_5)SFp$	Br^- salt, I (81%) ^1H NMR (CD_3COCD_3): 5.08 (s, CH_2), 5.33, 5.40 (Cp) reacts with I^- to give $FpCSSCH_2C_6H_5$ (see No. 28)

Table 8. Continued

No.	Fe=C(ER)$_2$	anion, method of preparation (yield) properties and remarks
34	Fe=C(SCH$_2$CH=CH$_2$)SFp	Br$^-$ salt, I (78%) ^{1}H NMR (CD$_3$COCD$_3$): 5.25, 5.42 (Cp) reacts with I$^-$ to give FpCSSCH$_2$CH=CH$_2$ (see No. 28)
*35	Fe=C(SC$_6$H$_5$)$_2$	PF$_6^-$ salt ^{1}H NMR (CD$_3$COCD$_3$): 5.43 (Cp)
*36	Fe=C(SFp)$_2$	BF$_4^-$ salt; black-brown, dec. p. 85 °C ^{1}H NMR (CDCl$_3$): 5.11 (5 H), 5.28 (10 H)
*37	Fe=C(SFp)(SRe(CO)$_5$)	BF$_4^-$ salt; red, air-stable crystals ^{1}H NMR (CD$_2$Cl$_2$): 5.4, 5.6 (Cp)
*38	Fe= S S	PF$_6^-$ salt, IIb (75%), IIIc (78%); dark yellow, m.p. 87 °C ^{1}H NMR: 4.13 (s, CH$_2$) in CD$_3$COCD$_3$; 3.90 (s, CH$_2$) in CD$_3$CN ^{13}C NMR (CD$_3$CN): 48.54 (CH$_2$), 209.92 (CO), 295.40 (C=Fe)
*39	Fe= S S	PF$_6^-$ salt, IIIc (60%); yellow, m.p. 166 to 168 °C ^{1}H NMR (CD$_3$COCD$_3$): 2.55 (m, CH$_2$), 3.43 (t, CH$_2$) ^{13}C NMR (CD$_3$COCD$_3$): 19.1 (CH$_2$), 39.6 (SCH$_2$), 210.9 (CO), 283.7 (C=Fe) BF$_4^-$ salt, oil (impure) ^{1}H NMR (CD$_3$COCD$_3$): 2.55 (m, CH$_2$), 3.43 (t, SCH$_2$) ^{13}C NMR (CD$_3$CN): 18.9 (CH$_2$), 39.6 (SCH$_2$), 210.5 (CO), 283.9 (C=Fe)
*40	Fe=C S S Pt P(C$_6$H$_5$)$_3$ P(C$_6$H$_5$)$_3$	BF$_4^-$ salt; yellow plates, dec. p. 170 °C
*41	Fe= S S CO Fe–C$_5$H$_5$ SH	CF$_3$SO$_3^-$ salt; olive green oil

No.	Properties and data
*42	$CF_3SO_3^-$ salt; deep green, m.p. 130 to 133 °C (dec.) ^{1}H NMR ($CDCl_3$): 3.15 (s, CH_3), 5.17, 5.32 (Cp) ^{13}C NMR (CD_2Cl_2): 29.82 (CH_3), 87.94, 88.21 (Cp), 209.24, 210.12, 210.21 (CO), 286.11 (C=Fe at FeCp(CO)), 312.12 (C=Fe at Fp)
*43	$CF_3SO_3^-$ salt; deep green, m.p. 119 to 121 °C (dec.) ^{1}H NMR (probably $CDCl_3$): 5.23, 5.38 (Cp) ^{13}C NMR (CD_2Cl_2): 12.14 (CH_3), 41.07 (CH_2), 88.00, 88.38 (Cp), 210.20, 210.29, 210.32 (CO), 286.21 (C=Fe at FeCp(CO)), 311.21 (C=Fe at Fp)
44	$CF_3O_3^-$ salt ^{1}H NMR (CD_3COCD_3): 5.13, 5.53, 5.58 (Cp)
*45	$CF_3SO_3^-$ salt; green, m.p. 140 to 145 °C ^{1}H NMR (CCl_3): 5.16, 5.54 (Cp)
46	$CF_3O_3^-$ salt, IV (59%) ^{1}H NMR ($CDCl_3$): 3.70, 3.79 (s's, CH_3), 4.80, 5.15 (Cp)
47	$CF_3SO_3^-$ salt, IV (48%) ^{1}H NMR ($CDCl_3$): 4.81, 5.17 (both Cp)

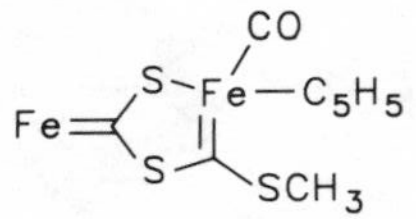

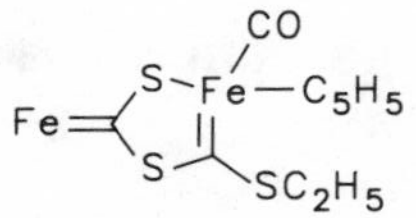

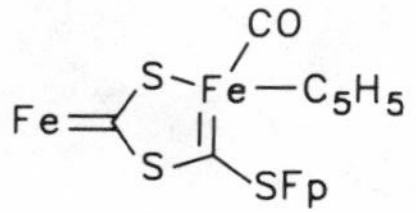

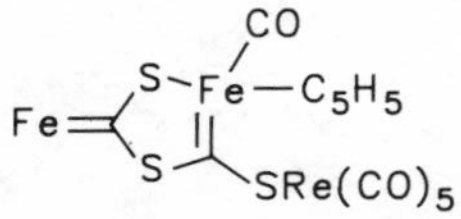

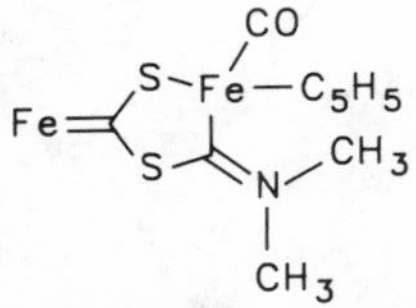

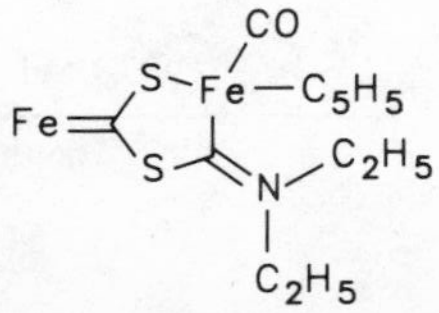

Table 8. Continued

No.	$Fe{=}C(ER)_2$	anion, method of preparation (yield) properties and remarks
48	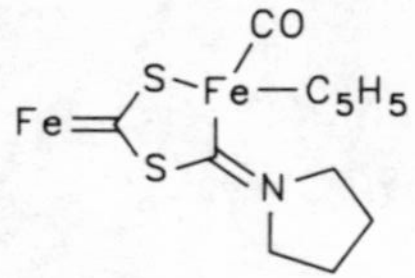	$CF_3SO_3^-$ salt, IV (61%) ^1H NMR ($CDCl_3$): 4.86, 5.17 (both Cp)
*49	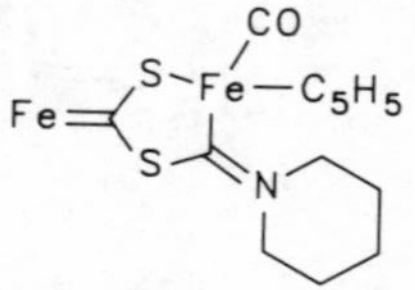	$CF_3SO_3^-$ salt, IV (66%); red orange, m.p. 136 to 138 °C ^1H NMR ($CDCl_3$): 4.79, 5.18 (both Cp) ^{13}C NMR (CD_2Cl_2): 59.24, 63.85 (CH_2N), 84.92, 87.71 (Cp), 210.71 (CO of Fp), 213.63 (CO of Fe(CO)Cp), 260.91 (CFeN), 289.42 (C=Fe)
heteroatoms E = S, Se		
*50	$Fe{=}C(SCH_3)SeC_6H_5$	PF_6^- salt, I (74%); yellow, m.p. 142 to 145 °C ^1H NMR (CD_3CN): 3.16 (s, CH_3) ^{13}C NMR (CD_3CN at 73 °C): 34.4 (CH_3), 209.8 (CO), 321.0 (C=Fe)
heteroatoms E = S, N		
*51	$Fe{=}C(SCH_3)N(CH_3)_2$	PF_6^- salt, IIIb (44%); yellow, m.p. 190 °C (dec.) ^1H NMR (CD_3COCD_3): 2.96 (s, CH_3S), 3.76, 4.02 (s's, CH_3N)
*52	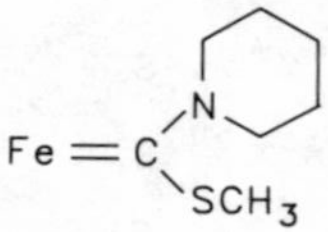	PF_6^- salt, IIIb (69%); yellow, m.p. 153 to 155 °C ^1H NMR (CD_3COCD_3): 2.91 (s, CH_3S) ^{13}C NMR (CD_3CN): 26.4 (CH_3S), 59.5, 64.9 (CH_2N), 211.4 (CO), 237.6 (C=Fe)

53 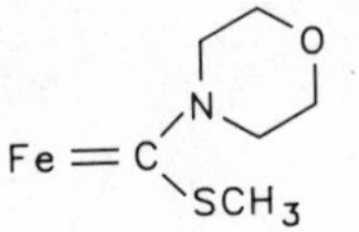 PF_6^- salt, IIIb (64%); yellow, m.p. 185 to 187 °C (dec.)
^{1}H NMR (CD_3COCD_3): 2.94 (s, CH_3), 3.94, 4.46 (m's, CH_2)

54 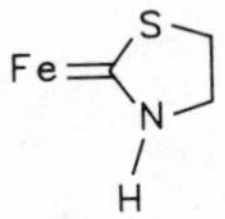 PF_6^- salt, IIIc (48%), IIb (83%); yellow-brown, m.p. 168 to 170 °C
^{1}H NMR (CD_3CN): 3.69 (t, CH_2N), 4.45 (t, CH_2S)
^{13}C NMR (CD_3CN): 33.45 (CH_2N), 56.77 (CH_2S), 211.31 (CO), 234.36 (C=Fe)

heteroatoms E = N

55 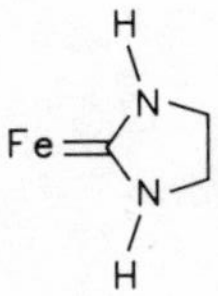 PF_6^- salt, IIIc (54%); pale yellow, m.p. 183 to 185 °C
^{1}H NMR (CD_3COCD_3): 3.78 (s, CH_2)
^{13}C NMR (CD_3COCD_3): 46.4 (CH_2), 200.0 (C=Fe), 212.2 (CO)

56 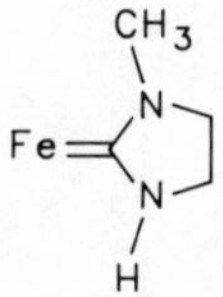 PF_6^- salt, IIIc (66%); yellow, m.p. 178 to 180 °C
^{1}H NMR (CD_3COCD_3): 3.32 (s, CH_3), 3.81 (s, CH_2)

57 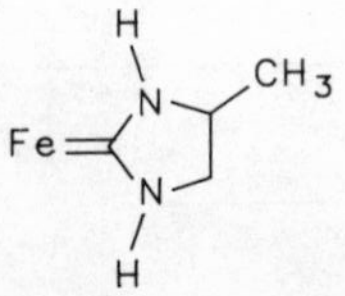 PF_6^- salt, IIIc (85%); yellow, m.p. 177 to 179 °C
^{1}H NMR (CD_3COCD_3): 1.31 (d, CH_3), 3.30, 4.40 (m's, CH_2CH)

Table 8. Continued

No.	$Fe{=}C(ER)_2$	anion, method of preparation (yield) properties and remarks
58	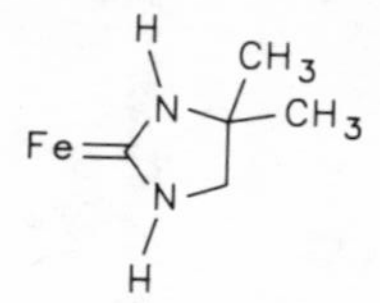	PF_6^- salt, III (88%); yellow, m.p. 246 °C (dec.) ^{1}H NMR (CD_3COCD_3): 1.39 (s, CH_3), 3.56 (s, CH_2)
59	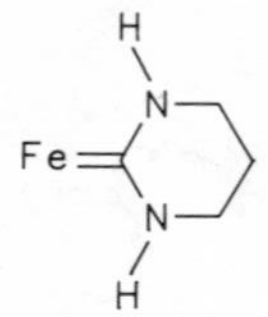	PF_6^- salt, IIIc (66%); yellow, m.p. 204 °C ^{1}H NMR (CD_3COCD_3): 2.00 (m, CH_2), 3.40 (t, CH_2N) ^{13}C NMR (CD_3COCD_3): 20.1 (CH_2), 42.7 (CH_2N), 192.2 (C=Fe), 212.7 (CO)
60	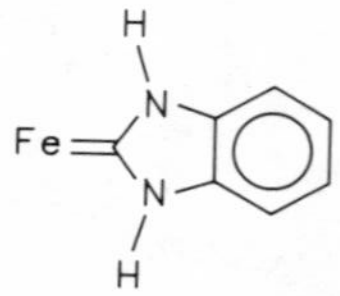	PF_6^- salt IIIc (21%); pale yellow, m.p. 230 to 233 °C

15% yield in the reaction of the $B(C_6H_5)_4^-$ salt of IV with BCl_3. The complex is only slightly soluble in CH_2Cl_2.

Hydrolysis produces $[FpCO]BF_4$. From $IrCl(CO)(C_6H_5P(CH_3)_2)_2$ and the BF_4^- salt (excess) in CH_2Cl_2 solution the adduct IV with a bridging CF_2 group is obtained in 66% yield.

F F CO
C P(CH3)2C6H5
Fe Ir X⁻
OC C P(CH3)2C6H5
O Cl

IV

$[Fp{=}CCl_2]BCl_4$ (Table **8**, No. **2**). To a frozen toluene solution of $FpCF_3$ are added three equivalents of BCl_3. Warming the mixture to room temperature results in the formation of a yellow precipitate. The complex decomposes on storage at room temperature even under N_2.

A structure determination was carried out and the data collected at 133 K. The salt crystallizes in the space group $P2_1/c$ with a = 12.238(4), b = 9.444(4), c = 13.440(4) A, β = 113.55(3)°; Z = 4 and d_c = 1.924 g/cm³. The carbene ligand lies within 3° of the mirror plane bisecting the OC–Fe–CO angle (see general remarks). The structure of the cation is depicted in Fig. 12.

Solutions form significant amounts of $FpCCl_3$ at room temperature after several days. Hydrolysis gives $[FpCO]^+$. With [PNP]Cl, the neutral complex $FpCCl_3$ is obtained; no reaction is observed with CO.

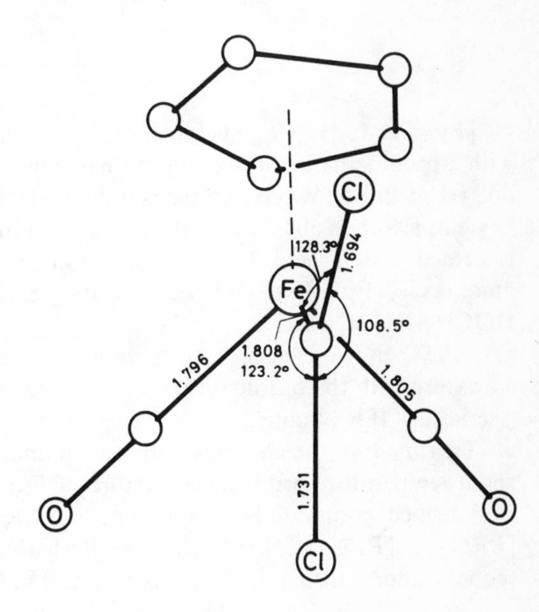

Fig. 12. Molecular structure of the cation of $[Fp{=}CCl_2]BCl_4$

$[Fp{=}C(OCH_3)_2]X$ (Table **8**, No. **4**, $X{=}PF_6$, CF_3SO_3). The PF_6^- salt is obtained as pale yellow crystals in 44% yield by the displacement of both substituents at the carbene carbon atom of $[Fp{=}C(OC_6H_5)SCH_3]PF_6$ (No. 10) with CH_3OH. The $CF_3SO_3^-$ salt is similarly prepared from No. 9 and dry CH_3OH (0.5 h, 89% yield).

The salts are fairly stable and show a tendency to decompose slowly to $[FpCO]^+$ on prolonged exposure to air in solution or in the solid state.

Reaction of the PF_6^- salt with CH_3O^- gives $FpCOOCH_3$ and Fp_2. The formation of the iron orthoester $FpC(OCH_3)_3$ in the first step was deduced from IR spectroscopic results. Reduction with $Li[BH(C_2H_5)_3]$ gives $FpCH(OCH_3)_2$, the starting material for the secondary carbene complex $[Fp{=}CHOCH_3]PF_6$.

$[Fp{=}C(OCH_3)OC_2H_5]CF_3SO_3$(Table **8**, No. **5**) is obtained according to the procedure above by use of C_2H_5OH (1 h, 57% yield) and shows the same properties.

$[Fp{=}C(OCH_3)OC_6H_{11}\text{-}c]BF_4$ (Table **8**, No. **6**) is formed by protonation of the metallocycle carbene complex V at $-78\,°C$ with $HBF_4{\cdot}O(C_2H_5)_2$ in the presence of CO (80% yield). In the first step the proton cleaves the Fe–C σ bond to generate the 16-electron carbene cation $[Cp(CO)Fe{=}C(OCH_3)OC_6H_{11}\text{-}c]^+$, which slowly disproportionates at $-20\,°C$ in CH_2Cl_2 in the absence of CO to give No. 6.

With $N(C_2H_5)_3$, demethylation occurs to give the neutral acyl complex $FpCOOC_6H_{11}\text{-}c$ in 83% yield.

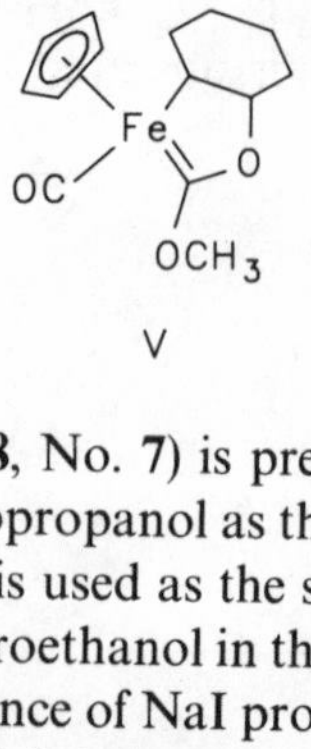

V

$[Fp{=}CO_2C_2H_4]PF_6$ (Table **8**, No. **7**) is prepared according to method IIa with 2-bromoethanol or 3-bromopropanol as the solvent; by-products are FpBr and $[(Fp)_2Br]Br$. When oxirane is used as the solvent the compound is formed in yields less than 30%. In 2-chloroethanol in the presence of NaCl, the complex is formed in 60% yield. The presence of NaI produces only 18% with significant amounts of FpI. It is obtained by the reaction of $[Fp(CO)]CF_3SO_3$ with $HOCH_2CH_2Br$, which is also the solvent, by addition of NaH in small portions at 0 °C. When the reaction in the presence of NaH is carried out without cooling, a mixture with the bis(dioxycarbene) derivative $[Cp(CO)Fe({=}CO_2C_2H_4)_2]PF_6$ (see Sect. 1.4) is obtained.

Treatment of the complex with an equimolar amount of NaI in acetone at room temperature produces a mixture of FpI and $FpCOOCH_2CH_2I$; similar ring-opened compounds of the type $FpCOOCH_2CH_2X$ are obtained with [PPN]Cl, [PPN]SCN, $[N(C_4H_9\text{-}n)_4]I$, $NaOCH_3$, and $NaSCH_2C_6H_5$; subsequent addition of CF_3SO_3H produces $[FpCO]CF_3SO_3$. With $(CH_3)_3NO$,

H_2O, or iodosobenzene the dioxocarbene ligand is converted into ethylene carbonate. Reduction with NaCp, sodium amalgam, or sodium naphthalenide in THF gives Fp_2 in 70% yield along with CO_2 and $CH_2{=}CH_2$; a mechanism involving radical intermediates is proposed. Treatment with $P(C_6H_5)_3$ or $PCH_3(C_6H_5)_2$ in CH_2Cl_2 produces a mixture of the corresponding monosubstituted carbene complex (70%), $[FpPR_3]PF_6$ (12%), and $[Cp(CO)Fe(PR_3)_2]PF_6$ (6%); the NaCp-catalyzed substitution with $(C_6H_5)_2PCH_3$ gives 72% of the substitution product along with 21% of $[Cp(CO)Fe(PR_3)_2]PF_6$. Prolonged reaction with the phosphine (18 h) gives $[Cp(CO)Fe(PR_3)_2]PF_6$ in 85%.

$[Fp{=}CO_2W(Cp)_2]BF_4$ (Table **8**; No. **8**) forms via a [2 + 2] cycloaddition of the WO bond of $(Cp)_2W{=}O$ across the CO bond of a carbonyl ligand of $[Fp(CO)]^+$ in CH_2Cl_2 solution at 22 °C (95% yield).

The ^{17}O NMR spectrum of the complex was measured with a sample prepared with $(Cp)_2W{=}^{17}O$ indicating that the ^{17}O isotope is also incorporated into the CO ligands. Crossover experiments show that completely free $(Cp)_2W{=}O$ is not liberated and the exchange must occur via dissociation and readdition within the solvent cage or by an intramolecular process.

$[Fp{=}C(OCH_3)SCH_3]X$ (Table **8**; No. **9**, X = PF_6, CF_3SO_3). The compounds should be stored at low temperature. At room temperature and more rapidly at elevated temperature (100 °C), conversion into the sulfido bridged complex $[(Fp)_2SCH_3]CF_3SO_3$ (61%; Formula VI) and $[FpCO]CF_3SO_3$ (12%) occurs. With ROH (R = CH_3, C_2H_5) the corresponding dioxocarbene cations $[Fp{=}C(OCH_3)OR]^+$ (Nos. 4 and 5) are obtained. The $CF_3SO_4^-$ salt reacts in refluxing CH_3CN (10 h) to give $[Fp(CH_3CN)]^+$ in 47% yield.

$$\left[(Fp)_2S{-}CH_3 \right]^+ [CF_3SO_3]^-$$

VI

$[Fp{=}C(OC_6H_5)SCH_3]X$ (Table **8**, No. **10**, X = PF_6, FSO_3). For the preparation, see No. 27. For thermal stability of the cation, see No. 9.

It reacts with CH_3OH to give the dimethoxycarbene complex No. 4. With piperidine, a mixture of No. 52 and 17 (48 and 18%, respectively) is obtained.

$[Fp{=}C(OCH_3)NHR]CF_3SO_3$ (Table **8**, No. **12**, R = CH_3; No. **14**; R = $CH_2C_6H_5$). Intramolecular removal of CH_3OH and partial conversion into the corresponding isocyanide occurs when the compounds are exposed to elevated temperatures in the solid state. No such conversion was found on heating the compounds in solution.

$[Fp{=}C(OC_6H_5)NC_5H_{10}]PF_6$ (Table **8**, No. **17**) is obtained similar to method IIIa (18%) as a mixture with $[Fp{=}C(SCH_3)NC_5H_{10}]PF_6$ (No. 52, 48%) by reaction of $[Fp{=}C(OC_6H_5)SCH_3]PF_6$ with piperidine.

[Fp=C_3H_5NO]X (Table **8**, No. **18**, X = PF_6, CF_3SO_3). The $CF_3O_3^-$ salt is obtained from [Fp=C(OCH_3)$NHCH_2CH_2OH$]CF_3SO_3 (No. 13) by loss of CH_3OH and ring closure on refluxing it in CH_3CN (12 h, 81%).

The complex (as the PF_6^- salt) can be reversibly deprotonated with excess NaH or K_2CO_3 in CH_2Cl_2 (2 h) to give the imidoyl complex VII; for subsequent reactions, see preparation of Nos. 19 and 20.

[Fp=C_4H_7NO]PF_6 (Table **8**, No. **19**). The neutral imidoyl complex VII is allowed to react with [O(CH_3)$_3$]PF_6 in CH_2Cl_2 solution for 15 min.

[Fp=C_6H_9NO]PF_6 (Table **8**, No. **20**) is similarly obtained from VII by reaction with CH_2=$CHCH_2Br$ under reflux (4 h); then excess KPF_6 is added (33%).

Photolysis of the complex in CH_2Cl_2 for 2 h gives the carbene cation VIII in 82% yield.

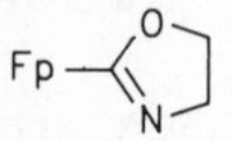

VII 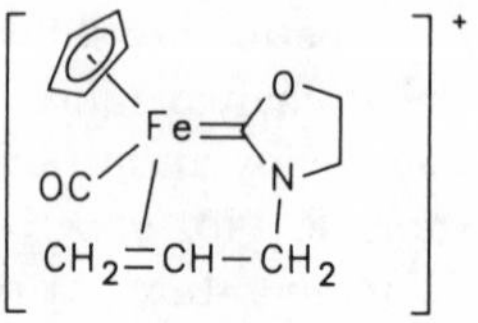 VIII 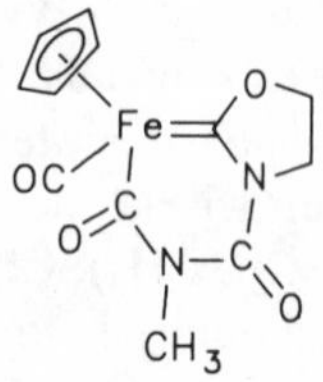 IX

[Fp=$COCH_2CH_2NCONHCH_3$]PF_6 (Table **8**, No. **21**) is obtained by addition of excess NH_4PF_6 to a solution of complex IX (obtained from VII and CH_3NCO) in CH_2Cl_2 (75%).

Addition of few drops of D_2O to an CH_3CN solution of the complex causes H/D exchange at the NH group and the NCH_3 doublet to collapse to a singlet and the NH signal to disappear. Addition of NaH in mineral oil to a CH_2Cl_2 solution of the complex regenerates the starting compound IX.

[Fp=C_4H_7NO-c]PF_6 (Table **8**, No. **22**) is prepared by modifying method IIa using azetidine instead of aziridine. Thus, a solution of [Fp(CO)]CF_3SO_3 in CH_3CN is allowed to react with azetidine in the presence of equimolar amounts of [Br(CH_2)$_3NH_3$]Br for 20 min. Further azetidine is added and the mixture stirred for 2 h (54% yield). The complex also forms in 30% yield when the reaction with [Br(CH_2)$_3NH_3$]Br is carried out in oxirane solution at 0 °C. 3-Bromopropylamine (obtained from ring opening of azetidine) is considered to be the reactive agent. The evidence for the intermediate carbamoyl complex, FpCONH(CH_2)$_3$Br, is derived from IR experiments.

[Fp=C(SH)SFe(CO)$_2$Cp]X (Table **8**, No. **23**, X = BF_4, CF_3SO_3). The compounds are obtained by addition of ethereal HBF_4 at 0 °C or stoichiometric amounts of CF_3SO_3H at room temperature to solutions of the neutral complex $FpCS_2Fp$ in CH_2Cl_2 and ether, respectively (73% yield). The conductivity of the BF_4^- complex in CD_3COCD_3 at 18 °C (5.7×10^{-5} mol/L) is 127 Ω^{-1} mol^{-1} cm^2.

Reaction of the $CF_3SO_3^-$ salt with CH_2N_2/ether in CH_2Cl_2 solution produces [Fp=C(SCH_3)SFp]CF_3SO_3 (No. 28) in 85% yield.

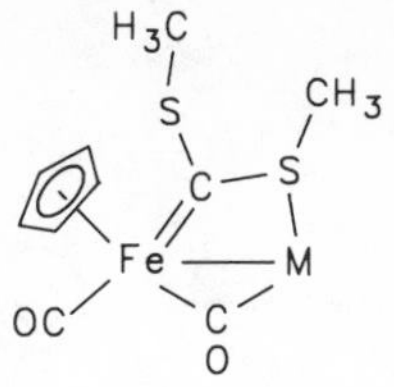

X

[Fp=C(SCH_3)$_2$]X (Table **8**, No. **24**, X = PF_6, BF_4, CF_3SO_3). The cation is also obtained as the only CO-containing product in about 75% yield by the oxidation of the binuclear compounds, formula X (M=Co(CO)$_2$, Fe(CO)(NO), Co(CO)P(C_2H_5)$_3$, Fe(NO)P(C_2H_5)$_3$), with two equivalents of [(Cp)$_2$Fe]$FeCl_4$ in CH_2Cl_2 solution. Other oxidizing agents are Br_2, I_2, [C(C_6H_5)$_3$]PF_6, and [C_7H_7]BF_4. It can be obtained in almost quantitative yield if FpC(SCH_3)$_3$ is treated with strong acids such as CF_3SO_3H, or as a mixture with the cation [Fp=C(SCH_3)SC_6H_5]$^+$ (No. 27) from the acid and FpC(SCH_3)$_2$$SC_6H_5$.

The PF_6^- salt is stable towards the influence of moisture and air at room temperature and solutions exposed to the air are stable for several days. It is soluble in CH_2Cl_2, CH_3CN, CD_3COCD_3, slightly soluble in H_2O, CH_3Cl, THF, and insoluble in hexane or ether. For the temperature dependent ^1H NMR spectrum, see also general remarks.

The PF_6^- salt crystallizes in the monoclinic space group C2/c-C_{2h}^6 (No. 15) with a = 32.91(1), b = 6.790(3), c = 14.509(4) Å, β = 105.48(3)°; Z = 8, d_c = 1.82 g/cm^3. The carbene ligand is planar with *syn*- and *anti*-SCH_3 groups; the plane of the carbene moiety is 90° from the plane of the Fp fragment as shown in Fig. 13. The relatively long Fe–$C_{carbene}$ distance and short S–$C_{carbene}$ distances suggest weak π bonding between carbon and the iron atom and significant π bonding between carbon and the sulfur atoms.

Irradiation of the PF_6^- salt in CH_3CN replaces one CO group to generate [Cp(CO)(CH_3CN)Fe=C(SCH_3)$_2$]PF_6 in 90% yield which can be used as a starting material of a variety of substituted cations of the type [Cp(CO)(^{2}D)Fe =C(SCH_3)$_2$]$^+$ (see Sect. 1.1.2.3). No reaction is observed with I^-, Cl^-, I_2, HCl, HF, $H_2PC_6H_{11}$, or P(C_6H_5)$_3$ at room temperature. The NaCp-catalyzed reaction of the PF_6^- salt with $PCH_3(C_6H_5)_2$ in THF solution gives the corresponding cations [Cp(CO)Fe(PR_3)$_2$]$^+$, [FpPR_3]$^+$, and [Cp(CO)(PR_3)Fe =C(SCH_3)$_2$]$^+$ in 32, 48 and 18% yields, respectively; the same compounds are produced without NaCp catalysis in CH_2Cl_2 solution in 64, 13 and 4% yields, respectively.

Reaction of the PF_6^- salt with SR^- (R = CH_3, C_6H_5; from SRH and NaH in THF) gives the corresponding adducts FpC(SCH_3)$_2$SR in about 75% yield. The complex FpC(SCH_3)$_3$ is also obtained in low yields if the title cation is treated with bases such as N(CH_3)$_3$ (probably via the unstable cationic ylide complex

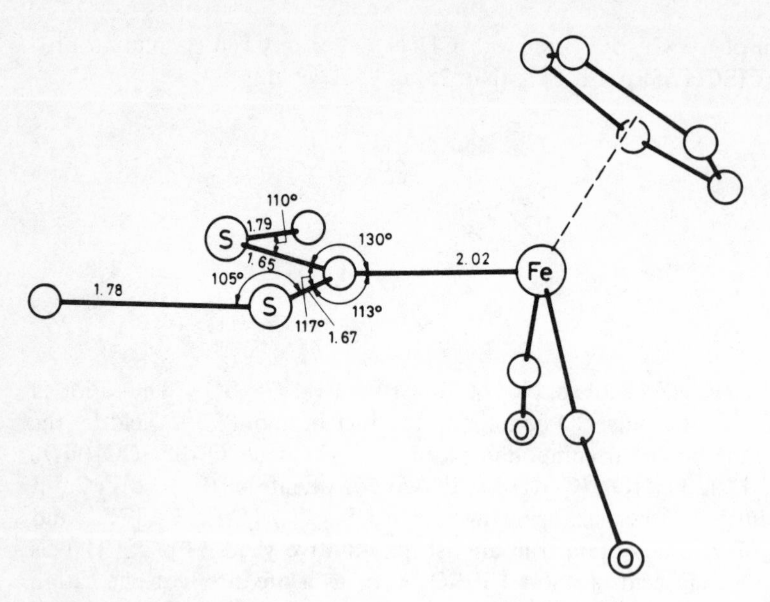

Fig. 13. Molecular structure of the cation of $[Fp{=}C(SCH_3)_2]PF_6$

$[FpC(SCH_3)_2N(CH_3)_3]^+$), CH_3O^-, $C_6H_5O^-$, $C_6H_5Se^-$, and aqeous OH^-. For the reaction with HSC_6H_5, see preparation of No. 35.

Primary amines, H_2NR, react in CH_2Cl_2 at room temperature to give the corresponding isocyanide complexes $[FpCNR]PF_6$ in 55 to 88% yield along with traces of $FpC(SCH_3)_3$. NH_3 yields FpCN which is also formed by the reaction of the carbene complex with NaN_3. Sterically less demanding secondary amines such as $HN(CH_3)_2$ or heterocyclic amines replace one SCH_3 group to produce the mixed carbene complexes $[Fp{=}C(SCH_3)NR_2]PF_6$. The second SCH_3 group is not replaced, even with large excess of amine. $HN(C_2H_5)_2$ and higher amines are too bulky and give $FpC(SCH_3)_3$. Aminothiocarbene complexes have also been identified with piperazine, pyrrolidine, or aziridine but not isolated. The reaction with diamines at room temperature generally provides a high yield of cyclic diaminocarbene complexes with loss of 2 moles of CH_3SH. In the case of ethylenediamine, a mixture of approximately equal amounts of the carbene complex No. 55 and the isocyanide complex $[FpCNCH_2CH_2NCFp][PF_6]_2$ is formed. While β-mercaptoethylamine gives the cyclic carbene complex No. 54, 3-aminopropanol behaves as a simple primary amine and produces the isocyanide complex $[FpCNCH_2CH_2OH]PF_6$.

$Li[AlH_4]$, $Li[BH(C_2H_5)_3]$, or $Li[AlH(C_4H_9\text{-}t)_3]$ in THF give $FpCH(SCH_3)_2$ in about 85% yield. Reduction of the cation with sodium naphthalenide leads to the formation of a mixture of Fp_2 (19%), $(Cp(CO)FeSCH_3)_2$ (5%), $FpC(SCH_3)_3$ (37%), and XI (37%).

The PF_6^- salt reacts with [PNP] $[Co(CO)_4]$ in THF to give X ($M{=}Co(CO)_2$) in 60% yield.

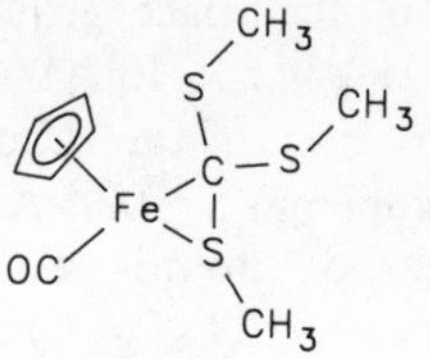

XI

[Fp=C(SCH$_3$)SC$_6$H$_5$]X (Table **8**, No. **27**; X = PF$_6$, CF$_3$SO$_3$). The starting complex, FpCSSC$_6$H$_5$, for Method I is prepared in situ from [FpCS]$^+$ and [SC$_6$H$_5$]$^-$; see No. 10. The PF$_6^-$ salt is produced from a solution of the CF$_3$SO$_3^-$ salt in acetone by anion exchange. The CF$_3$SO$_3^-$ salt is also obtained (40%) as a mixture with [Fp=C(SCH$_3$)$_2$]CF$_3$SO$_3$ (No. 24, 56%) by the reaction of FpC(SCH$_3$)$_2$SC$_6$H$_5$ with CF$_3$SO$_3$H. The cation also forms as a mixture with No. 35 by the reaction of [Fp=C(SCH$_3$)$_2$]$^+$ with C$_6$H$_5$SH.

The reaction of the PF$_6^-$ salt with benzylamine produced the isocyanide complex [FpCNCH$_2$C$_6$H$_5$]PF$_6$ and with piperidine the carbene complex No. 52 is obtained.

[Fp=C(SCH$_3$)SFp]X (Table **8**, No. **28**; X = CF$_3$SO$_3$, PF$_6$, I) is also obtained by dropwise treatment of a CH$_2$Cl$_2$ solution of No. 23 with an ethereal CH$_2$N$_2$ solution (85%).

The compounds react with [N(C$_4$H$_9$-*t*)$_4$]I in CH$_3$COCH$_3$ to give FpCSSCH$_3$ and FpI. The alkylation of FpCSSFp with various RBr or RSO$_3$CF$_3$ compounds, followed by treatment of the stable carbene cation [Fp=C(SR)SFp]$^+$ with I$^-$ is a general method for the preparation of other FpCSSR compounds.

[Fp=C(SCH$_3$)SRe(CO)$_5$)]X (Table **8**, No. **30**, X = CF$_3$SO$_3$, PF$_6$, BF$_4^-$). The BF$_4^-$ salt is obtained as a yellow powder in 90% yield by slow addition of FpCSSCH$_3$ in portions to a CH$_2$Cl$_2$ solution of (CO)$_5$ReFBF$_3$. The compounds are sparingly soluble in THF and CH$_2$Cl$_2$, soluble (with decomposition) in acetone, and insoluble in H$_2$O. The compounds are stable in the solid state upon exposure to air, but solutions of the complex exposed to air decompose completely within one day.

With CH$_3$NH$_2$ the PF$_6^-$ salt is split into the neutral compounds FpCSSCH$_3$, Re(CO)$_4$(CH$_3$NH$_2$)CONHCH$_3$, and [CH$_3$NH$_3$]PF$_6$, probably via the unstable neutral carbene complex Fp=C(SCH$_3$)SRe(CO)$_4$CONHCH$_3$. LiBr (5-fold excess) converts the complex into FpCSSCH$_3$ and Re(CO)$_5$Br.

[Fp=C(SC$_6$H$_5$)$_2$]PF$_6$ (Table **8**, No. **35**) as the major component is obtained together with complex No. 27 by refluxing a mixture of excess C$_6$H$_5$SH and [Fp=C(SCH$_3$)$_2$]PF$_6$ (No. 24) in acetone or CH$_2$Cl$_2$.

[Fp=C(SFe(CO)$_2$Cp)$_2$]BF$_4$ (Table **8**, No. **36**) can be prepared in 72% yield by successive reaction of Na[Fp] with CS$_2$ (to give the anion [FpCS$_2$]$^-$), FpI (to give FpCS$_2$Fp), and FpBF$_4$.

[Fp=C(SFe(CO)$_2$Cp)(SRe(CO)$_5$]BF$_4$ (Table **8**, No. **37**) FpCSSFp is added to a suspension of (CO)$_5$ReFBF$_3$, probably in CH$_2$Cl$_2$ (80% yield).

The complex crystallizes in the space group $P2_12_12_1\text{-}D_2^4$ (No. 19) with the unit cell parameters a = 8.589(3), b = 13.299(4), c = 24.06(1) Å; Z = 4, d_c = 2.04 g/cm^3. The molecular structure of the cation is depicted in Fig. 14. The three metal atoms are nearly coplanar ($\pm$ 0.05 Å) with the CS_2 plane. The sum of the angles at the carbene carbon atom is 360°.

[Fp=$CS_2C_2H_4$-*c*]X (Table **8**, No. **38**, X = PF_6, $FeCl_4$). The cation forms by the oxidation of XII ($n = 2$, M = Fe(CO)(NO), $Co(CO)_2$, $Co(CO)P(C_2H_5)_3$) with $[Cp_2Fe][FeCl_4]$ in 75, 87 and 53% yield, respectively.

The PF_6^- salt reacts with gaseous NH_3 to give FpCN. With benzylamine, loss of $HSCH_2CH_2SH$ occurs to give $[FpCNCH_2C_6H_5]PF_6$ in 85% yield. With secondary amines, no aminothiocarbene complexes are produced. $(CH_3)_2NH$ and $N(CH_3)_3$ give new unidentified species which finally decompose to form Fp_2. Addition of $NaSCH_3$ to a solution of the complex in THF generates the neutral complex XIII (R = SCH_3, $n = 2$) in 65% yield; addition of $HBF_4{\cdot}O(C_2H_5)_2$ quantitatively regenerats the starting carbene cation. Excess

XII XIII

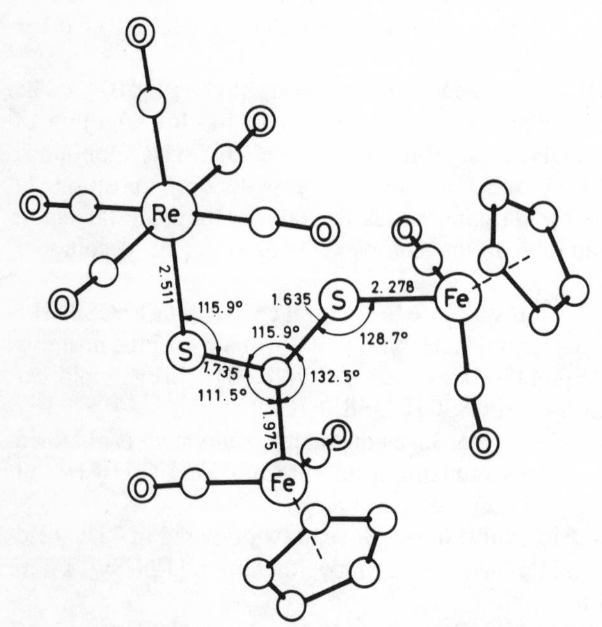

Fig. 14. Molecular structure of the cation $[Fp{=}CSFp(SRe(CO)_5)]^+$

$(C_6H_5)_2PCH_3$ in CH_2Cl_2 solution (15 min) produces the corresponding mono-substituted carbene complex $[Cp(CO)((C_6H_5)_2PCH_3)Fe{=}CS_2C_2H_4\text{-}c]PF_6$ in 85% yield. Irradiation at 254 nm (2.5 h) in CH_3CN solution similarly replaces one CO ligand to give $[Cp(CO)(CH_3CN)Fe{=}CS_2C_2H_4\text{-}c]PF_6$ in 64% yield.

$[Fp{=}CS_2C_3H_6\text{-}c]X$ (Table **8**, No. **39**, X = PF_6, BF_4). The BF_4^- complex is obtained as an impure oil by H^- abstraction from the neutral complex XIII (R = H, $n = 3$) with $[C(C_6H_5)_3]BF_4$ in CH_2Cl_2 solution. Oxidation of the binuclear complexes XII ($n = 3$, M = Fe(CO)(NO), $Co(CO)_2$) with $[Cp_2Fe][FeCl_4]$ produces the cation in nearly quantitative yields.

The reaction of the PF_6^- salt with bases proceeds in a manner similar to that described for No. 38, and irradiation in the presence of CH_3CN analogously gives $[Cp(CO)(CH_3CN)Fe{=}CS_2C_3H_6\text{-}c]PF_6$ (89%).

$[Fp{=}CS_2Pt(P(C_6H_5)_3)_2]BF_4$ (Table **8**, No. **40**). A THF solution of Fp^- (from Fp_2 and $NaK_{2.8}$) is first treated at −70 °C with one equivalent of CS_2 and than with 0.5 equivalent of $[(P(C_6H_5)_3)_2PtCl]_2[BF_4]_2$ (76% yield). It can be stored indefinitely at −20 °C.

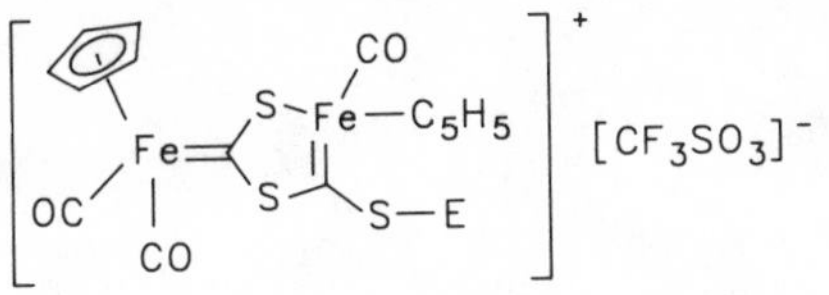

$[CF_3SO_3]^-$

XIV 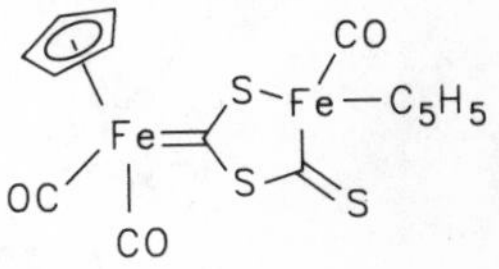 XV

$[Fp{=}C_2S_3RFe(CO)Cp]CF_3O_3$ (E = H, CH_3, C_2H_5, Formula XIV; Table **8**, Nos. **41**, **42**, and **43**) are obtained by reaction of the neutral complex XV with $HOSO_2CF_3$, $CH_3OSO_2CF_3$, or $C_2H_5OSO_2CF_3$, respectively. No. 41 is prepared in ether, resulting in an olive green oil that decomposes on attempted crystallization. Nos. 42 and 43 are obtained in CH_2Cl_2 solution at room temperature during 15 min in 88 and 91% yield. Alkylation of XV with CH_3I also gives No. 42, but longer reaction times are required. For the carbene character of the bond between the ring carbon atom and the endocyclic iron atom, see General Remarks.

No. 43 crystallizes in the monoclinic space group $P2_1/n\text{-}C_{2h}^5$ (No. 14) with the parameters a = 14.565(4), b = 10.369(2), c = 16.176(2) Å, and β = 101.73(1)°; Z = 4 and d_c = 1.73 g/cm³. The central unit of the cation is the planar five-membered ring (maximal deviation from the average plane = 0.04 Å) containing the carbene carbon atom, the endocyclic iron atom, two sulfur atoms, and the CSR carbon atom. The ring plane forms an angle of 73° with the symmetry plane of the Fp group. The outer S atom of the C_2H_5S group deviates from the ring plane by 0.13 Å. All CO group are on the same side of the ring. The molecular structure of the cation is depicted in Fig. 15.

No. 42 reacts with a variety of nucleophiles that add at the sp^2 ring carbon atom. Thus, H^-, CN^-, or SR^- produce the neutral complexes XVI with E=H, CN, SR (R=CH_3, C_2H_5), respectively. The action of $[HS(CH_2)_3S]^-$ replaces the SCH_3 group to produce the spirocomplex XVII. Secondary amines,

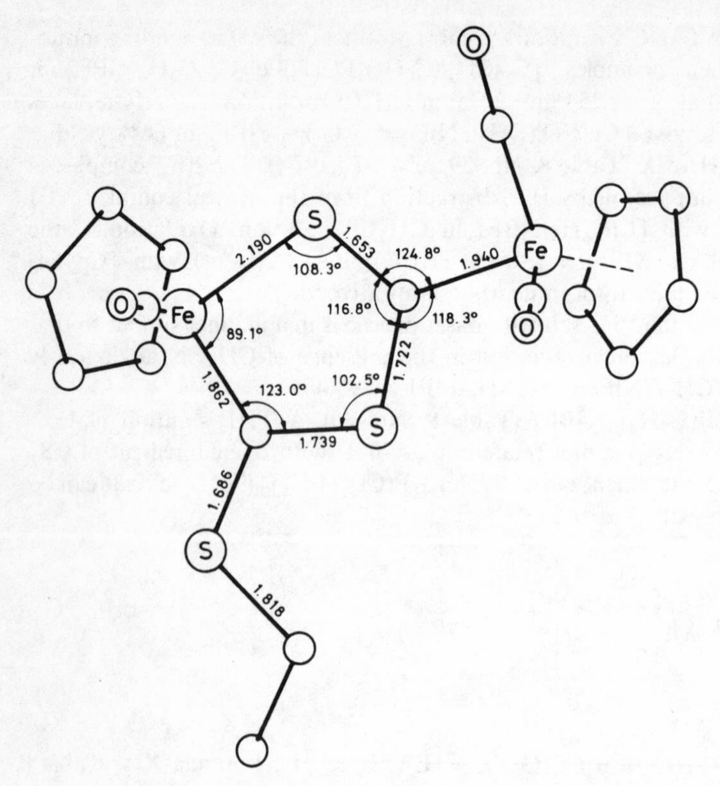

Fig. 15. Molecular structure of the cation $[Fp{=}CC_3H_5S_3Fe(CO)Cp]^+$

NHR_2, generate the carbene cations Nos. 46 to 49, whereas primary amines, H_2NR, give the neutral isocyanides FpCSSFe(CNR)(CO)Cp (R = CH_3, C_6H_{11}, C_4H_9-*n*). The reactions are carried out in CH_2Cl_2 at room temperature with yields in the range of 40 to 60%.

CO, S–Fe–C_5H_5, Fe=, S, E, OC, CO, SR

XVI

CO, S–Fe–C_5H_5, Fe=, S, S, S, OC, CO

XVII

$[Fp{=}C_{13}H_5FeO_6ReS_3]CF_3SO_3$ (Table **8**, No. **45**, Formula XIV, E = $Re(CO)_5$) is prepared by the reaction of $[FpCS]CF_3SO_3$ with $FpCSSRe(CO)_5$ in CH_2Cl_2 solution (69% yield).

$[Fp{=}C_7H_{10}NS_2Fe(CO)Cp]CF_3SO_3$ (Table **8**, No. **49**) crystallizes in the triclinic space group $P\bar{1}$-C_i^1 (No. 2) with the parameters a = 11.570(2), b = 12.242(2), c = 10.552(2) Å, α = 104.60(2)°, β = 87.51(1)°, γ = 117.24(2)°;

Z = 2 and d_c = 1.68 g/cm³. The central five-membered ring is slightly puckered and the deviations from the plane defined by the two S atoms and the carbene carbon atom are − 0.28 (Fe-2), + 0.19 (C of CN), and − 0.13 Å (Fe-1). The dihedral angle between the plane of the Fp moiety and the CS_2 carbene plane is 41°, thus being midway between the two favored configurations (0° and 90°) for Fe–C-π bonding. The planar geometry at the N atom is typical for an iminium ion and the positive charge formally resides on this atom. The carbenic character of the Fe=C bond is more pronounced at the chiral endocyclic Fe atom than at Fe of the Fp group; see General Remarks concerning resonance structures of this compounds. The structure of the cation is depicted in Fig. 16.

[Fp=C(SCH_3)SeC_6H_5]PF_6 (Table 8, No. **50**). The action of benzylamine gives [Fp($CNCH_2C_6H_5$)]PF_6 (58%) and piperidine produces the carbene complex No. 52.

[Fp=C(SCH_3)N(CH_3)$_2$]PF_6 (Table **8**, No. **51**) does not react with $(C_6H_5)_2PCH_3$ at 25 °C; in the presence of sodium napthalenide a small quantity of Fp_2 is formed along with other unidentified products.

[Fp=C(SCH_3)C_5H_{10}N-*c*]PF_6 (Table **8**, No. **52**) is also obtained in 66% yield by treatment of complex No. 50 or 27 with piperidine, and it forms as a mixture with No. 17 (48 and 18%, respectively) by the reaction of piperidine with [Fp=C(SCH_3)OC_6H_5]PF_6 (No. 10).

Reaction with excess (CH_3)NH_2 occurs at a CO ligand to give an equilibrium between the neutral complex Cp(CO)(CH_3NHCO)Fe=C(SCH_3)NC_5H_{10} and starting material in CH_2Cl_2.

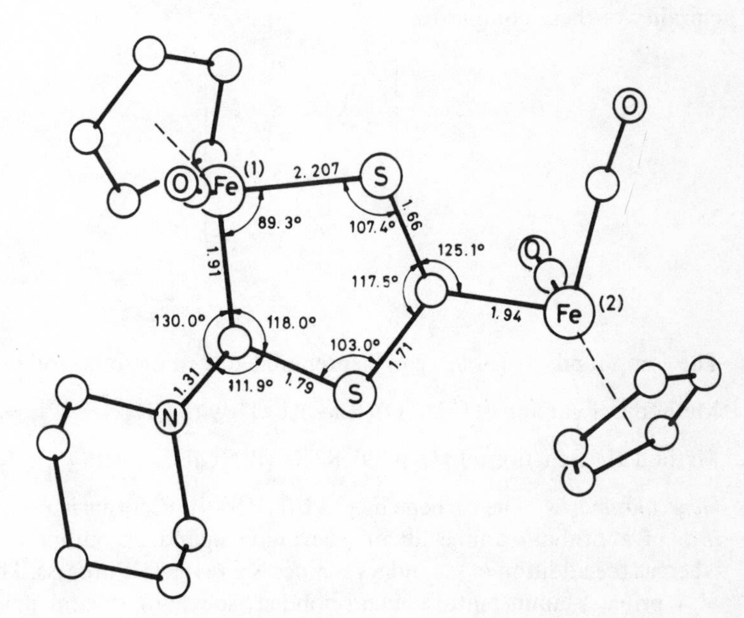

Fig. 16. Molecular structure of the cation of [Fp=$C_7H_{10}NS_2$Fe(CO)Cp]CF_3SO_3

Compounds with ^5L ligands other than Cp

$[C_5(CH_3)_5(CO)_2Fe{=}C(SCH_3)_2]PF_6$ reacts with NaH in THF in the presence of CH_3SH (2 h) to give the neutral complex $C_5(CH_3)_5(CO)_2FeC(SCH_3)_3$ in 92% yield.

1.1.4 Cationic Carbene Complexes with Additional Isocyanide Ligands

In the following sections carbene complexes are described which contain one or two isocyanide ligands in place of CO in the Fp group. In contrast to the compounds with ^2D ligands, the scope of compounds is restricted in both cases to cationic complexes with two heteroatoms at the carbene ligand. Whereas in the series $[^5L(RNC)_2Fe{=}C(ER)_2]^+$ only two complex are described, a variety of compounds with four different ligands at the iron atom, $[^5L(RNC)(CO)Fe{=}C(ER)_2]^+$, have been prepared.

For theoretical considerations and comparison, see Sect. 1 and Sect. 1.1.2.3 containing compounds of the type $[^5L(CO)(^2D)Fe{=}C(ER)_2]^+$. No corresponding compounds of the type $[^5L(^2D)_2Fe{=}C(ER)_2]^+$ with two heteroatoms at the carbene ligand are known.

1.1.4.1 Complexes of the Type $[^5L(CO)(RNC)Fe{=}C(ER)_2]^+$

The compounds in this section (Formula I) contain Cp (compounds Nos. 1 to 15) and C_9H_7 (indenyl; compounds Nos. 16 to 21) as ^5L ligands. Although the iron atom is surrounded by four different ligands, nothing is reported about the chirality of these complexes.

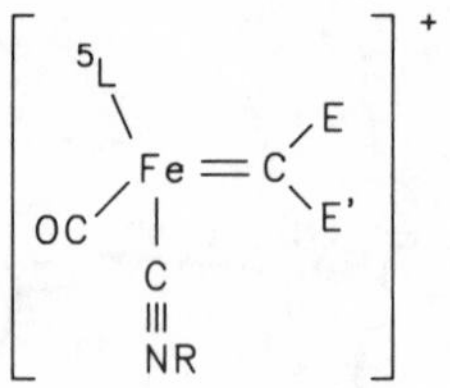

I

The compounds in Table 9 can be prepared according to the following methods:

Method I: Reaction of $[^5L(CO)(RNC)_2Fe]^+$ with CH_3NH_2 (^5L = Cp, indenyl).

Method II: Reaction of $[Cp(CO)(RNC)(R'NC)Fe]^+$ with $(CH_3)_2NH$.

General remarks. The carbene ligand C(NHR)NHR′ forms irreversibly by addition of a primary amine at an isocyanide ligand as outlined in Method I, whereas the addition of secondary amines is a reversible process. The conversion of a primary amine into a metal-bonded isocyanide ligand upon prolonged reaction, according to the general equation: $[^5L(CO)(CH_3NC)_2Fe]^+$

$+ RNH_2 \rightarrow [^5L(CO)(RNC)Fe{=}C(NHCH_3)]^+$ (see, e.g., complexes Nos. 5, 6), probably occurs in a series of steps involving chelating intermediates which could not be isolated.

The preparation of the indenyl complexes follows that of the corresponding Cp derivatives and occurs at a slightly faster rate. Yields are similar and the products are recovered as oils.

Compounds containing two NHR groups at the carbene carbon atom have a structure in solution as depicted in Formula II. In the case of R = R′ (A and B are identical) two signals are observed for R in the NMR spectra according to *cis* and *trans* positions of R to Fe and restricted rotation about the $C_{carbene}$–N bond. At elevated temperatures, averaging of the signals occurs and first-order rate constants have been calculated for this process; ΔG‡ values for the rotational free energy barriers are between 16.9 and 17.9 kcal/mol.

If R and R′ are different ($R = CH_3$; $R' = C_2H_5$, C_3H_7-*i*) the formation of the isomers A and B is observed. The relative isomer population is not very sensitive to solvent change and is given in Table 9 with the individual compounds. In the case of the isocyanide complexes, preference for configuration B is found in which the larger group is *cis* to the iron atom; for other rotamer preference with ^2D ligands, see Sect. 1.1.2.3.

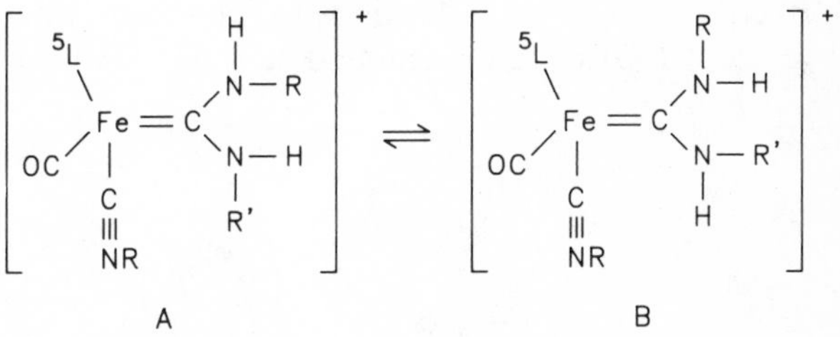

A B

II

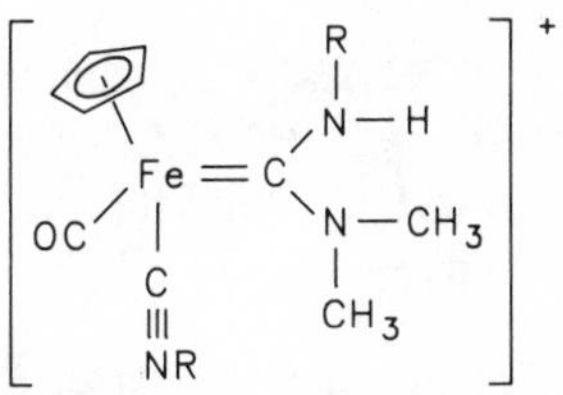

III

Compounds with the $N(CH_3)_2$ group (Formula III, Nos. 11 and 12, $R = CH_3$; No. 15, $R = C_3H_7$-*i*) form only one isomer in a reversible reaction as shown by NMR spectroscopy, with the preferred orientation of R cis to Fe (Formula III). They decompose on standing in solution into $(CH_3)_2NH$ and the corresponding isocyanide complexes, and the half-life values decrease with increasing number of the bulky C_3H_7-*i* group in R of III.

The $N(CH_3)_2$ protons in these compounds appear as two singlets (*cis* and *trans* methyl groups) in ^1H NMR and ^{13}C NMR low-temperature spectra and as averaged signals at high temperature (e.g room temperature); energy barriers of rotation are in the range of 13.1 to 14.1 kcal/mol and are less than those of $NHCH_3$ groups, see above.

^1H NMR/^{13}C NMR chemical shifts for CH_3 protons when the $NHCH_3$ group is *cis* to Fe range from 3.10 to 3.25/35.1 to 36.8 ppm and those for *trans* to Fe range from 2.88 to 2.96/31.4 to 32.6 ppm (Formulas II and III). Similarly, CH carbon atoms of the C_3H_7-*i* group appear in the 51.4 to 52.6 range (*cis*) or at about 45.8 ppm (*trans*).

NH protons appear in the ^1H NMR spectra as broad signals in the region 7.0 to 8.5 ppm and the singlet of the Cp group is found in the region 5.00 to 5.20 (in CD_3COCD_3, $CDCl_3$), and 4.75 to 4.95 (in CD_3CN, CD_3SOCD_3) ppm.

The CN carbon atoms of the isocyanide ligands appear in the ^{13}C NMR spectra as broad signals at 154 to 155 ppm; for the Cp group shifts between 84 and 87 ppm are recorded.

The carbon atoms of the indenyl ligand C_9H_7 (Formula IV) have been assigned in the ^{13}C NMR spectra as follows for complex No. 16: 70.8 (C-3), 71.9 (C-1), 92.0 (C-2), 102.7 (C-3a), 104.0 (C-7a), 125.5 (C-5,6), 128.3 (C-4,7). Similar values are reported for the other indenyl complexes.

In the IR spectra, the ν (C≡N) vibrations are reported between 2155 and 2200 cm^{-1}. A single ν (CO) vibration is found at about 1980 cm^{-1}.

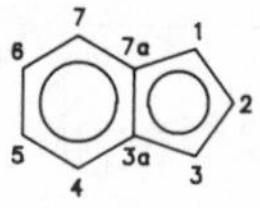

7 7a 1 6 2 5 3a 3 4

IV

Further Information

$[Cp(CO)(CH_3NC)Fe{=}C(SCH_3)_2]PF_6$ (Table **9**, No. **1**) forms by the reaction of the carbene complex $[Cp(CO)(CH_3CN)Fe{=}C(SCH_3)_2]PF_6$ with equimolar amounts of CH_3NC in CH_2Cl_2 for 5 h (65% yield); see the formation of the bis-isocyanide complex in Sect. 1.5.1. The complex is stable at room temperature.

From temperature-dependent ^1H NMR experiments, the coalescence temperature relating to syn-anti isomerization, T_c, was found to be − 28 °C and the free energy of activation, $\Delta G^{\ddagger}$, to be 12.2 kcal/mol.

For the reaction with piperidine; see complex No. 2. With $C_6H_5CH_2NH_2$ at room temperature in CH_3CN solution (4 h) the bis-isocyanide complex $[Cp(CO)(CH_3NC)(C_6H_5CH_2NC)Fe]^+$ is obtained in 58% yield.

$[Cp(CO)(CH_3NC)Fe{=}C(SCH_3)NC_5H_{10}]PF_6$ (Table **9**, No. **2**) is obtained in 73% yield as stable irregular glass-like crystals by stirring compound No. 1 with a 10-fold excess of piperidine in CH_3CN solution for 6 h.

Table 9. Cationic Carbene Compounds of the Type $[^5L(CO)(RNC)Fe{=}C(ER)_2]^+$; ER = OR, SR, NR_2. An asterisk indicates further information at the end of the table

No.	$Fe{=}C(ER)_2$ CNR	Properties and remarks method of preparation (yield)
Compounds with Cp as the ^5L ligand		
heteroatoms E = S		
*1	$Fe{=}C(SCH_3)_2$ CH_3NC	PF_6^- salt; orange crystals, m.p. 128 to 130 °C ^1H NMR (CD_3COCD_3): 3.20 (s, CH_3S) ^{13}C NMR (CD_3CN): 29.6 (CH_3S), 214.9 (CO), 315.8 (C=Fe)
heteroatoms E = S, N		
*2	Fe=C(N-piperidino)(SCH_3) CH_3NC	PF_6^- salt; yellow crystals, m.p. 75 to 78 °C ^1H NMR (CD_3CN): 1.77 (m, $3CH_2$), 2.78 (s, CH_3S), 4.16 (t, CH_2N), 4.35 (t, CH_2N) ^{13}C NMR (CD_3CN): 24.4 (CH_2-4), 25.1 (CH_3S), 58.3, 64.5 (CH_2-2, 6), 216.7 (CO), 248.1 (C=Fe)
heteroatoms E = N, N		
*3	$Fe{=}C(NHCH_3)_2$ CH_3NC	$B(C_6H_5)_4^-$ salt, I (44%); bright yellow ^1H NMR (CD_3SOCD_3): 2.72, 3.01 (d's, CH_3N), 7.40 to 7.80 (br, NH) I^- salt, I (60%) ^1H NMR $CDCl_3$: 2.95, 3.20 (d's, CH_3NH) ^{13}C NMR ($CDCl_3$): 32.4, 35.9 (CH_3NH), 206.4 (C=Fe), 216.9 (CO)
4	$Fe{=}C(NHCH_3)_2$ CNC_2H_5	I^- salt for preparation, see No. 7
*5	$Fe{=}C(NHCH_3)_2$ CNC_3H_7-*i*	I^- salt; m.p. 182 °C ^1H NMR (CD_3COCD_3): 2.90, 3.18 (d's, CH_3NH) ^{13}C NMR ($CDCl_3$): 32.6, 35.9 (CH_3NH), 206.4 (C=Fe), 217.1 (CO)
*6	$Fe{=}C(NHCH_3)_2$ CNC_4H_9-*t*	I^- salt ^1H NMR (CD_3COCD_3): 2.92, 3.19 (d's, CH_3N)

Table 9. Continued

No.	Fe=C(ER)$_2$ CNR	Properties and remarks method of preparation (yield)
*7	Fe=C(NHCH$_3$)NHC$_2$H$_5$ CNCH$_3$	I$^-$ salt; isomer mixture A:B = 7:13 (Formula II), m.p. 58 °C (dec., impure, containing 15% of isomer No. 4) ^{1}H NMR (CD$_3$COCD$_3$; A/B): 1.18/1.27 (d's, CH$_3$C), 3.18/2.96 (d's, CH$_3$NH), 3.42/3.62 (m's, CH$_2$N) ^{13}C NMR (CDCl$_3$; A/B): 14.2/16.1 (CH$_3$C), 35.1/31.4 (CH$_3$NH), 39.3/43.6 (CH$_2$N), 204.0, 204.9 (C=Fe), 216.4 (CO)
8	Fe=C(NHCH$_3$)NHC$_2$H$_5$ CNCH$_2$CH$_3$	I$^-$ salt, I (60%); isomer mixutre A:B = 7:13 (Formula II), m.p. 129 °C ^{1}H NMR (CD$_3$COCD$_3$; A/B): 1.16/1.24 (t's, CH$_3$C), 3.17/2.9 (d's, CH$_3$N), 3.41/3.66 (m's, CH$_2$NH) ^{13}C NMR (CDCl$_3$; A/B): 14.3/16.3 (CH$_3$C), 35.7/31.9 (CH$_3$N), 39.7/44.2 (CH$_2$NH), 204.5, 205.3 (C=Fe), 216.6 (CO)
9	Fe=C(NHCH$_3$)NHC$_3$H$_7$-*i* CNCH$_3$	I$^-$ salt; isomer mixture A:B = 3:7 (Formula II) ^{1}H NMR (CD$_3$COCD$_3$, A/B): 1.32 (m, CH$_3$C), 3.19/2.90 (d, CH$_3$NH), 4.40 (m, CH) ^{13}C NMR (CDCl$_3$, A/B): 23.4, 23.6 (CH$_3$C), 35.9/32.5 (CH$_3$NH), 45.6/51.4 (CH), 203.3, 204.5 (C=Fe), 216.3 (CO)
10	Fe=C(NHCH$_3$)NHC$_3$H$_7$-*i* CNC$_3$H$_7$-*i*	I$^-$ salt, I (60%); isomer mixture A:B = 1:4 (Formula II), m.p. 68 to 70 °C ^{1}H NMR (CD$_3$COCD$_3$, A/B): 1.40 (m, 2CH$_3$C), 3.20/2.96 (d, CH$_3$NH), 4.40 (m, 2CH) ^{13}C NMR (CDCl$_3$, A/B): 22.9, 23.6 (CH$_3$C), 35.6/31.7 (CH$_3$N), 45.8/51.7 (CH), 202.5, 204.3 (C=Fe), 216.2 (CO)

No.	Compound	Data
*11	Fe=C(NHCH$_3$)N(CH$_3$)$_2$ CNCH$_3$	PF_6^- salt, II; yellow solid ^{1}H NMR (CD$_3$COCD$_3$): 3.29 (d, CH$_3$NH), 3.37 (s, (CH$_3$)$_2$N) ^{13}C NMR (CDCl$_3$ at − 50 °C): 36.2 (CH$_3$NH), 38.4, 48.1 ((CH$_3$)$_2$N), 207.4 (C=Fe), 215.7 (CO)
*12	Fe=C(NHCH$_3$)N(CH$_3$)$_2$ CNC$_3$H$_7$-*i*	PF_6^- salt, II; amber oil ^{1}H NMR (CD$_3$COCD$_3$): 3.23 (d, CH$_3$NH), 3.38 (s, (CH$_3$)$_2$N) ^{13}C NMR (CDCl$_3$ at − 50 °C): 37.3 (CH$_3$NH), 41.2, 48.0 ((CH$_3$)$_2$N), 205.2 (C=Fe), 216.8 (CO)
13	Fe=C(NHC$_2$H$_5$)$_2$ CNC$_2$H$_5$	I^- salt, I (60%); m.p. 146 °C ^{1}H NMR (CD$_3$COCD$_3$): 1.19, 1.28 (t's, CH$_3$C), 3.49, 3.71 (m's, CH$_2$N) ^{13}C NMR (CDCl$_3$): 14.3, 16.4 (CH$_3$C), 41.7, 44.2 (CH$_2$N), 203.9 (C=Fe), 216.7 (CO)
14	Fe=C(NHC$_3$H$_7$-*i*) CNC$_3$H$_7$-*i*	I^- salt, I (60%); m.p. 179 °C ^{1}H NMR (CD$_3$COCD$_3$): 1.27, 1.35 (d's, CH$_3$C) ^{13}C NMR (CDCl$_3$): 22.3. 23.4 (CH$_3$C), 45.8, 52.6 (CH), 201.6 (C=Fe), 216.4 (CO)
*15	Fe=C(NHC$_3$H$_7$-*i*)N(CH$_3$)$_2$ CNC$_3$H$_7$-*i*	PF_6^- salt, II; amber oil ^{1}H NMR (CD$_3$COCD$_3$): 1.35 (d, CH$_3$C), 3.41 (s, (CH$_3$)$_2$N), 4.40 (m, CHNH) ^{13}C NMR (CDCl$_3$ at − 50 °C): 23.0 (CH$_3$C), 40.4, 48.1 ((CH$_3$)$_2$N), 203.0 (C=Fe), 215.7 (CO)

Compounds with C_9H_7 (indenyl) as the ^{5}L ligand

heteroatoms E = N, N

No.	Compound	Data
16	Fe=C(NHCH$_3$)$_2$ CNCH$_3$	I^- salt, I; m.p. < 25 °C ^{1}H NMR (CD$_3$COCD$_3$): 2.92, 2.96 (d's, CH$_3$N) ^{13}C NMR (CDCl$_3$): 32.7, 35.3 (CH$_3$N), 206.3 (C=Fe), 217.1 (CO)
17	Fe=C(NHCH$_3$)$_2$ CNC$_3$H$_7$-*i*	I^- salt; m.p. < 25 °C ^{1}H NMR (CD$_3$COCD$_3$): 2.92, 2.94 (d's, CH$_3$N) ^{13}C NMR (CDCl$_3$): 33.2, 35.5 (CH$_3$N), 206.5 (C=Fe), 217.5 (CO) for preparation, see procedure outlined for No. 5
18	Fe=C(NHCH$_3$)$_2$ CNC$_4$H$_9$-*t*	I^- salt; m.p. < 40 °C (dec.) ^{1}H NMR (CD$_3$COCD$_3$): 2.97 (m, CH$_3$N) ^{13} C NMR (CDCl$_3$): 33.3, 35.6 (CH$_3$N), 206.4 (C=Fe), 217.4 (CO) for preparation, see procedure outlined for No. 6

Table 9. Continued

No.	$Fe=C(ER)_2$ CNR	Properties and remarks method of preparation (yield)
19	$Fe=C(NHCH_3)NHC_2H_5$ $CNCH_3$	I^- salt; m.p. $< 25\,°C$ (B is the major isomer, Formula II) ^1H NMR (CD_3COCD_3, A/B): 1.10/1.19 (t's, CH_3C), 2.92 (m, CH_3N), 3.36 (m, CH_2N) ^{13}C NMR ($CDCl_3$, A/B): 15.2/16.3 (CH_3C), 35.1/32.4 (CH_3N), 40.0/43.5 (CH_2N), 204.3/205.1 (C=Fe), 216.9 (CO) for preparation, see procedure outlined for No. 7
20	$Fe=C(NHCH_3)NHC_2H_5$ $CNCH_2CH_3$	I^- salt, I; m.p. $< 25\,°C$ (B is the major isomer, Formula II) ^1H NMR (CD_3COCD_3, A/B): 0.99/1.16 (t's, CH_3C), 2.91 (m, CH_3N), 3.44 (m, CH_2N) ^{13}C NMR ($CDCl_3$, A/B): 15.4/16.5 (CH_3C), 35.6/32.3 (CH_3N), 40.2/43.7 (CH_2N), 204.8/205.5 (C=Fe), 217.0 (CO)
21	$Fe=C(NHCH_3)NHC_3H_7$-*i* $CNCH_3$	I^- salt; m.p. $< 25\,°C$ (A and B isomer) ^1H NMR (CD_3COCD_3, A, B): 1.30 (m, CH_3C), 2.92 (m, CH_3N), 4.40 (m, CH) for preparation, see procedure outlined for No. 9

[Cp(CO)(CH$_3$NC)Fe=C(NHCH$_3$)$_2$]X (Table **9**, No. **3**; X = B(C$_6$H$_5$)$_4$, I). The temperature-dependent ^{1}H NMR spectrum in CD$_3$SOCD$_3$ of the cation between 30 and 80 °C was measured. At elevated temperature, loss of coupling between CH$_3$ and H disappears shown by NH proton exchange. At about 72 °C, the signals of the CH$_3$ protons (Formula II, R = R′=CH$_3$) at the carbene ligand coalesce to a singlet, indicating rapid rotation around the C$_{carbene}$–N bond; see also general remarks.

[Cp(CO)(*i*-C$_3$H$_7$NC)Fe=C(NHCH$_3$)$_2$]I (Table **9**, No. **5**) forms by refluxing a CH$_2$Cl$_2$ solution of [Cp(CO)(CH$_3$NC)$_2$Fe]I with an excess of *i*-C$_3$H$_7$NH$_2$ for 48 h. If equimolar quantities of the reactants are stirred at room temperature for 200 h, a 1:4 mixture with No. 9 is obtained.

[Cp(CO)(*t*-C$_4$H$_9$NC)Fe=C(NHCH$_3$)$_2$]I (Table **9**, No. **6**) is obtained in about 35% yield by refluxing a mixture of excess *t*-C$_4$H$_9$NH$_2$ and [Cp(CO)(CH$_3$NC)$_2$Fe]I in CH$_2$Cl$_2$ for 10 d. Purification proceeds by chromatography.

[Cp(CO)(CH$_3$NC)Fe=C(NHCH$_3$)NHC$_2$H$_5$]I (Table **9**, No. **7**). The reaction of [Cp(CO)(CH$_3$NC)$_2$Fe]I with an equimolar quantity of C$_2$H$_5$NH$_2$ in CH$_2$Cl$_2$ solution at 0 °C for 4 h produces a mixture of 85% of the title complex along with 15% of No. 4. With excess amine a mixture with increased portion of No. 4 is formed.

[Cp(CO)(CH$_3$NC)Fe=C(NHCH$_3$)NHC$_3$H$_7$-*i*]I (Table **9**, No. **9**) forms according to Method I with an equimolar quantity of *i*-C$_3$H$_7$NH$_2$ for 24 h. Unreacted starting material is removed by chromatography. Prolonged reaction time (200 h) leads to a 4:1 mixture with No. 5.

With excess amine, quantitative conversion into No. 5 is observed.

[Cp(CO)(RNC)Fe=C(NHCH$_3$)N(CH$_3$)$_2$]PF$_6$ (Table **9**, No. **11**, R = CH$_3$; No. **12**, R = C$_3$H$_7$-*i*) lose (CH$_3$)$_2$NH at 300 K with the rate constants k = 4.3 × 10^{-4} and 5.3 × 10^{-3} s^{-1}; half-lives of the compounds are 1610 and 130 min, respectively.

The free energy of rotation, ΔG‡ (see also No. 15), around the C$_{carbene}$–N bond is 14.1 and 13.6 ± 0.3 kcal/mol, respectively.

[Cp(CO)(*i*-C$_3$H$_7$NC)Fe=C(NHC$_3$H$_7$-*i*)N(CH$_3$)$_2$]PF$_6$ (Table **9**, No. **15**) is rapidly converted into [Cp(CO)(*i*-C$_3$H$_7$NC)$_2$Fe]PF$_6$ on standing in solution at room temperature with the rate constant k = 2.7 × 12^{-2} s^{-1} at 300 K and a half-life of 26 min.

From variable-temperature ^{1}H NMR spectra in CD$_3$COCD$_3$, the free energy of rotation around the C$_{carbene}$–N bond, ΔG‡, is estimated as 13.1 ± 0.3 kcal/mol.

1.1.4.2 Complexes of the Type [^{5}L(RNC)$_2$Fe=C(ER)$_2$]$^+$

The two known carbene compounds in this section contain two isocyanide ligands and are prepared by different routes. Whereas the first compound forms by ligand exchange at the iron atom of a carbene complex, the second complex is obtained by addition of CH_3NH_2 to an isocyanide ligand; see method I in Sect. 1.4.2. Similar reactions have also been carried out with $C_2H_5NH_2$.

[Cp(CH$_3$NC)$_2$Fe=C(SCH$_3$)$_2$]PF$_6$ was not obtained analytically pure. It forms as a byproduct by the reaction of the acetonitrile carbene complex $[Cp(CO)(CH_3CN)Fe{=}C(SCH_3)_2]PF_6$ (see Sect. 1.2.3) with a slight excess CH_3NC to give mainly $[Cp(CO)(CH_3NC)Fe{=}C(SCH_3)_2]PF_6$; see Sect. 1.4.2.

The NMR spectra in CD_3CN solution exhibits the following signals (in ppm): ^1H NMR: 3.05 (s, CH_3S), 3.44 (s, CH_3NC), 4.65 (s, Cp). ^{13}C NMR (CD_3CN): 29.0 (CH_3S), 31.7 (CH_3N), 84.3 (Cp), 159.9 (CN), 325.5 (C=Fe). The IR spectrum in CH_2Cl_2 shows $\nu(C{\equiv}N)$ bands at 2161 s, 2187 s, and 2225 m cm^{-1}.

[Cp(4-CH$_3$OC$_6$H$_4$NC)$_2$Fe=C(NHCH$_3$)NHC$_6$H$_4$OCH$_3$-4]X (X = BF_4, PF_6). Both salts are obtained by bubbling CH_3NH_2 into a suspension of the corresponding salt $[Cp(4\text{-}CH_3OC_6H_4NC)_3Fe]X$ in ether for 1 h. The mixture is stirred for about 8 h to give a yellow precipitate. The BF_4^- salt was purified by chromatographing it on silica gel. $CHCl_3$ elutes a yellow impurity first and acetone elutes the product.

The ^1H NMR spectrum in $CDCl_3$ of the PF_6^- salt shows the presence of two rotational isomers (A and B, Formula I) with an A/B ratio of 1.6 (in ppm): 3.03 (br, CH_3N of A), 3.23 (d, CH_3N of B), 3.72/3.62 (s's, CH_3O at carbene), 3.76 (s, CH_3O at isocyanide), 4.84/4.62 (s's, Cp), 6.7 to 7.5 (m, C_6H_4), 8.06/7.85 (br, HN).

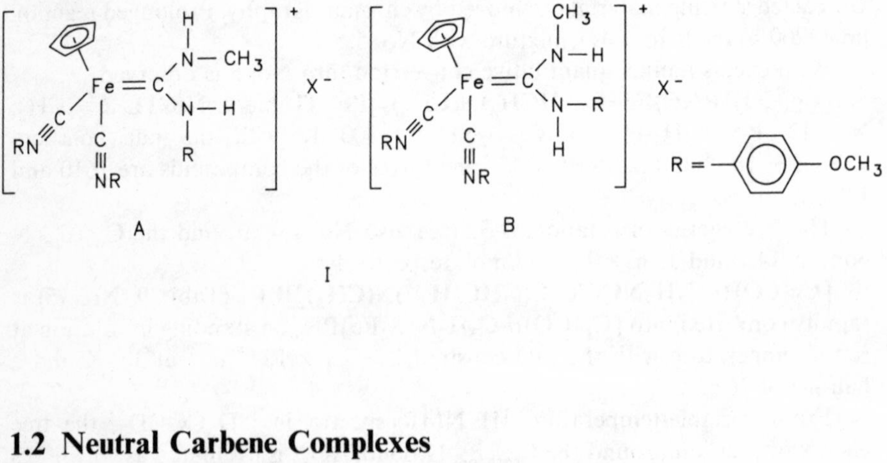

I

1.2 Neutral Carbene Complexes

This section comprises all neutral complexes of the general types $^5L(CO)XFe{=}CR(ER)$, $^5L(CO)XFe{=}C(ER)_2$, $^5L(^2D)_2Fe{=}CR(ER)$, $^5L(CO)(^2D)Fe{=}CR(ER)$, $^5L(CO)_2Fe{=}C(ER)_2$, and $^5L(CO)RFe{=}C(ER)_2$; two 19-electron carbene compounds with the ^5L ligand $C_5(CH_3)_5$ are included. E represents a heteroatom such as O, S, N, and R a set of organic groups to achieve the rare gas shell. To give a general survey of this field of chemistry we have also included compounds in which a second or a third iron atom is part of the carbene ligand (parts of the general groups R or ER).

1.2.1 Carbene Complexes of the Type $^{5}L(CO)(X)Fe{=}C(R)ER$ and $^{5}L(CO)(X)Fe{=}C(ER)_2$

The Fe atom in these compounds is surrounded by four different ligands in a pseudotetrahedral arrangement, thus being chiral at iron, as shown in Formula I (both ER groups can be different or equal). Only a few compounds have been prepared containing carbene ligands with one and two heteroatoms and the σ-bonded ligand X is restricted to CN, I, $Ge(C_6H_5)_3$, and $Sn(C_6H_5)_3$. The most recent results (see Nos. 2 and 3) and the compounds $Cp(CO)XFe{=}C(C_6H_5)ER$ ($ER = OC_2H_5$, NHC_2H_5; $X = I$, $Sn(C_6H_5)_3$), $Cp(CO)IFe{=}CC(CH_3)_2C_2H_4O$, and $C_5(CH_3)_5(CO)IFe{=}CC_3H_6O$ are too new to be included in Table 10.

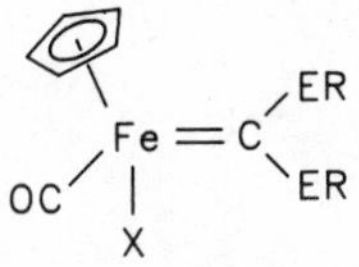

I

Compounds of the type Cp(CO)Fe (^{1}L-^{2}D) in which the chelating ligand ^{1}L-^{2}D is a conjugated system as shown in Formula II ($Z^1 = O$, NR; $Z^2 = CR$, N) can also be described as carbene complexes, according to resonance Formula III. These compounds are not treated here but two compounds of this type can be found in Organoiron Compounds B 11, Table 26, Nos. 15 and 22. For the latter compound (Formulas II and III, $R^1 = C_6H_5$, $R^2 = CH_3$, $Z^1 = O$, $Z^2 = CC_6H_5$) the ^{13}C value of the carbene carbon atom is recorded and the shift to low field (257.97 ppm) agrees with the carbene formulation III. Similarly, the structure of No. 15 in "Organoiron Compounds" B 11, 1983, p. 240 (Formulas II and III, $R^1 = R^2 = CF_3$, $Z^1 = NH$, $Z^2 = N$) shows a very short Fe=C bond typical for carbene complexes; see also discussion of adducts of Lewis acids at conjugated acyl systems in Sect. 1.1.

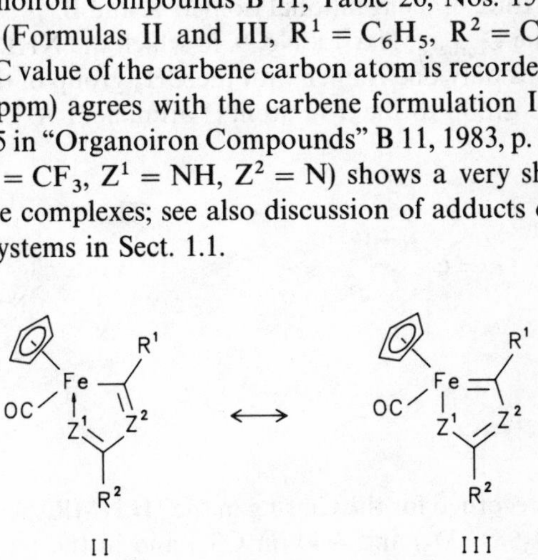

II ⟷ III

Concerning compounds which contain a similar ligand arrangement as described in this section, see the complexes Nos. 23 to 35 in Table 11 and Formula IV in the subsequent Sect. 1.2.3. In these compounds, the CpFe fragment is part of a five-membered ring system.

Several intermediates of the general types $^5L(CO)(X)Fe{=}CR(ER)$ and $^5L(CO)(X)Fe{=}C(ER)_2$ have been formulated in the literature to explain isomerization phenomena. Thus, in the C-1 racemization process during photochemical substitution of CO by PR_3 in optically pure $FpCH(OCH_3)C_6H_5$, the complex $Cp(CO)(H)Fe{=}C(C_6H_5)OCH_3$ probably equilibrates with the intermediate 16-electron species $Cp(CO)FeCH(OCH_3)C_6H_5$ via H migration from C-1 to Fe. The carbene mercaptide $Cp(CO)(CH_3S)Fe{=}C(SCH_3)_2$, a short-lived intermediate, is probably in rapid equilibrium with IV, as deduced by variable-temperature ^1H NMR studies and chemical reactions of IV. An intermediate as depicted in Formula V would account for the equivalence of diastereotopic carbons of compound No. 27 in Table 11 in Sect. 1.2.3.

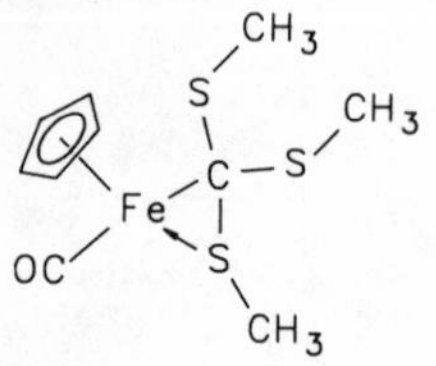

IV 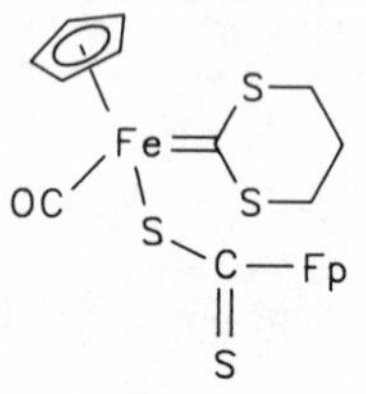 V

The compounds described in this section with NHR groups at the carbene carbon atom exist exclusively in the "*amphi*" configuration with one group R in *cis* and one in *trans* position with respect to the Fe atom (see Formula VI). The complexes No. 6 to 8 exhibit two different ligands at the two N atoms and exist therefore as a mixture of the rotational isomers A and B, as a result of restricted rotation about the $C_{carbene}$–N axes. No. 9, in which the NHR group is replaced by $N(CH_3)_2$, forms only one isomer with the CH_3 group of the $NHCH_3$ ligand probably in *cis* position to the iron atom (Formula VI A, $NHR^1 = N(CH_3)_2$).

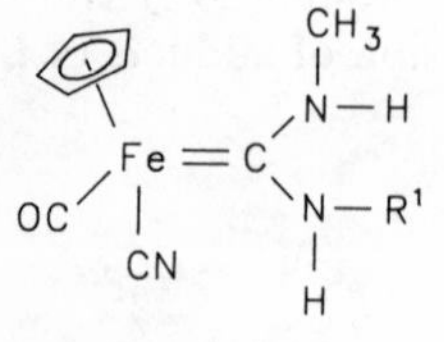

A 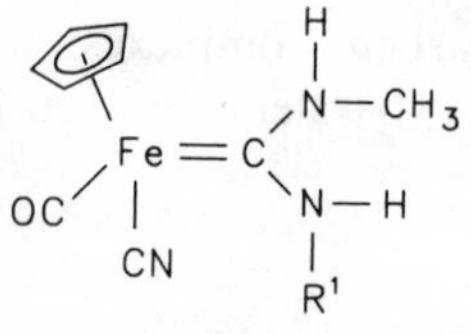 B

VI

Singlets were recorded for the Cp ring in the ^1H NMR spectrum in the range 4.65 ppm (in CD_3COCD_3) and 4.40 (in CS_2) and in the ^{13}C NMR spectra at about 83 ppm in $CDCl_3$. The strong ν (CO) absorption in the IR spectra gives two bands if mixtures of isomers are present.

Further Information

$Cp(CO)(Ge(C_6H_5)_3)Fe{=}C(CH_3)OCH_3$ (Table **10**, No. **1**) is obtained in a very low yield by treatment of the complex $Cp(CO)_2FeGe(C_6H_5)_3$ with $LiCH_3$, generating the intermediate anionic acyl complex $[Cp(CO)(Ge(C_6H_5)_3)FeCOCH_3]^-$, followed by alkylation with $[O(CH_3)_3]BF_4$.

$Cp(CO)(I)Fe{=}C(SCH_3)_2$ (Table **10**, No. **4**) is produced in low yields by the reaction of the cationic carbene complex $[Cp(CO)(CH_3CN)Fe{=}C(SCH_3)_2]PF_6$ with an excess $[N(C_2H_5)_4]I$ in CH_2Cl_2. The complex is not very stable.

$Cp(CO)(NC)Fe{=}C(NHCH_3)NHR$ (Table **10**, Nos. **5** to **8**, R = CH_3, C_2H_5, C_3H_7-*n*, C_3H_7-*i*) are prepared by bubbling CH_3NH_2 through a solution of the corresponding isocyanide complexes $Cp(CO)(CNR)FeCN$ in CH_2Cl_2 (80% yield). Nos. 6 to 8 are mixtures of the rotational isomers A and B (see Formula VI).

In the ^1H NMR spectrum the NH protons are observed as broad signals in the region of 7.0 to 8.5 ppm. CN gives broad signals at 145 to 155 ppm in the ^{13}C NMR spectrum. Using variable-temperature NMR spectroscopy, the rotational free energy barrier was measured to be about 17 kcal/mol.

$Cp(CO)(CN)Fe{=}C(NHCH_3)N(CH_3)_2$ (Table **10**, No. **9**) could be obtained only as a 1:1 mixture with the starting complex $Cp(CO)(CNCH_3)FeCN$ by bubbling $(CH_3)_2NH$ through a solution of this complex in CH_2Cl_2.

$Cp(CO)(Sn(C_6H_5)_3)Fe{=}CN_2C_2H_6$ (Table **10**, No. **10**) is produced by the reaction of the thiocarbonyl complex $Cp(CO)(CS)FeSn(C_6H_5)_3$ with $NH_2CH_2CH_2NH_2$ in the presence of pyridine in ether. The solvent is evaporated and the residue extracted with CS_2.

1.2.2 Carbene Compounds of the Type $^5L(CO)(^2D)Fe{=}C(R)ER$ and $^5L(^2D)_2Fe{=}C(R)ER$

Compounds of these types are limited to structures I, II and the zwitterionic complex V with the ^5L ligands Cp and $C_5(CH_3)_5$, respectively. Concerning the formulation of I and II as acyl oxygen donors, coordinating at the Mn atom ($Fe{-}C(CH_3){=}O{\rightarrow}Mn$) or at BH_3 ($Fe{-}CH{=}O{\rightarrow}BH_3$), see discussion in Sect. 1.1. Compounds of type II have been only spectroscopically characterized and could not be isolated.

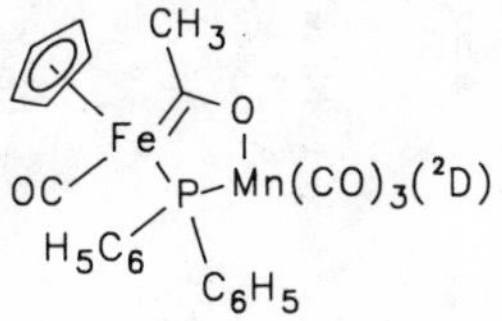

I

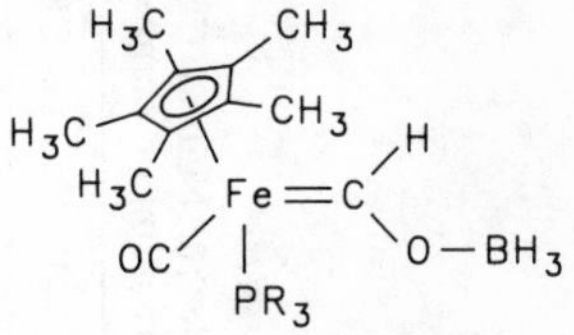

II

Table 10. Neutral Carbene Complexes of the Types $Cp(CO)(X)Fe{=}CR(ER)$ and $Cp(CO)(X)Fe{=}C(ER)_2$ ($Cp(CO)(X)Fe$ is abbreviated as Fe). Further information on numbers preceded by an asterisk is given at the end of the table

No.	$Fe{=}CR(ER)$ or $Fe{=}C(ER)_2$	X	properties and remarks
carbene ligand with one heteroatom (Fe=CR)(ER)			
*1	$Fe{=}C(CH_3)OCH_3$	$Ge(C_6H_5)_3$	^1H NMR: 2.28 (s, CH_3C), 3.52 (s, CH_3O)
2	$Fe{=}C(C_6H_5)OC_2H_5$	I	m.p. 55 °C (dec.)
3	Fe= (2-oxacyclopentylidene ring)	I	brown
carbene ligand with two hetroatoms (Fe=C(ER)$_2$)			
*4	$Fe{=}C(SCH_3)_2$	I	^1H NMR (CS_2): 3.15 (s, CH_3)
*5	$Fe{=}C(NHCH_3)_2$	CN	m.p. 186 °C ^1H NMR (CD_3COCD_3): 2.90 (d, CH_3), 3.16 (d, CH_3) ^{13}C NMR ($CDCl_3$): 30.4, 35.9 (CH_3), 214.4 (C=Fe), 220.3 (CO)
*6	$Fe{=}C(NHCH_3)NHC_2H_5$	CN	m.p. 97 °C, 1:1 mixture of the isomers A and B (Formula VI) ^1H NMR (CD_3COCD_3): 1.22 (t, CH_3C, A, B), 2.90 (d, CH_3N, B) 3.13 (d, CH_3N, A), 3.25 (m, CH_2, A), 3.70 (m, CH_2, B) ^{13}C NMR ($CDCl_3$): 14.0 (CH_3C, A), 16.1 (CH_3C, B), 30.3 (CH_3N, B), 36.2 (CH_3N, A), 28.4 (CH_2N, A), 44.3 (CH_2N, B), 214.2 (C=Fe), 220.5 (CO)

*7	Fe=C(NHCH$_3$)NHC$_3$H$_7$-*n*	CN	m.p. 26 to 28 °C, 1:1 mixture of the isomers A and B (Formula VI) ^{1}H NMR (CD$_3$COCD$_3$): 2.91 (d, CH$_3$N, B), 3.12 (d, CH$_3$N, A), 3.18 (m, CH$_2$N, A), 3.56 (m, CH$_2$N, B) ^{13}C NMR (CDCl$_3$): 30.0 (CH$_3$N, B), 35.7 (CH$_3$N, A), 45.2 (CH$_2$N, A), 51.2 (CH$_2$N, B), 215.8 (C=Fe), 220.8 (CO)
*8	Fe=C(NHCH$_3$)NHC$_3$H$_7$-*i*	CN	m.p. 26 to 28 °C 1:3 mixture of the isomers A and B (Formula VI) ^{1}H NMR (CD$_3$COCD$_3$): 1.24, 1.28 (d's, CH$_3$C, A, B), 2.88 (d, CH$_3$N, B), 3.13 (d, CH$_3$N, A), 4.25 (m, CHN, A, B) ^{13}C NMR (CDCl$_3$): 23.0, 24.7 (CH$_3$C, A, B), 30.5 (CH$_3$N, B), 36.0 (CH$_3$N, A), 43.9 (CH, A), 52.0 (CH, B), 215.6 (C=Fe), 220.6 (CO)
*9	Fe=C(NHCH$_3$)N(CH$_3$)$_2$	CN	^{1}H NMR (CD$_3$COCD$_3$): 3.2 to 3.3 (m's, 3CH$_3$)
*10	H N Fe= N H	Sn(C$_6$H$_5$)$_3$	yellow brown, m.p. 125 to 130 °C (dec.) ^{1}H NMR (CS$_2$): 3.05 (br, CH$_2$), 5.45 (br, HN)

Cp(CO)Fe=C(CH$_3$)OMn(CO)$_4$P(C$_6$H$_5$)$_2$ (Formula I, $^2D = CO$). To a solution of $(CO)_5MnCH_3$ in THF, an equimolar amount of $FpP(C_6H_5)_2$ in THF is added dropwise and the mixture is stirred for 48 h at room temperature. Evaporation of the solvent and extraction of the residue with CH_2Cl_2, followed by preparative thin-layer chromatography on silica gel with CH_2Cl_2/hexane (3:7) gives two yellow bands ($Mn_2(CO)_{10}$ and an unidentified product), a purple band ($Cp(CO)FeMn(\mu\text{-}H)(\mu\text{-}P(C_6H_5)_2)(CO)_4$) (Formula IV), and a large orange band containing the title complex which is obtained as a red oil (17% yield). A ^{13}C-enriched complex was prepared from ^{13}CO-enriched $(CO)_5MnCH_3$. Monitoring the reaction by ^{31}P NMR, ^{13}C NMR, and IR spectroscopy indicates the formation of some reasonably stable intermediates.

^1H NMR spectrum (C_6D_6, in ppm): 2.5 (s, CH_3), 4.1 (s, Cp). ^{13}C NMR spectrum (C_6H_6/C_6D_6, in ppm): 51.3 (CH_3), 86.1 (Cp), 212.2, 213.3 (br, CO), 218.7 (d, CO; J(P,C) = 19.0), 220.1 (br, CO), 319.3 (d, C=Fe; J(P,C) = 11.9). ^{31}P NMR spectrum in C_6D_6: 106.4 ppm. The IR spectrum in hexane shows ν (CO) bands at 1950 (FeCO), 1993, 2002, and 2080 (all MnCO) cm^{-1}.

The complex crystallizes in the triclinic space group $P\bar{1}-C_i^1$ (No. 2) with two crystallographically independent molecules per unit cell with the parameters a = 11.485(4), b = 12.989(4), c = 16.932(5) Å, $\alpha = 68.22(2)°$, $\beta = 87.17(2)°$, $\gamma = 88.88(2)°$; Z = 4 and $d_c = 1.542$ g/cm^3. The molecular structure, which is depicted in Fig. 17, shows no Fe–Mn bond (Fe ... Mn = 3.718 Å and 3.713 in molecules 1 and 2, respectively).

$LiCH_3$ adds at one of the CO groups at the Mn atom to give the anionic acyl complex III; see Section 1.3. Thermolysis in refluxing THF (4 h) gives low yields of seven products from which four have been identified as $MnH(CO)_3(PCH_3(C_6H_5)_2)$, Fp_2, $Cp(CO)Fe(\mu\text{-}H)(\mu\text{-}P(C_6H_5)_2)Mn(CO)_4$ (Formula IV), and I ($^2D = PCH_3(C_6H_5)_2$). The complex does not react with HBF_4, $[C(C_6H_5)_3]BF_4$, $[O(CH_3)_3]BF_4$, or H_2. For the reaction with $PCH_3(C_6H_5)_2$, see below.

Cp(CO)Fe=C(CH$_3$)OMn(CO)$_3$(PCH$_3$(C$_6$H$_5$)$_2$P(C$_6$H$_5$)$_2$ (Formula I, $^2D = PCH_3(C_6H_5)_2$) is prepared from the tetracarbonyl complex I ($^2D = CO$), see above, by dropwise addition of $PCH_3(C_6H_5)_2$ in THF solution (24 h). The workup procedure for the starting complex gives an orange band from which the title complex is obtained as an orange powder in 62% yield. The complex forms also by thermolysis of I ($^2D = CO$) probably via coupling of the $\mu\text{-}P(C_6H_5)_2$ ligand with the acetyl methyl group.

^1H NMR (C_6D_6, in ppm): 1.81 (d, CH_3P; J(P,H) = 8.0), 2.55 (s, CH_3C), 4.18 (s, Cp). ^{13}C NMR (C_6H_6/C_6D_6, in ppm): 50.7 (CH_3), 86.1 (Cp), 217 (CO), 219 (d, CO; J(P,C) = 18.6), 315 (d,C=Fe; J(P,C) = 6.6). ^{31}PNMR (C_6H_6/C_6D_6, in ppm): 44 (s, br), 112.7 (d; J(P,P) = 34.5). The ν (CO) bands in the IR spectrum in hexane solution appear at 1202, 1927, 1950, and 2020 cm^{-1}.

C$_5$(CH$_3$)$_5$(CO)(PR$_3$)Fe=C(H)OBH$_3$ (Formula II, R = CH_3, C_4H_9-*n*, or C_6H_5). The formation of these compounds along with the corresponding formyl complexes, $C_5(CH_3)_5(CO)(PR_3)FeCHO$, was detected by ^1H NMR spectro-

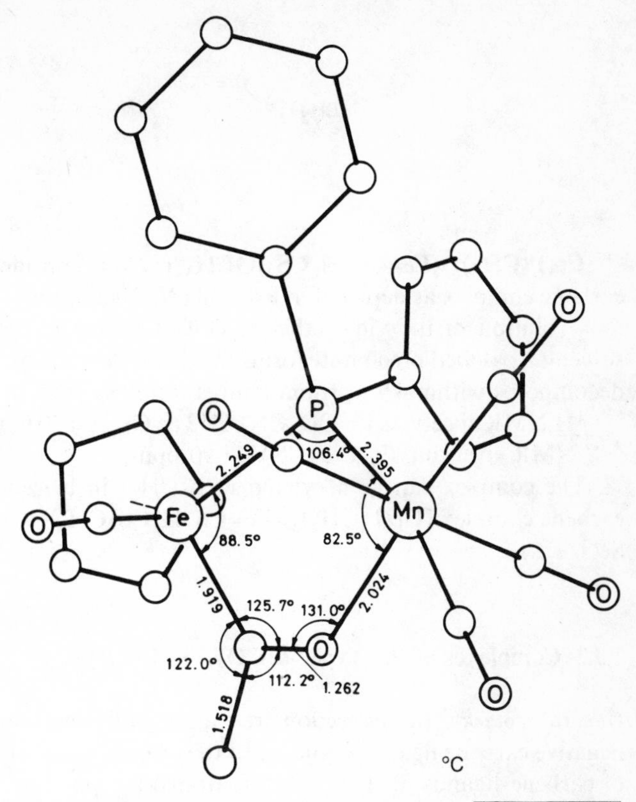

Fig. 17. Molecular structure of one of the two independent molecules of $Cp(CO)Fe{=}C(CH_3)OMn(CO)_4P(C_6H_5)_2$

CH_3 O
Fe O C—CH_3
OC P–Mn(CO)$_3$ Li
H_5C_6 C_6H_5

III

H_5C_6 C_6H_5
P
Fe—Mn(CO)$_4$
OC H

IV

scopy when the corresponding cations, $[C_5(CH_3)_5(PR_3)Fe(CO)_2]^+$, were treated with $NaBH_4$ at $-80\,°C$ in THF-d_8 solution. The formyl-to-adduct ratio was 1:1 for R = CH_3, C_4H_9 and 1:2.5 for R = C_6H_5.

^{1}H NMR THF-d_8, in ppm) for R = CH_3: 1.20 (d, CH_3P; J(P,H) = 30), 1.91 (s, CH_3C), 12.83 (d, CH; J(P,H) = 1.2); for R = C_4H_9-*n*: 1.41 (m, C_4H_9), 1.80 (s, CH_3C), 12.80 (d, br, CH; J(P,H) = 1.3); for R = C_6H_5: 12.48 (d, br, CH; J(P,H) = 1.5).

At about $-20\,°C$, conversion of the mixture into the corresponding methyl complexes, $C_5(CH_3)_5(CO)(PR_3)FeCH_3$, was observed.

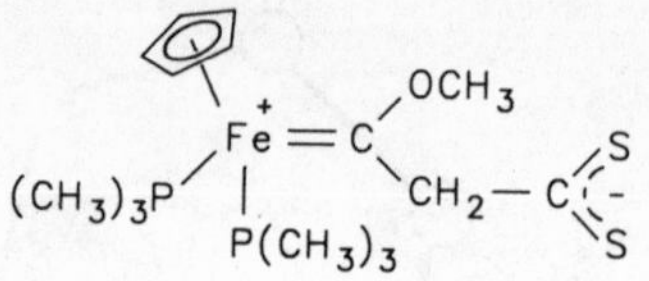

V

$Cp(P(CH_3)_3)_2Fe{=}C(CH_2CS_2)OCH_3$ can be considered as a zwitterionic carbene complex as depicted in Formula V. A solution of CS_2 in C_6H_6 is added to a solution of the vinyl ether $Cp(P(CH_3)_3)_2FeC({=}CH_2)OCH_3$ in the same solvent; a pale red precipitate forms in 78% yield, m.p. 58 °C (dec.). The complex decomposes within 4 h at room temperature.

^1H NMR spectrum (in CD_3CN): 2.13 (s, CH_2), 4.30 (t, CH_3O; J(P,H) = 0.50). ^{31}P NMR spectrum (in CD_3CN): 31.90 ppm.

The complex can be alkylated with CH_3I in benzene to give the cationic carbene complex $[Cp(P(CH_3)_3)_2Fe{=}C(CH_2CS_2CH_3)OCH_3]I$ in 68% yield; see Sect. 1.1.1.2.

1.2.3 Complexes of the Type $^5L(CO)_2Fe{=}C(ER)_2$

General remarks. In this section are collected all compounds in which a formally negative carbene ligand is bonded to the 16-electron $[Fp]^+$ moiety. Three types of carbene ligands (0, 1, or 2 heteroatoms) are possible. Among these neutral compounds, with exception of No. 1, only compounds in which two sulfur heteroatoms stabilize the carbene carbon atom are known; see also the analogous cationic species in Sect. 1.1.3.3.

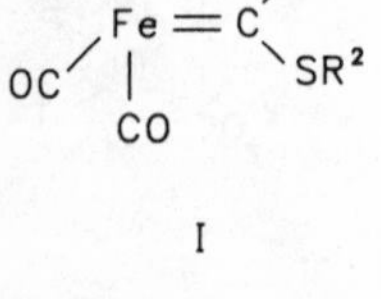

I 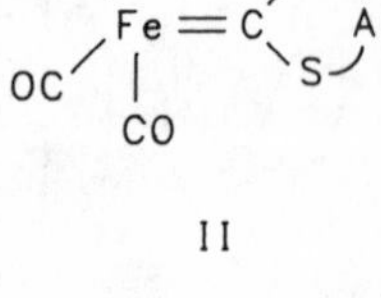 II

With E = S, two types of compounds are described. In compounds of the general Formula I, R^1 is represented by CH_3 or a 17-electron fragment, while R^2 is a 16-electron fragment. According to the two resonance forms shown in III, these complexes can be viewed as neutral carbene complexes (Formula IIIa) or as adducts of the neutral thioacyl compounds $FpCSSR^1$ to the 16-electron fragments ML_n (Formula IIIb). C-S and S-M distances from a structure determination of complex No. 10 are indicative of some contribution from resonance form IIIb. In Formula II the carbene carbon atom is part of a 4- or 5-membered ring system.

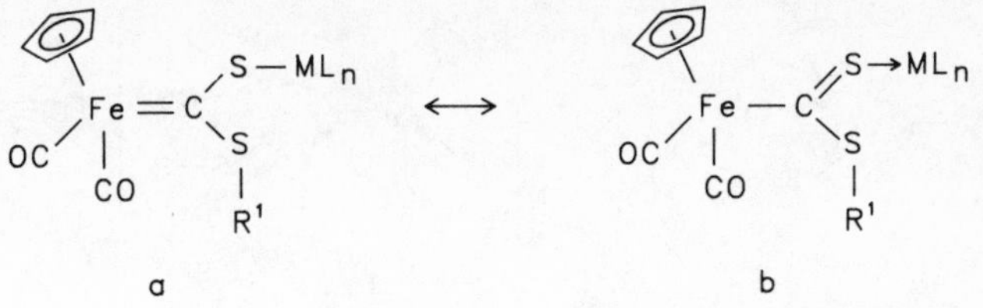

III

Compounds containing additional iron atoms have also been included. They contain the 17 electron fragment $C_5H_5(CO)_2Fe$ (abbreviated as Fp) as a part of the carbene ligand (Nos. 7 to 10 and 15) or a Cp(CO)Fe fragment (Cp(CS)Fe in No. 17) as a member of a heterocyclic carbene ligand (Nos. 16, 17 and 23 to 35). As shown in the resonance formulas IVa and IVb, the iron atom in the heterocycle similarly contains a carbenoid carbon atom and compounds of the type IV formally belong to the general type $^5L(CO)XFe{=}C(ER)_2$ as well; cf. Sect. 1.2.3.

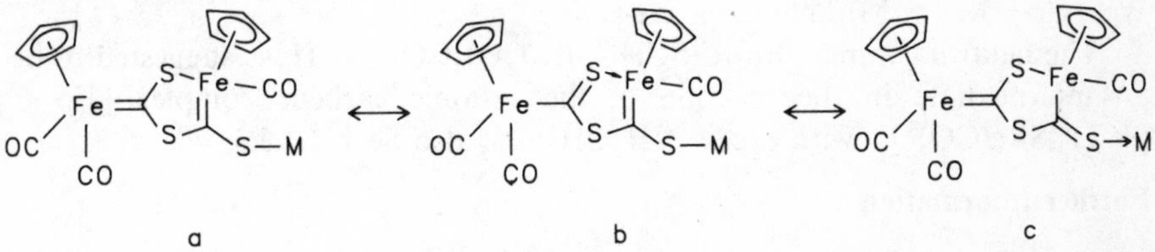

IV

The complexes listed in Table 11 have been prepared according to the following methods:

Method I From $FpCSSCH_3$ and $M(CO)_5(THF)$ (M = Cr, Mo, W) or $CP^*Mn(CO)_2(THF)$ (Cp^*=Cp, $C_5H_4CH_3$) in THF.

Method II From $FpCSSML_n$ (ML_n = Fp, $Re(CO)_5$) and the photochemically generated adducts $M(CO)_5(THF)$ (M=Cr, Mo, W) or CpMn $(CO)_2(THF)$.

Method III Reaction of the neutral complex No. 28 (Formula V) with $(CO)_5M$ (THF), HgX_2, or BF_3.

Method IV Reaction of the cationic complex VI with R^- (R = SCH_3, SC_2H_5, $S(CH_2)_3S^-$, or CN).

The ^{13}C NMR chemical shift of the carbene carbon atom (E = S) is found between 280 and 310 ppm, similar to that of the analogous cationic species of the type $[Fp{=}C(SR)_2]^+$: see Sect. 1.1.3.3.

Doubling of the ν (CO) frequencies of the Fp group in cyclohexane solution, observed in some complexes (Nos. 16, 20, 28, and 30 to 32), is attributed to the presence of different conformers in unequal concentrations. Isomers in which

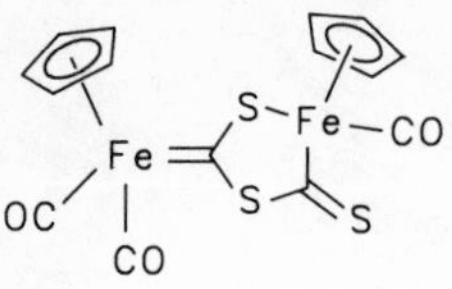

V

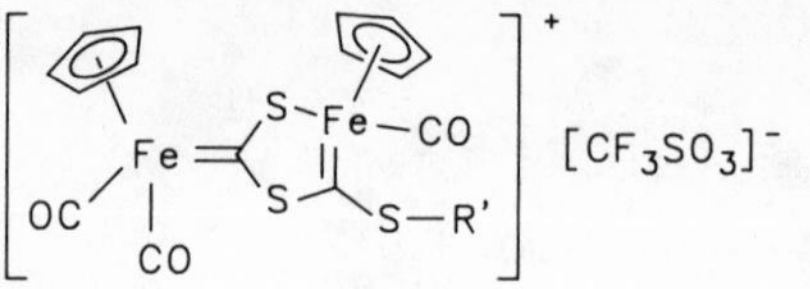

VI

the CS_2 plane coincides with the symmetry plane of the Fp group or 90° to the symmetry plane and the low energy differences between both structures are discussed; see also theoretical studies presented in Sect. 1.

Compounds No. 23 and 24 contain two chiral centers. Two diastereomers are obtained for No. 23, while only one isomer of No. 24 is formed. In the prevalent isomer of No. 23 and in No. 24 the position of the CO group of the endocyclic Fe atom is trans to the ring H atom or the CN group, respectively.

Compounds No. 7 to 14 are air-stable in the solid state, but decompose in chlorinated solvents within a few hours. The stability is in the order W > Cr > Mo > Mn. No. 24 is air-stable.

The neutral complex $Fp{=}C(SCH_3)SRe(CO)_4CONHCH_3$ is suggested to be an intermediate in the reaction of the cationic carbene complex $[Fp{=}C(SCH_3)SRe(CO)_5]^+$ with excess NH_2CH_3; see also Sect. 1.1.3.3.

Further Information

$(Fp{=}(C_3H_4NO\text{-}c)PdCl_2)_2$ (Table **11**, No. **1**, Formula VII) is obtained by metallation the imine complex VIII with $PdCl_2(CH_3CN)_2$ in CH_2Cl_2 solution (28% yield). The complex is moderately soluble in CH_2Cl_2 or CH_3CN, slightly soluble ether, and insoluble in hexane; it is unstable in solution. This dimeric complex can also be viewed as a carbene complex.

Bubbling HCl into a CH_2Cl_2 solution of the complex provides the cationic carbene complex IX along with $[PdCl_4]^{2-}$.

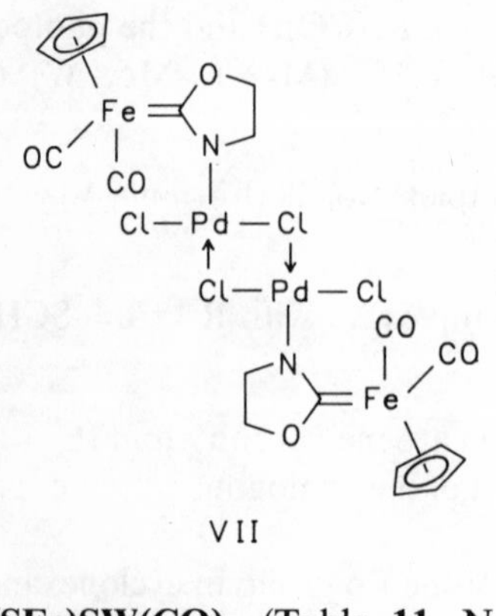

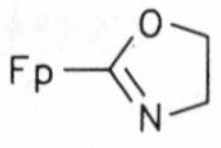

VIII

VII

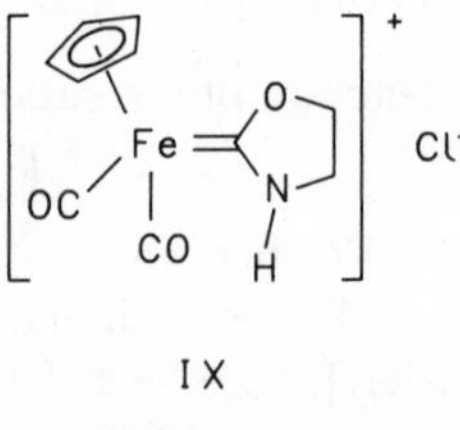

IX

$Fp{=}C(SFp)SW(CO)_5$ (Table **11**, No. **10**) decomposes when refluxed in CH_2Cl_2 to give $W(CO)_6$, No. 16 (70%), and small amounts of Fp_2 [5].

Table 11. Complexes of the Type Fp=C(ER)$_2$. Further information on numbers given by an asterisk is given at the end of the table

No.	Fe=C(ER)$_2$	method of preparation (yield) properties and remarks
the heteratoms E are O and N		
*1	Fe=C (ring: O–CH$_2$–CH$_2$–N), N–PdCl$_2$	dimeric (Formula VII), orange-brown powder, m.p. 126 °C (dec.) ^{1}H NMR (CD_3CN): 3.59 (m, CH_2N), 4.05 (m, CH_2O) ^{13}C NMR (C_6D_6): 56.0 (CH_3N), 67.0 (CH_3O), 193.6 (C=Fe, ?), 212.5 (CO)
the heteroatoms E are S		
2	Fe=C(SCH_3)SMn(CO)$_2$Cp	I (43%), red brown, m.p. 115 to 116 °C (dec.) ^{1}H NMR ($CDCl_3$): 2.81 (s, CH_3), 4.53, 5.01 (s's, Cp)
3	Fe=C(SCH_3)SMn(CO)$_2$$C_5H_4CH_3$	I (52%); red brown, m.p. 96 to 97 °C (dec.) ^{1}H NMR ($CDCl_3$): 1.87 (s, CH_3C), 2.83 (s, CH_3S), 4.40 (m, C_5H_4)
4	Fe=C(SCH_3)SCr(CO)$_5$	I (77%); orange, m.p. 93 to 98 °C (dec.) ^{1}H NMR ($CDCl_3$): 2.09 (s, CH_3)
5	Fe=C(SCH_3)SMo(CO)$_5$	I (48%); orange, m.p. 96 to 101 °C (dec.) ^{1}H NMR ($CDCl_3$): 2.91 (s, CH_3)
6	Fe=C(SCH_3)SW(CO)$_5$	I (82%); orange, m.p. 118 to 122 °C ^{1}H NMR ($CDCl_3$): 2.93 (s, CH_3)
7	Fe=C(SFp)SMn(CO)$_2$Cp	II (38%); red-brown, m.p. 114 to 115 °C (dec.) ^{1}H NMR (C_6D_6): 4.10, 4.50, 4.65 (s's, Cp)
8	Fe=C(SFp)SCr(CO)$_5$	II (58%); red, m.p. 122 to 123 °C (dec.) ^{1}H NMR (C_6D_6): 4.05, 4.40 (s's, Cp)
9	Fe=C(SFp)SMo(CO)$_5$	II (40%); red, m.p. 105 to 106 °C (dec.) ^{1}H NMR (C_6H_6): 4.09, 4.43 (s's, Cp)
*10	Fe=C(SFp)SW(CO)$_5$	II (75%); red, m.p. 141 to 142 °C (dec.) ^{1}H NMR (C_6D_6): 4.06, 4.45 (s's, Cp)

Table 11. Continued

No.	$Fe{=}C(ER)_2$	method of preparation (yield) properties and remarks
11	$Fe{=}C(SRe(CO)_5)SMn(CO)_2Cp$	II (18%); red-brown, m.p. 108 to 109 °C (dec.)
12	$Fe{=}C(SRe(CO)_5)SCr(CO)_5$	II (43%); orange, m.p. 114 to 115 °C (dec.) ^1H NMR ($CDCl_3$): 4.99 (s, Cp)
13	$Fe{=}C(SRe(CO)_5)SMo(CO)_5$	II (32%); orange, m.p. 109 to 110 °C (dec.) ^1H NMR ($CDCl_3$): 5.01 (s, Cp)
14	$Fe{=}C(SRe(CO)_5)SW(CO)_5$	II (65%); orange, m.p. 129 to 130 °C (dec.) ^1H NMR (C_6D_6): 4.28 (s, Cp)
15	$Fe{=}C(S)_2Pt(S)_2C{=}Fp$	see "Organoiron Compounds" C 4, 216
*16	$Fe{=}C(S)_2Fe(Cp)CO$	red needles, m.p. 114 to 115 °C (dec.) ^1H NMR ($CDCl_3$): 4.51 (s, CpFe (CO)), 4.87 (s, Fp) ^{13}C NMR ($CDCl_3$): 79.7, 82.2 (Cp), 212.2 (CO of Fp), 218.6 (CO of CpFe(CO)), 306.3 (C=Fe)
*17	$Fe{=}C(S)_2Fe(Cp)CS$	orange-red, m.p. 117 to 120 °C (dec.) ^1H NMR ($CDCl_3$): 4.77, 4.95 (s's, Cp)
*18	$Fe{=}C(S)_2Mn(CO)_4$	yellow crystals, m.p. 128 to 129 °C (dec.) ^1H NMR ($CDCl_3$): 4.90 (s, Cp)

No.	Compound	Data
*19	Fe=C, S, S, $Mn(CO)_3$, $P(C_6H_5)_3$	yellow crystals, m.p. 168 °C (dec.) ^{1}H NMR ($CDCl_3$): 4.63 (s, Cp)
*20	Fe=C, S, S, Mn, NO	yellow, m.p. 96 to 98 °C ^{1}H NMR (C_6D_6): 4.11, 4.50 (s's, Cp)
*21	Fe=C, S, S, $Re(CO)_4$	yellow crystals, m.p. 155 °C (dec.) ^{1}H NMR ($CDCl_3$): 4.98 (s, Cp)
*22	Fe=C, S, S, $Re(CO)_3$, $P(C_6H_5)_3$	yellow crystals, m.p. 220 °C (dec.)
23	Fe=, S, S, Fe—CO, H, SCH_3	IV (67%); dark green crystals, m.p. 122 to 124 °C (dec), 5:1 mixture of diastereomers, *cis* and *trans*) ^{1}H NMR ($CDCl_3$, *cis/trans*): 4.55/4.45 (s, Cp), 4.88 (s, Cp), 5.24/6.46 (s, CH) ^{13}C NMR (CD_2Cl_2, *cis/trans*): 66.8/67.10 (CH), 82.71/87.16 (Cp at the endocyclic Fe), 82.98 (Cp at the exocyclic Fe), 212.30, 212.67, 221.70 (CO), 281.91 (C=Fe)
*24	Fe=, S, S, Fe—CO, CN, SCH_3	IV (78%); black crystals, m.p. 158 to 160 °C (dec.) ^{1}H NMR ($CDCl_3$): 4.57, 4.93 (s's, Cp) ^{13}C NMR (CD_2Cl_2): 82.62, 83.84, 87.17 (all Cp), 211.32, 211.56, 219.44 (CO), 281.20 (C=Fe)
25	Fe=, S, S, Fe—CO, SCH_3, SCH_3	IV (88%); black crystals, m.p. 116 to 118 °C (dec.) ^{1}H NMR ($CDCl_3$): 4.57, 4.90 (s's, Cp)

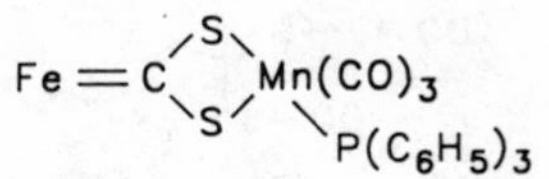

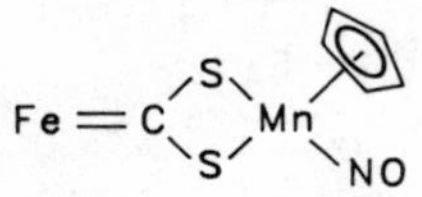

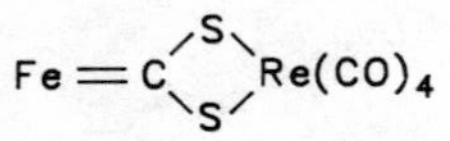

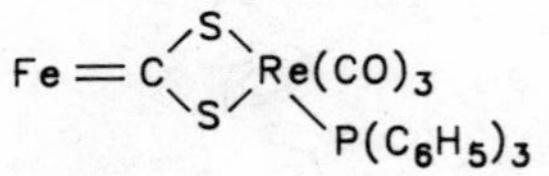

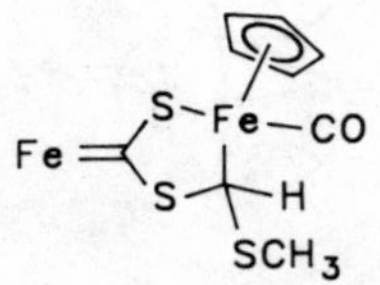

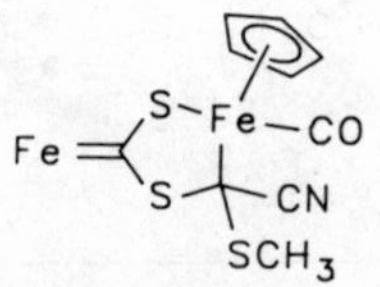

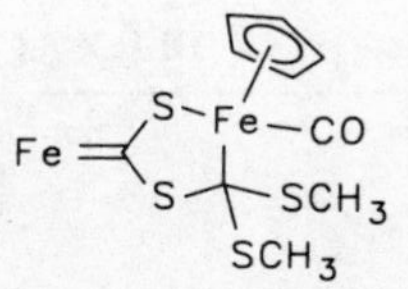

Table 11. Continued

No.	$Fe{=}C(ER)_2$	method of preparation (yield) properties and remarks
26	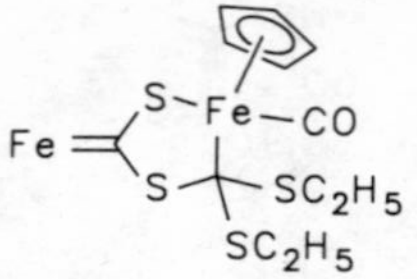	IV (83%); m.p. 112 to 114 °C (dec.) ^1H NMR ($CDCl_3$): 4.57, 4.89 (s's, Cp)
*27	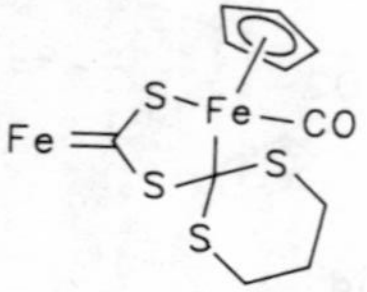	IV (74%); black crystals, m.p. 134 to 136 °C (dec.) ^1H NMR ($CDCl_3$): 4.62, 4.89 (s's, Cp) ^{13}C NMR (CD_2Cl_2): 79.63 (CS_2), 83.19, 87.10 (Cp), 212.11, 212.29, 220.49 (CO), 273.28 (C=Fe)
*28	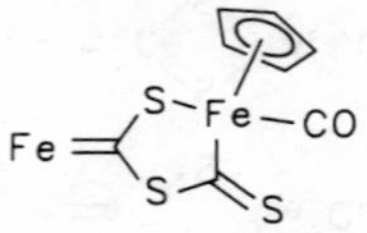	deep green crystals, m.p. 144 to 145 °C (dec.) ^1H NMR ($CDCl_3$): 4.75, 5.13 (s's, Cp) ^{13}C NMR (CD_2Cl_2): 87.70, 87.83 (Cp), 211.23 (2CO), 213.51 (1CO), 279.24 (C=Fe), 329.49 (C=S)
29	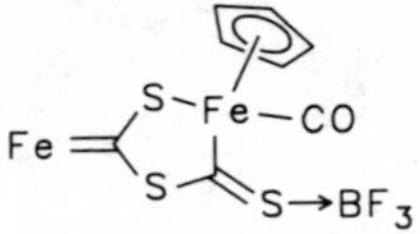	III; olive-green decomposes rapidly
30	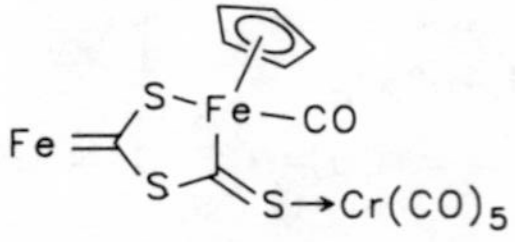	III (54%); dark purple crystals, m.p. 135 to 139 °C (dec.) ^1H NMR ($CDCl_3$): 4.78, 5.13 (s's, Cp)

No.	Compound	Data
31	$Fe{=}\overline{CS\text{-}Fe(Cp)(CO)\text{-}C(=S{\to}Mo(CO)_5)S}$	III (40%); purple crystals, m.p. 137 to 140 °C (dec.) ^{1}H NMR ($CDCl_3$): 4.78, 5.10 (s's, Cp)
*32	$Fe{=}\overline{CS\text{-}Fe(Cp)(CO)\text{-}C(=S{\to}W(CO)_5)S}$	III (78%); brown crystals, m.p. 155 °C ^{1}H NMR ($CDCl_3$): 4.83, 5.15 (s's, Cp) ^{13}C NMR (CD_2Cl_2): 86.56, 86.86 (Cp), 198.20 (CO-*trans* at W), 198.54 (CO-*cis* at W), 210.08, 210.17, 211.54 (all COFe), 280.29 (C=Fe), 321.51 (C=S)
33	$Fe{=}\overline{CS\text{-}Fe(Cp)(CO)\text{-}C(=S{\to}HgCl_2)S}$	III (52%); dark green crystals, m.p. 132 to 134 °C (dec.) ^{1}H NMR (CD_3COCD_3): 5.29, 5.55 (s's, Cp)
34	$Fe{=}\overline{CS\text{-}Fe(Cp)(CO)\text{-}C(=S{\to}HgBr_2)S}$	III (81%); dark green crystals, m.p. 150 to 152 °C (dec.) ^{1}H NMR (CD_3COCD_3): 5.18, 5.50 (s's, Cp) ^{13}C NMR (CD_2Cl_2): 87.78, 89.19 (Cp), 210.74 (2CO), 211.18 (1CO), 285.08 (C=Fe), 326.95 (C=S)
35	$Fe{=}\overline{CS\text{-}Fe(Cp)(CO)\text{-}C(=S{\to}HgI_2)S}$	III (77%); dark green crystals, m.p. 141 to 143 °C (dec.) ^{1}H NMR (CD_3COCD_3): 4.89, 5.14 (s's, Cp)

The complex crystallizes in the triclinic space group $P\bar{1}$-C_i^1 (No. 2) with a = 6.647(1), b = 12.145(2), c = 14.795(2) Å, α = 82.33(1)°, β = 85.73(1)°, γ = 79.73(1)°; Z = 2, d_c = 2.15 g/cm³. As depicted in Fig. 18 the $W(CO)_5$ group is *cis* and the Fp group *trans* to the carbene iron atom. The two $C_{carbene}$–S and S-metal distances are slightly different, indicating some thioacyl donor character (resonance formula IIIb) to the $W(CO)_5$ fragment.

Fp=CS$_2$Fe(CO)Cp (Table **11**, No. **16**) is obtained in 17% yield by irradiation of a THF solution of FpCSSFp. The complex can be prepared in 70% yield by refluxing compound No. 10 in CH_2Cl_2 or in 31% yield from K[$FpCS_2$] and [Cp(CO)Fe($CH_3CN)_2$]PF_6 at − 78 °C. It also forms in trace amounts upon refluxing FpCSSFp in CH_2Cl_2.

Fp=CS$_2$Fe(CS)Cp (Table **11**, No. **17**) forms in about 11% yield upon irradiation of complex No. 28 in THF.

Fp=CS$_2$Mn(CO)$_4$ and **Fp=CS$_2$Re(CO)$_4$** (Table **11**, Nos. **18** and **21**). To a solution of the anion $[FpCS_2]^-$ in THF the appropriate solid $M(CO)_5Br$ (M = Mn, Re) is added and the mixture stirred at room temperature for about 40 min; chromatography on silica gel gives $Mn_2(CO)_{10}$ and $FpMn(CO)_5$, followed by the title complex No. 18 (46%) and Fp_2. Chromatography in the case of

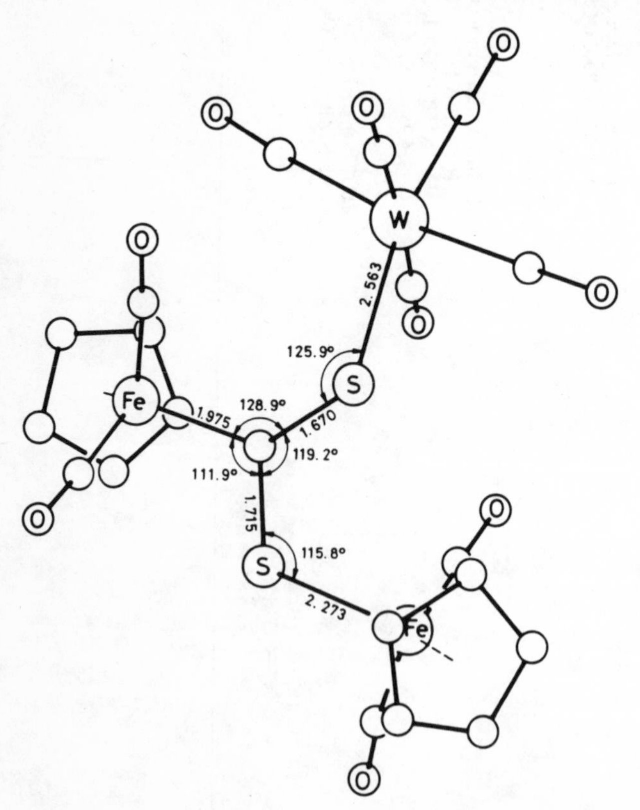

Fig. 18. Molecular structure of Fp=C(SFp)SW(CO)$_5$

the Re compound gives small amounts of $Re_2(CO)_{10}$, Fp_2, and No. 21 (48% yield).

For the reaction with $P(C_6H_5)_3$, see preparation of Nos. 19 and 22.

$Fp{=}CS_2Mn(CO)_3P(C_6H_5)_3$ and **$Fp{=}CS_2Re(CO)_3P(C_6H_5)_3$** (Table **11**, Nos. **19** and **22**) are obtained by the reaction of Nos. 18 and 21, respectively, with a twofold excess of $P(C_6H_5)_3$ in CH_2Cl_2 solution; 85 and 79% yields, respectively.

$Fp{=}CS_2Mn(NO)P(C_6H_5)_3$ (Table **11**, No. **20**) is prepared by addition of $[Cp(CO)_2Mn(NO)]PF_6$ to a solution of $[FpCS_2]^-$. Chromatography on Al_2O_3 gives small amounts of No. 16 followed by the title complex (41% yield) and $FpCS_2Fp$.

$Fp{=}CS_2C(CN)(SCH_3)Fe(Cp)(CO)$ (Table **11**, No. **24**). The addition of CN^- to the cationic starting complex takes place stereoselectively to form only one diastereoisomer with CN trans to the CO group of the endocyclic Fe atom.

The complex crystallizes in the triclinic space group $P\bar{1}$-C_i^1 (No. 2) with a = 6.894(2), b = 10.057(1), c = 14.520(2) Å, α = 96.00(1)°, β = 96.17(1)°, γ = 106.31(1)°; Z = 2, d_c = 1.70 g/cm³. The dihedral angle of the CS_2 plane and the symmetry plane of the Fp group is 89°. The structure of the molecule is depicted in Fig. 19.

$Fp{=}CS_2C(S_2C_3H_6)Fe(CO)Cp$ (Table **11**, No. **27**, Formula X). The broad signal of the CH_2S carbon atoms in the ^{13}C NMR spectrum at room temperature in CH_2Cl_2 splits at −60 °C into a pair of lines at 28.4 and 32.4 ppm, while the signals of the other carbon atoms remain unchanged. Coalescene occurs at 19 °C to give a single line at 31.0 ppm at 50 °C in $CDCl_3$ (32.6 ppm at 90 °C in

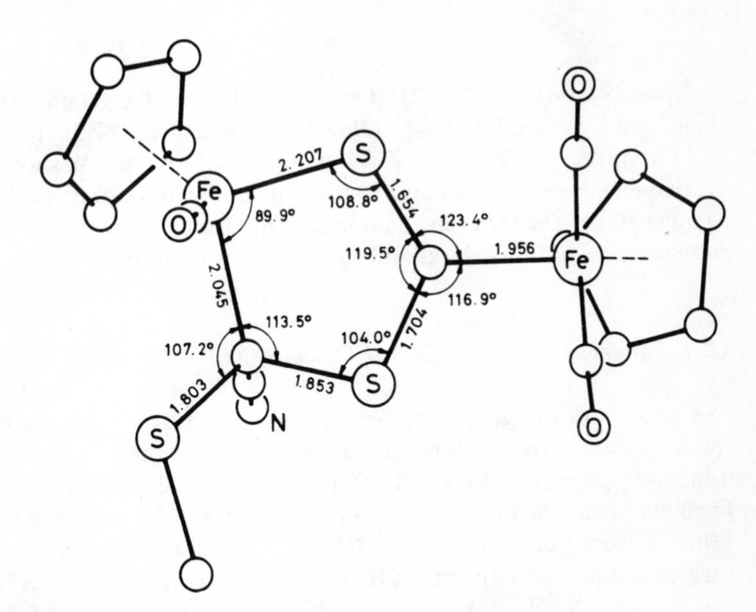

Fig. 19. Molecular structure of $Fp{=}CS_2C(CN)(SCH_3)Fe(Cp)(CO)$

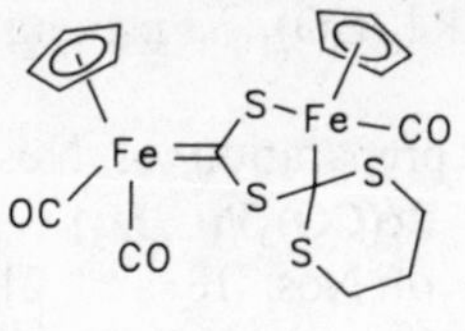

X

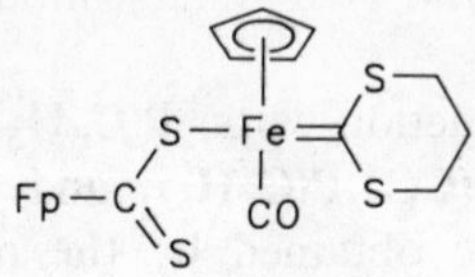

XI

toluene-d_8). The equilibration of the diastereotopic CH_2 groups is explained by C–S bond cleavage, affording the dithiocarbene intermediate XI, followed by rotation about the Cp(CO)Fe=C bond and ring closure.

Fp=C_2S_3Fe(CO)Cp (Table **11**, No. **28**) is prepared by the reaction of the cationic adducts XII (X = CF_3SO_3, M = $Re(CO)_5$, Fp; see also Formula IV) with $[(C_2H_5)_4N]Br$ in refluxing CH_2Cl_2 or with NaI in boiling acetone (67% yield).

Irradiation of the complex in THF gives the thiocarbonyl complex No. 17 along with other, unidentified products. The compound gives adducts with $(CO)_5M(THF)$, HgX_2, or BF_3 (Method III). Protonation with HSO_3CF_3, alkylation with RSO_3CF_3, or reaction with $AgBF_4$ produces the corresponding cationic adducts XII (M = H, CH_3, C_2H_5, or $Ag_{0.5}$).

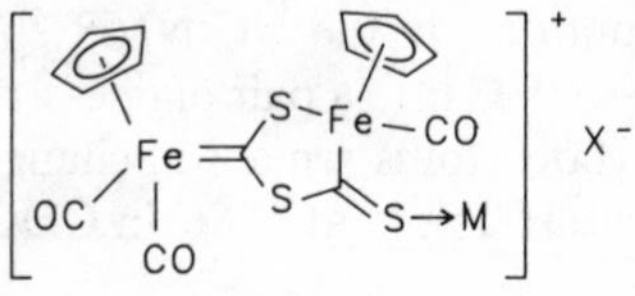

XII

Fp=C_2S_3{W(CO)$_5$}Fe(CO)(Cp) (Table **11**, No. **32**) crystallizes in the monoclinic space group $P2_1/a\text{-}C^5_{2h}$ (No. 14) with a = 13.162(2), b = 13.156(1), c = 14.092(3) Å, β = 93.66(1)°; Z = 4 and d_c = 2.11 g/cm^3. As shown in Fig. 20, the five-membered ring is planar with a maximum deviation of 0.11 Å from the average plane. The exocyclic S atom deviates from the ring plane by 0.3 Å. The endocyclic Fe atom is chiral and the crystals are racemic.

1.2.4 Complexes of the Type $^5L(CO)RFe=C(ER)_2$

Compounds of the general type $^5L(CO)RFe=C(ER)_2$, including R,E-bridging species with a carbene ligand and an additional α-bonded carbon atom at the iron atom, can be divided into the types of Formula I with separate ligands and Formula II in which the iron atom is part of a 3- to 6-membered ferracycle. Whereas from I only one species (III) is known and is described in this section, various compounds of Formula II are known and are described in "Organoiron Compounds" B 16b, 1990, Sect. 1.5.3.3.7, in which the compounds are arranged in the order of increasing size of the ferracycle, regardless of the bonding

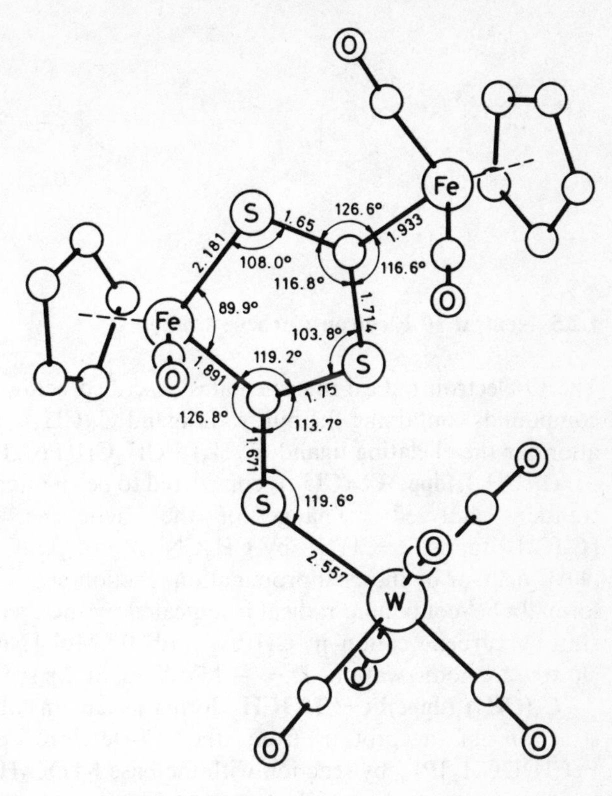

Fig. 20. Molecular structure of $Fp{=}C_2S_3\{W(CO)_5\}Fe(CO)(Cp)$

properties at the iron atom. In Table 16 on pp. 117–132 of the said volume Nos. 10 to 13, 32, 37 to 39, and 41 are relevant compounds with $^5L = Cp$, No. 61 bears $^5L = (C_6H_5)_3CC_5H_4$.

Fe=C(ER)(ER), OC, R — I

Fe=C(ER)(E), OC, R — II

Cp(CO)(CH$_3$NHCO)Fe=C(SCH$_3$)NC$_5$H$_{10}$-*c* (Formula III) forms an equilibrium with the cationic carbene complex $[Fp{=}C(SCH_3)NC_5H_{10}\text{-}c]PF_6$ (Formula IV) when the latter complex is treated with a large excess of CH_3NH_2: IV + $CH_3NH_2 \leftrightharpoons$ III + $[CH_3NH_3]PF_6$. Complex III exhibits one CO absorption in the IR spectrum in CH_2Cl_2 solution at 1909 cm^{-1}. Upon evaporation of the solvent, the starting material is recovered.

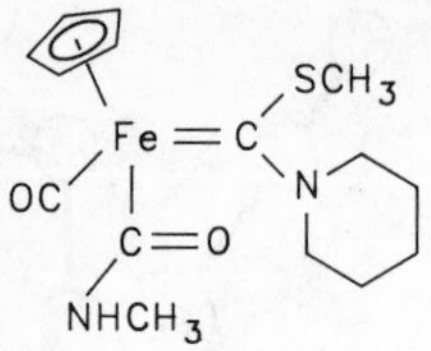

III

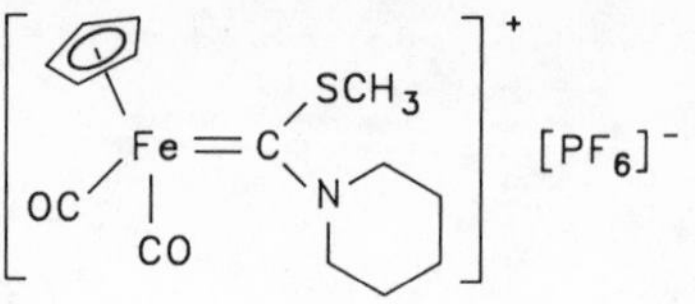

IV

1.2.5 Neutral 19 Electron Carbene Complexes

The 19-electron radical intermediates described below are represented only by compounds containing the bulky ^{5}L ligand $C_5(CH_3)_5$; dppe is used as abbreviation for the chelating ligand $(C_6H_5)_2PCH_2CH_2P(C_6H_5)_2$.

$C_5(CH_3)_5$(dppe)Fe=CH_2 is considered to be an intermediate by the electron-transfer catalyzed exchange of the methylene ligand in the cation $[C_5(CH_3)_5(dppe)Fe{=}CH_2]^+$ by CH_3CN to give $[C_5(CH_3)_5(dppe)Fe(NCCH_3)]^+$ (90% yield) or by the cyclopropanation reaction of this cation with styrene. The formally Fe^I-containing radical is indicated by the cyclic voltammogram of the starting carbene cation in CH_2Cl_2 with 0.1 Mol $[N(C_4H_9\text{-}n)]PF_6$ at the Pt electrode (anodic wave at E = − 1.70 V vs. SCE); see also Section 1.1.1.1.

$C_5(CH_3)_5$(dppe)Fe=$CHOCH_3$ forms as an unstable intermediate by abstraction of a proton from the 17-electron species $[C_5(CH_3)_5(dppe)FeCH_2OCH_3]PF_6$ by reaction with the base $K[OC_4H_9\text{-}t]$ in THF at − 80 °C for 1 h. In a second step, electron transfer occurs in the 17-electron species to form $C_5(CH_3)_5(dppe)FeCH_2OCH_3$ and the cationic carbene complex $[C_5(CH_3)_5(dppe)Fe{=}CHOCH_3]PF_6$.

The ESR spectrum (133 K) of a sample generated in THF at − 100 °C by the procedure above and followed by immediate quenching at 77 K allows the observation of the signals of the radical at g = 2.08 and 2.40 along with the signals of the starting 17-electron cation.

With $[(Cp)_2Fe]PF_6$ or the 17-electron cation $[C_5(CH_3)_5(dppe)FeCH_2O\text{-}CH_3]PF_6$ at 0 °C, oxidation to the cationic carbene complex $[C_5(CH_3)_5(dppe)Fe{=}CHOCH_3]^+$ occurs.

1.3 Anionic Carbene Complexes

The single complex described in this section can be viewed a an anionic carbene complex of the type $[Cp(CO)(^2D)Fe{=}C(R)OR']^-$, according to the resonance form Ia, or as a compound in which the acyl oxygen atom of the $FeCOCH_3$ group coordinates at the Mn atom (Ib). No other anionic carbene complexes containing the ^{5}LFe group are described, but the ferracyclic compounds No. 35,

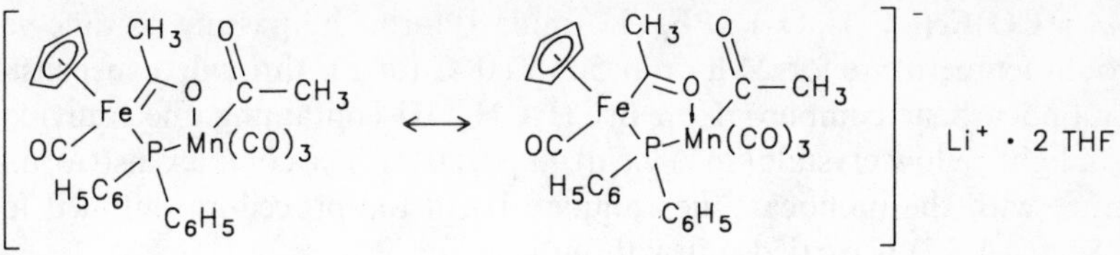

I

36, and 41 in Table 16 on p. 125 in "Organoiron Compounds" B 16b, 1990, can be understood as such.

$[Cp(CO)FeC_{19}H_{16}MnO_5P]Li \cdot 2$ THF (Formula I) can be prepared by addition of $[CH_3]^-$ to the neutral complex II (see Formula I, $^2D = CO$, in Sect. 1.2.2). Thus, to a solution of II in THF at $-78\,°C$, a solution of $LiCH_3$ in hexane is added dropwise via syringe; the complex is obtained as a red oil.

The NMR spectra were measured in CD_3COCD_3 solution at 22 °C (in ppm). ^1H NMR: 2.5, 2.9 (s's, CH_3), 4.4 (s, Cp). ^{13}C NMR: 87.2 (s, Cp), 207.4, 208.5 (s's, CO), 210 (s, br, CO), 219.9 (d, CO; J(P,C) = 20.0), 309.7 (d, C=O at Mn; J(P,C) = 9.8), 323.8 (d, C=Fe; J(P,C) = 12.2). ^{31}P NMR: 93.3 (s).

Protonation with 20% H_3PO_4 probably occurs at the terminal acyl oxygen to generate an intermediate hydroxy carbene complex that decomposes to the starting complex II.

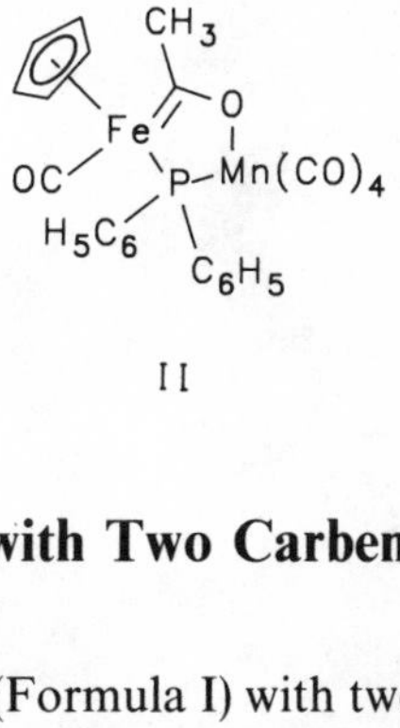

II

1.4 Carbene Compound with Two Carbene Ligands

Only one cationic compound (Formula I) with two carbene ligands is described with CO as additional ligand. For similar neutral compounds of Formula II (X = H, F; various groups R,R'), see "Organoiron Compounds" B 16b, 1990, Nos. 42 to 59 in Table 16 on pp. 127–131.

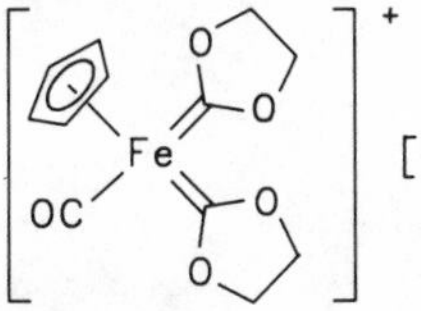

I

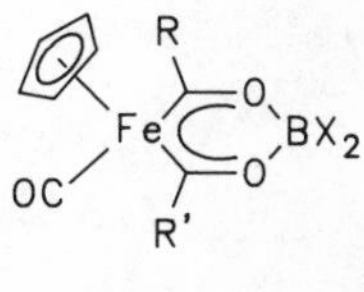

II

$[Cp(CO)Fe(=C_3H_4O_2)_2]PF_6$ (Formula I) forms by passing oxirane vapors at room temperature for 24 h or at 50 to 60 °C for 2 h through a suspension of the monocarbene complex III in $BrCH_2CH_2OH$ containing one equivalent of NaBr; light yellow crystals (26% yield); m.p. 180 °C. The complex also forms as a mixture with the monocarbene complex III, if the procedure outlined for III (see Sect. 1.1.3.3) is carried out without cooling.

The NMR spectra were recorded in CD_3CN solution (in ppm). ^1H NMR: 4.62 (m, CH_2), 5.06 (s, Cp). ^{13}C NMR: 72.54 (CH_2), 87.70 (Cp), 216.51 (CO), 251.00 (C=Fe).

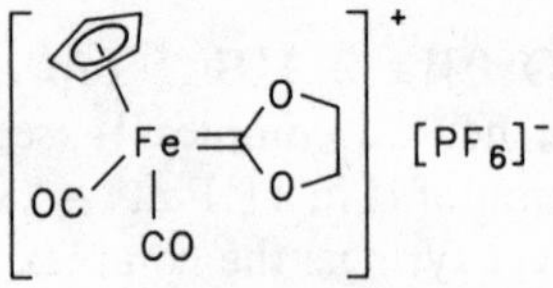

$[PF_6]^-$

III

2 Vinylidene Complexes

Only cationic vinylidene complexes of the types I to III with one vinylidene ligand are known. The vinylidene ligand can be considered as a carbene complex in which a carbon atom is interposed between the metal atom and the terminal CRR′ group. In contrast to the related carbene complexes, no heteroatoms are introduced and R or R′ represent alkyl, aryl, or hydrogen. The cations can also be considered as metal-stabilized vinyl cations, $M-C^+=CRR'$. The thermal stability of the complexes increases with the number of CO groups replaced by ligands of the type ^2D with greater σ donor/π acceptor ratios, thus increasing in the order I < II < III.

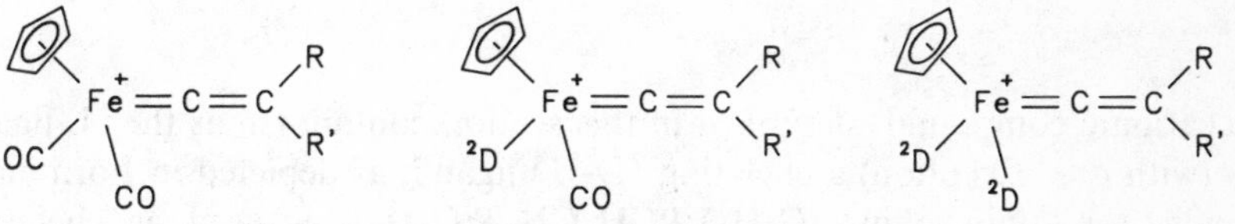

I II III

Low-temperature NMR experiments fail to freeze out rotation of the $=C=C(CH_3)_2$ ligand around the Fe=C bond axis of $[Cp(CO)\{P(C_6H_5)_3\}Fe=C=C(CH_3)_2]^+$, where the ground state should have nonequivalent CH_3 groups. The rotation is considered to be fast even at − 100 °C.

Nonparametrized molecular orbital calculations were carried out on the model vinylidene cations $[Cp(PH_3)_2Fe=C=CH_2]^+$ and $[Fp=C=CH_2]^+$. The metal-vinylidene bond is between double and triple, which permits facile rotation of the ligand. Whereas the preferred orientation of the CRR′ plane in carbene complexes is within the plane of the Fp group, the introduction of a π bond with the interposed carbon atom changes the preferred orientation by a 90° twist, as shown by extended Hückel calculations and illustrated in Formula IV. The rotation barrier for the simplest vinylidene ligand $=C=CH_2$ is calculated to be 3.6 kcal/mol, as compared with 6.2 kcal/mol for the $=CH_2$ carbene ligand, and should further decrease with the number of carbon atoms introduced.

In contrast to the results of the calculations, the X-ray determination of the cation $[Cp(dppe)Fe=C=C(CH_3)CS_2CH_3]^+$ ($(C_6H_5)_2PCH_2CH_2P(C_6H_5)_2$ = dppe) shows that the orientation of the vinylidene plane with respect to the

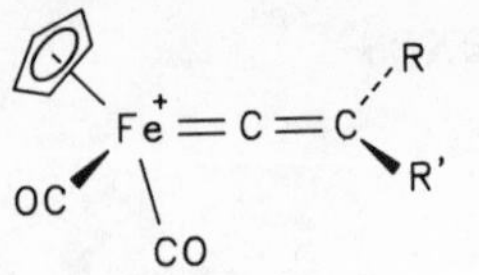

IV

mirror plane of the [Cp(dppe)Fe] fragment is not 90°, as predicted, but 49.7° (130.3°). This deviation is explained by electronic and steric factors.

The action of bases on the vinylidene cations depends on the nature of both systems. Thus, proton abstraction can occur from C-2 (reverse of formation reaction) to give alkynes, or addition of bases at the electrophilic vinylidene carbon C-1 takes place to generate adducts; see Sect. 2.4. If the bases are primary or secondary amines, H transfer from the resulting adduct to C-2 gives carbene complexes.

2.1 Cationic Complexes of the Type $[^5L(^2D\text{-}^2D)Fe{=}C{=}CRR']^+$ or $[^5L(^2D)_2Fe{=}C{=}CRR']^+$

The cationic compounds described in this section contain Cp as the ^5L ligand plus (with one exception) a chelating $^2D\text{-}^2D$ ligand, as depicted in Formula I. Usually, the diphosphine $(C_6H_5)_2PCH_2CH_2P(C_6H_5)_2$ is used as chelating ligand and is abbreviated as dppe. Similar substituted derivatives (Formula IV and V) are abbreviated as shown in the explanation for Table 12. One complex of the $[Cp(^2D)_2Fe{=}C{=}CRR']^+$ type is mentioned (Formula I, the $^2D\text{-}^2D$ chelating ligand is replaced by two $P(CH_3)_3$ groups) but not characterized further.

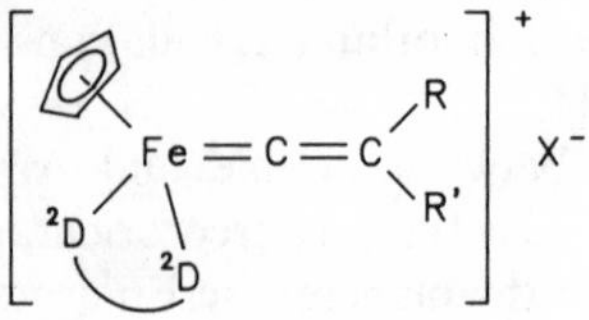

I

General remarks. Vinylidene cations of the $[Cp(^2D\text{-}^2D)Fe{=}C{=}CRR']^+$ type are chiral if R and R′ are different and the molecule does not contain a symmetry plane. Thus, if the molecule contains a further center of chirality, it gives rise to diastereomers as shown in compounds No. 3 to 5, 7, and 8. The equilibrium diastereomeric composition (dc) is indicated for each individual complex. Additionally, the five-membered ring formed by Fe and the chelating molecule dppe is puckered to give the chiral δ and λ conformations (Formula II), but rapid interconversion occurs from one chiral conformation to the other. However, when a chiral carbon atom is introduced by substitution of dppe by (*S*,*S*)-chiraphos and *trans*-cypenphos (Formula IV and V, respectively), the chelate ring is fixed in a single static chiral conformation, and in chiraphos the δ conformation with the CH_3 groups in the equatorial positions (e in Formula II) is preferred (Formula III).

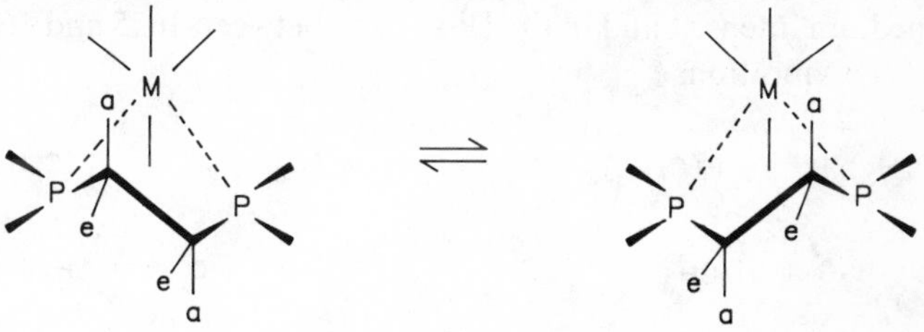

II

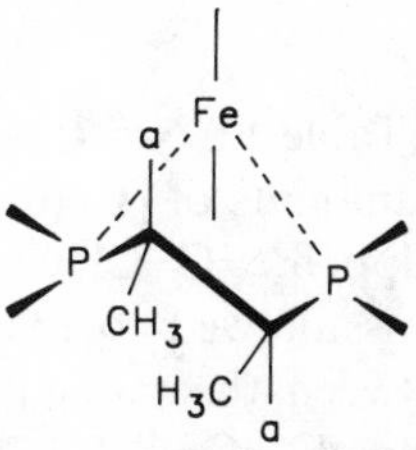

III

At room temperature Nos. 3 to 5, 7, and 8 therefore exhibit two doublets in the ^{31}P NMR spectrum (P-1 and P-2) with J(P,P) values of about 50 to 60 Hz. Temperature-dependent spectra indicate rapid rotation of the vinylidene ligand with barriers of rotation of about 9 to 10 kcal/mol. On lowering the temperature to about 60 K a diastereomeric equilibrium generates two pairs of doublets; the diastereomeric composition (dc) of both rotamers depends on the bulkiness of the R′ (R is H) and on the type of chiral ^2D-^2D ligand; for the dc values, shifts of the signals and explanations, see the individual compounds.

In the ^{13}C NMR spectra the signals of the vinylidene carbon atoms (assigned as C-1) appear at 355 to 370 ppm and that of the adjacent carbon atom (C-2) is found in the region of the phenyl carbon atoms.

The signal for the PF_6^- anion in the ^{31}P NMR spectrum (in $CD_2Cl_2/CDCl_3$ solution) was recorded at − 143.8 (sept; J(F,P) = 711 Hz) ppm.

The compounds collected in Table 12 can be prepared according to the following methods:

Method I Alkylation (CH_3SO_3F, RI, or $[C_7H_7\text{-}c]PF_6$) or protonation ($HPF_6 \cdot O(C_2H_5)_2$) of Cp(dppe)FeC≡CR compounds.

Method II From $Cp(^2D\text{-}^2D)FeBr$ and $RC{\equiv}CH$ in the presence of KPF_6 or NH_4PF_6.

Explanation for Table 12: The chelating $^2D\text{-}^2D$ ligands, (*S*,*S*)-chiraphos and *rac-trans*-cypenphos, are shown in Formula IV and V, respectively. ^1H NMR and ^{13}C NMR data of the chelating ligands and unassigned signals are omitted. The ^{13}C NMR shifts for the Cp ligand are in the range 88 to 90 ppm. The C-1 carbon atoms in dppe complexes appear as triplets with J(P,C) values of about 33 Hz. A medium intense band in the IR spectra between 1625 and 1690 belongs to the ν (C=C) vibration.

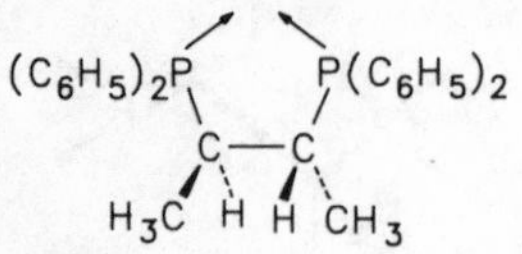

Further Information

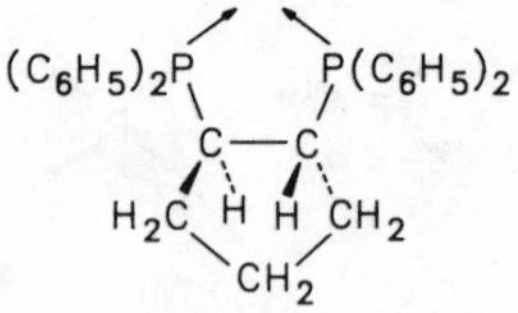

$[Cp(dppe)Fe{=}C{=}CH_2]X$ (Table **12**, No. **1**, X = PF_6, FSO_3). The FSO_3^- salt forms along with equimolar amounts of $[Cp(dppe)Fe{=}C{=}C(CH_3)_2]^+$ (No. 10) and small amounts of $[Cp(dppe)Fe{=}C{=}CHCH_3]^+$ (No. 2) when the alkyne complex $Cp(dppe)FeC{\equiv}CH$ is allowed to react with CH_3SO_3F in benzene solution. An equilibrium between the pairs $[Cp(dppe)Fe{=}C{=}CHCH_3]^+$/Cp(dppe)FeC≡CH and $[Cp(dppe)Fe{=}C{=}CH_2]^+$/$Cp(dppe)FeC{\equiv}CCH_3$ is considered to be responsible for the product distribution. The cation is thermally stable.

The equivalence of the vinylidene protons in the variable-temperature ^1H NMR spectrum in the range between − 60 and + 60 °C is interpreted in terms of rapid ligand rotation or, more likely, as evidence for the vinylidene plane being perpendicular to the mirror plane of the Cp(dppe)Fe fragment.

The PF_6^- salt is unreactive towards alcohols and H_2O. It is oxidized by iodosobenzene in CH_3CN or copper(II) acetate in CH_3OH similarly to No. 2, but gives a mixture of products. Treatment of the cation with bases regenerates $Cp(dppe)FeC{\equiv}CH$.

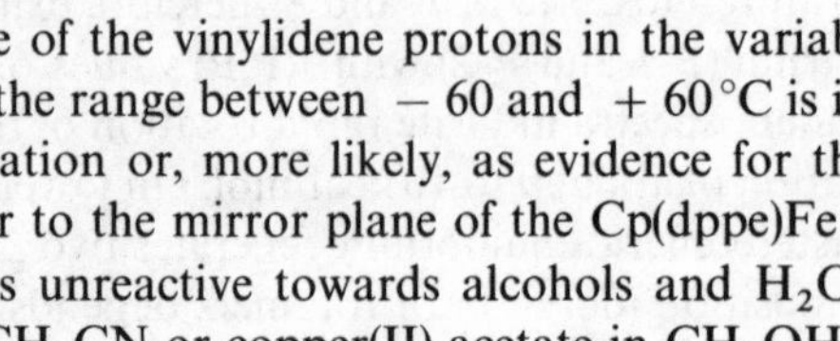

Table 12. Cationic Vinylidene Complexes of the Type $[Cp(^2D\text{-}^2D)Fe{=}C{=}CRR']^+$ and $[Cp(^2D)_2Fe{=}C{=}CRR']^+$. Further information on compounds with numbers preceded by an asterisk is given at the end of the table

No.	Fe=C=CRR′ ^{2}D-^{2}D or 2 ^{2}D	anion, method of preparation (yield) properties and remarks
*1	$Fe{=}C{=}CH_2$ dppe	PF_6^- salt, I ^{1}H NMR ($CDCl_3$): 3.91 (t, $CH_2{=}C$) FSO_3^- salt ^{13}C NMR ($CDCl_3$): 106.8 (C-2), 354.5 (m, C-1)
*2	$Fe{=}C{=}CHCH_3$ dppe	PF_6^- salt, I, orange crystals ^{1}H NMR (CD_3COCD_3): 1.02 (d of t, CH_3: $^5J(P,H) = 0.7$), 4.37 (q of t, HC; $^4J(P,H) = 3.0$) ^{13}C NMR (CD_3COCD_3): 4.68 (s, CH_3), 118.0 (C-2), 358.3 (t, C-1) BF_4^- salt, II (80%) FSO_3^- salt
*3	$Fe{=}C{=}CHCH_3$ (*S*,*S*)-chiraphos	PF_6^- salt, II (82%), yellow orange ^{1}H NMR (CD_2Cl_2): 1.35 (m, $CH_3C{=}$), 4.70, (t, 1H) ^{13}C NMR ($CD_2Cl_2/CDCl_3$, 7/3 ratio, at 289 K): 358.6 (C-1; $^2J(P,C) = 26$) ^{31}P NMR ($CD_2Cl_2/CDCl_3$, 7/3 ratio): 100.9, 96.1 (d's, P-1, 2; J(P,P) = 50) at 289 K; 105 (d, P-1), 96.6/95.2 (d's, 55/45 ratio, P-2) at 153 K
*4	$Fe{=}C{=}CHCH_3$ *trans*-cypenphos	PF_6^- salt, II (80%), yellow ^{1}H NMR (CD_2Cl_2): 1.60 (m, CH_3) ^{13}C NMR ($CD_2Cl_2/CDCl_3$, 7/3 ratio, at 289 K): 361.5 (dd, C-1, $^2J(P,C) = 28, 37$) ^{31}P NMR ($CD_2Cl_2/CDCl_3$, 7/3 ratio): 74.6, 72.1 (d's, P-1, 2; J(P,P) = 61) at 289 K; similar values at 153 K
5	$Fe{=}C{=}CHC_4H_9$-*t* *trans*-cypenphos	PF_6^- salt, II (49%), red ^{13}C NMR ($CD_2Cl_2/CDCl_3$, 7/3 ratio, at 289 K): 364.0 (dd, C-1; $^2J(P,C) = 28, 36$) ^{31}P NMR ($CD_2Cl_2/CDCl_3$, 7/3 ratio): 73.8, 68.2 (d's, P-1, 2; J(P,P) = 60) at 298 K for the rotamer population at 168 K, see further information for No. 4
*6	$Fe{=}C{=}CHC_6H_5$ dppe	PF_6^- salt; II ^{13}C NMR ($CD_2Cl_2/CDCl_3$, 7/3 ratio, at 298 K): 365 (C-1) ^{31}P NMR ($CD_2Cl_2/CDCl_3$, 7/3 ratio): 93.5 (s) at 298 K; 97.4, 91.4 (AB q, P-1, 2; J(P,P) = 42) at 180 K

Table 12. Continued

No.	Fe=C=CRR′ ^2D-^2D or 2 ^2D	anion, method of preparation (yield) properties and remarks
*7	Fe=C=CHC_6H_5 (*S,S*)-chiraphos	PF_6^- salt, II (> 90%), light brown microcrystals ^{13}C NMR (CD_2Cl_2/$CDCl_3$, 7/3 ratio, at 298 K): 355 (dd, C-1; $^2J(P,C)$ = 29.3, 35.4) ^{31}P NMR (CD_2Cl_2/$CDCl_3$, 7/3 ratio): 94.8, 93.2 (d's, P-1, 2; J(P,P) = 52) at 298 K; 103.2/84.0 (d's, 86/14 ratio P-1; J(P,P) = 48/60), 92.1/93.7 (d's, 86/14 ratio, P-2; J(P,P) = 48/60) at 168 K
*8	Fe=C=CHC_6H_5 *trans*-cypenphos	PF_6^- salt, II (87%), brown ^{13}C NMR (CD_2Cl_2/$CDCl_3$, 7/3 ratio, at 298 K): 368 (C-1; $^2J(P,C)$ = 29.4, 37.2) ^{31}P NMR (CD_2Cl_2/$CDCl_3$, 7/3 ratio): 71.5, 69.0 (d's, P-1, 2; J(P,P) = 61) at 298 K
*9	Fe=C=C$(CH_3)_2$ 2P$(CH_3)_3$	no anion and no properties reported
*10	Fe=C=C$(CH_3)_2$ dppe	FSO_3^- salt; I (> 90%), pale orange solid, m.p. 230.5 to 232.5 °C ^1H NMR ($CDCl_3$/CD_2Cl_2): 0.96 (t, CH_3) ^{13}C NMR ($CDCl_3$/CD_2Cl_2): 12.0 (CH_3), 363.3 (t, C-1)
*11	Fe=C=C(CH_3)C_6H_5 dppe	I^- salt, I PF_6^- salt prepared by anion exchange from the I^- salt (62%)
*12	Fe=C=C(C_2H_5)C_6H_5 dppe	I^- salt; I PF_6^- salt prepared by anion exchange from the I^- salt (70%)

*13	$Fe{=}C{=}C(CH_2C_6H_5)C_6H_5$ dppe	I^- salt, I PF_6^- salt prepared by anion exchange from the I^- salt (87%)
*14	$Fe{=}C{=}C(CH_3)COCH(C_6H_5)_2$ dppe	BF_4^- salt, bright orange crystals, m.p. 182 to 183 °C ^1H NMR ($CDCl_3$): 1.13 (s, CH_3) ^{13}C NMR ($CDCl_3$): 9.24 (CH_3), 127.1 (C-2), 366.7 (t, C-1) $CF_3CO_2^-$ salt
*15	$Fe{=}C{=}C(CH_3)CS_2CH_3$ dppe	I^- salt, CH_3OH solvate, deep yellow brown crystals, m.p. 171 to 177 °C (dec.) ^1H NMR ($CDCl_3$): 1.45 (s, br, CH_3C) ^{13}C NMR ($CDCl_3$): 13.4 (s, CH_3C), 144.7 (C-2), 364.5 (t, C-1)
*16	H $Fp{=}C{=}C$ C_6H_5 Fp=Cp(dppe)Fe	PF_6^- salt, I (78%), buff-orange powder, m.p. >155 °C (dec.) ^1H NMR ($CDCl_3$): 1.91 (t, CH of C_7H_7), 4.99 (m, 2CH= of C_7H_7), 5.93 (m, 2CH= of C_7H_7), 6.19 (m, 2CH= of C_7H_7)

[Cp(dppe)Fe=C=CHCH$_3$]X (Table **12**, No. **2**; X = PF_6, FSO_3, BF_4). For the formation of the alkyne complex Cp(dppe)FeC≡CH by alkylation with CH_3SO_3F, see above.

The vinylidene proton is acidic and the cation forms a conjugate acid/base pair with Cp(dppe)FeC≡CCH$_3$; pK = 7.74 ± 0.05 (determined by titration of the base Cp(dppe)FeC≡CCH$_3$ with HCl or the acid [Cp(dppe)Fe=C =CHCH$_3$]$^+$ with aqueous KOH in THF under N_2; deprotonation also occurs with $(CH_3)_3NO$. The BF_4^- salt is oxidized by iodosobenzene in CH_3CN (20 °C, 4 h) and involves carbon-carbon coupling to generate the dinuclear dicationic complex VI (R = CH_3, ^2D-^2D = dppe) in 77% yield. This one-electron oxidation can also be effected with copper(II) acetate in CH_3OH solution.

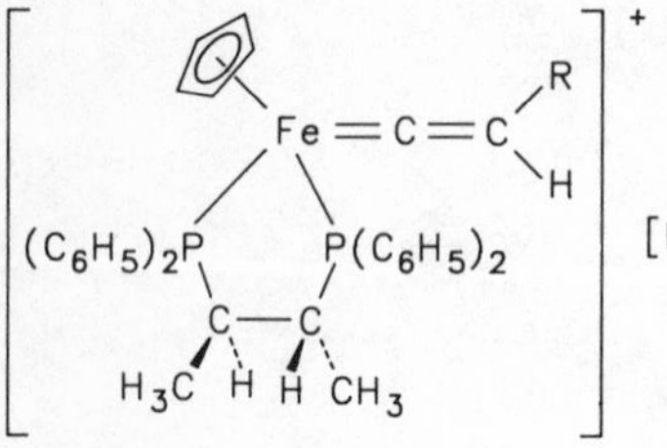

VIIa

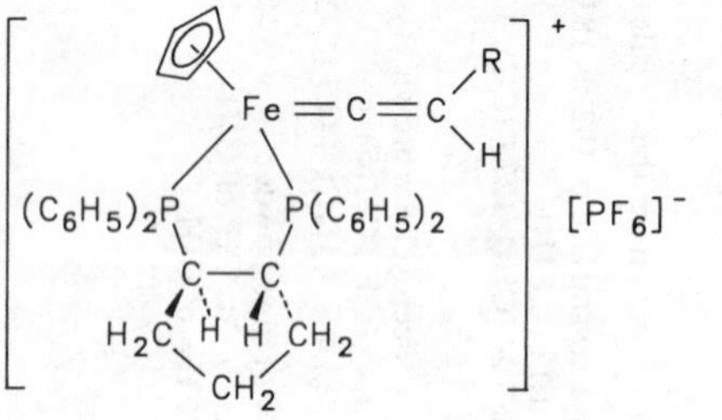

VIIb

[Cp{(C$_6$H$_5$)$_2$PCH(CH$_3$)CH(CH$_3$)P(C$_6$H$_5$)$_2$}Fe=C=CHCH$_3$]PF$_6$ (Table **12**, No. **3**). The complex (Formula VIIa, R = CH_3) exhibits only a low asymmetric induction and the population of the two possible diastereomeric conformations are found to be 55:45 at 160 K. Coupling constants between P-1 and P-2 at this temperature have not been determined because of incomplete resolution of the signals; see also general remarks.

[Cp{(C$_6$H$_5$)$_2$P(C$_5$H$_8$)P(C$_6$H$_5$)$_2$}Fe=C=CHCH$_3$]PF$_6$ (Table **12**, No. **4**). The complex (Formula VIIb, R = CH_3) shows only one set of signals in the low-temperature ^{31}P NMR spectrum, which is interpreted in terms of one diastereomer predominating (> 9:1). The signal of P-1 retain its shift, whereas that of P-2 goes downfield on lowering the temperature. The signals are isochronous at 210 K. This shift is probably caused by the concentration of the less stable rotamer decreasing below the detection point as a consequence of decreasing temperature.

[Cp(dppe)Fe=C=CHC$_6$H$_5$]PF$_6$ (Table **12**, No. **6**). The ^{31}P NMR at 180 K exhibits an AB quartet, indicating that the plane of the vinylidene ligand is perpendicular to the plane of the Cp(dppe)Fe fragment. The coalescence temperature to give a singlet is 210 K and the energy barrier for rotation of the vinylidene ligand is about 9.4 kcal/mol.

The complex undergoes oxidative coupling similar to No. 2 to generate the dication VI with R = C_6H_5.

[Cp{(C$_6$H$_5$)$_2$PCH(CH$_3$)CH(CH$_3$)P(C$_6$H$_5$)$_2$}Fe=C=CHC$_6$H$_5$]PF$_6$ (Table **12**, No. **7**). The set of doublets of the room-temperature ^{31}P NMR spectrum of the complex (Formula VIIa, R = C_6H_5) gives rise to two sets of doublets of 86:14 relative intensity at 168 K, reflecting the diastereomeric composition; see general remarks. Different coalescence temperatures for P-1 (about 220 K) and P-2 (about 205 K) are recorded; energy barrier of rotation is about 9.7 kcal/mol. The difference in free energy of formation for the two diastereomers is estimated to be about 0.55 kcal/mol.

$[Cp\{(C_6H_5)_2P(C_5H_8)P(C_6H_5)_2\}Fe{=}C{=}CHC_6H_5]PF_6$ (Table **12**, No. **8**). The complex (Formula VIIb, R = C_6H_5) forms only one diastereomer at low temperature (> 9:1), similar to No. 4; isochronous temperature for the ^{31}P NMR signals is 210 K; see No. 4.

$[Cp(dppe)Fe{=}C{=}C(CH_3)_2]FSO_3$ (Table **12**, No. **10**) forms also with equimolar amounts of No. 1 by the alkylation of Cp(dppe) FeC≡CH as described for No. 1. Thus, the Cp(dppe)FeC≡CCH_3 formed during the equilibrium process is alkylated to give the title complex.

The complex is thermally stable and does not react with alcohol and H_2O. With $Na[BH(OCH_3)_3]$ in THF as a hydride source a 4:1 mixture of Cp(dppe)FeCH=$C(CH_3)_2$ and Cp(dppe)FeH is obtained. The action of KOH in THF containing traces of water, or $Na[N(Si(CH_3)_3)_2]$ in ether, smoothly deprotonates the cation to generate exclusively complex VIII (containing a ferradiphospha[2.1.1.]bicyclohexane ring) in 66% yield. The zwitterionic intermediate IX is postulated to be formed in the first step. No product arises from an expected nucleophilic attack at C-1. No. 10 is not attacked by $(CH_3)_3NO$ or iodosobenzene.

φ = C_6H_5

VIII IX

$[Cp(dppe)Fe{=}C{=}C(R)C_6H_5]X$ (Table **12**, Nos. **11** to **13**, R = CH_3, C_2H_5, $CH_2C_6H_5$; X = I, PF_6). The I^- salts are obtained from Cp(dppe)FeC≡CC_6H_5 and the corresponding RI under mild conditions. Anion exchange gives the PF_6^- salts in 62, 70, and 87% yields, respectively.

$[Cp(dppe)Fe{=}C{=}C(CH_3)COCH(C_6H_5)_2]X$ (Table **12**, No. **14**, X = BF_4, CF_3CO_2). The BF_4^- salt is prepared by addition of diphenylketene to a solution of Cp(dppe)FeC≡ CCH_3 in dry CH_2Cl_2 at − 78 °C followed by treatment of the mixture with a freshly prepared solution of $Ti(OCH(CH_3)_2)_4$. Treatment with HBF_4 gives the complex in 47% yield. The $CF_3CO_2^-$ salt slowly forms on addition of 1 equivalent of CF_3COOH to the iron complex X (^2D-^2D = dppe).

$[Cp(dppe)Fe{=}C{=}C(CH_3)CS_2CH_3]X$ (Table **12**, No. **15**, X = I, BF_4, PF_6). The I^- salt is prepared by methylation of XI with CH_3I in CH_2Cl_2 solution (3 h).

X XI

The BF_4 or PF_6^- salts are obtained by exchange using aqueous $[NH_4]BF_4$ or $[NH_4]PF_6$, respectively, in CH_3OH or C_2H_5OH.

[Cp(dppe)Fe=C=C(CH$_3$)CS$_2$CH$_3$]I·CH_3OH crystallizes in the monoclinic space group $P2_1/c\text{-}C_{2h}^5$ (No. 14) with the parameters a = 11.625(3), b = 20.609(4), c = 15.490(2) Å, β = 103.25(2)°; Z = 4 and d_c = 1.487 g/cm³. The geometry at the iron atom is normal with average bond lengths of 2.13(1) Å between Fe and the carbon atoms of the Cp ring. The positive charge and the electron withdrawing character of the vinylidene ligand are considered to be responsible for a slight elongation of the Fe–P bonds relative to other complexes containing the Cp(dppe) Fe fragment. The vinylidene ligand is planar, which means that the Fe, C-1, C-2 atoms and the carbon atoms of the CH_3 and CS

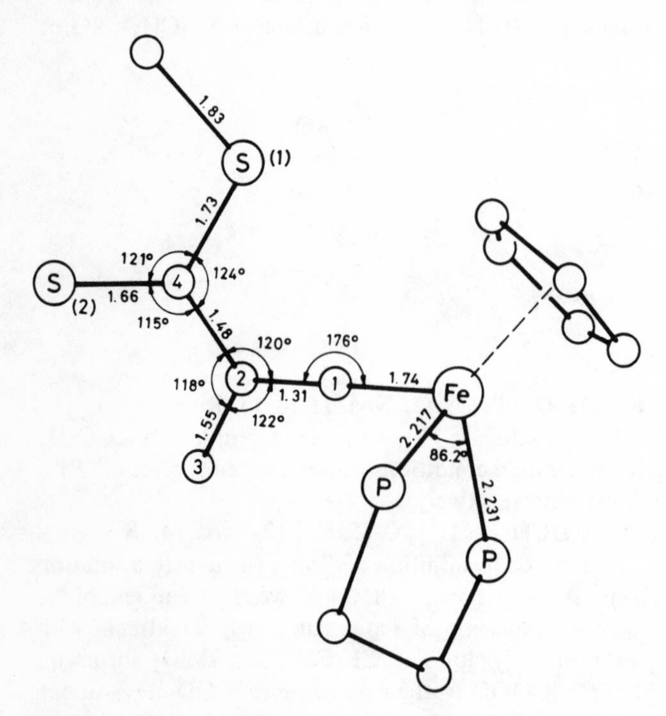

Fig. 21. Molecular structure of the cation of [Cp(dppe)Fe=C=C(CH$_3$CS$_2$CH$_3$]I; the phenyl rings are omitted for clarity

groups all lie within 0.05 Å of an idealized plane. The vinylidene plane and the mirror plane of the Cp(dppe)Fe fragment form a dihedral angle of 49.7° (130.3°). The Fe–C-1 bond length is shorter than the corresponding bond length in carbene complexes and supports the view of a formal double bond between Fe and C-1. The C-1–C-2 bond length is typical for an allene double bond. The vinylidene linkage is nearly linear. The methyl group on S-1 is *cis* to S-2 and lies in the dithiocarbomethoxy plane, which forms a dihedral angle of 17.9° with the plane of the vinylidene ligand (containing the Fe, C-1, C-2, C-3, and C-4 atoms). The bond angles at C-2 range between 118° and 122° and are consistent with sp^2 hybridization. The structure of the cation, showing only the central coordination sphere at the iron atom, is depicted in Fig. 21.

[Cp(dppe)Fe=C=C(C_6H_5)C_7H_7]PF_6 (Table **12**, No. **16**). The complex is insoluble in light petroleum, ether, and H_2O; slightly soluble in CH_3OH and C_2H_5OH, more soluble in THF, acetone, and $CHCl_3$; and very soluble in CH_2Cl_2 and $(CH_3)_2SO$.

Reactions with $NaOCH_3$, K[HB(C_4H_9-*n*)$_3$], CH_3OH, and H_2O were carried out with the corresponding Ru complex and were mainly characterized by loss of C_7H_7.

2.2 Cationic Complexes of the Type $[^5L(CO)(^2D)Fe{=}C{=}CRR']^+$

The compounds of this section (Formula I) have only Cp as the ^5L ligand and ^2D is restricted to phosphines. Nothing is reported about the orientation of the vinylidene plane with respect to the ligands at Fe.

Like the corresponding carbene complexes the cations are chiral at iron because they have four different ligands at the iron atom. If R and R′ are different (Nos. 5 to 7, 10), two isomers are expected. In the case of R = R′ two sets of signals should appear in the NMR spectra; however, the rotation of the vinylidene ligand around the Fe–C-1 axis of No. 8 could not be frozen out even at − 100 °C.

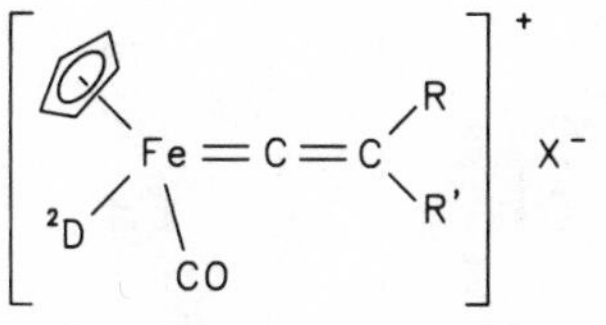

$$\left[(C_5H_5)(^2D)(CO)Fe{=}C{=}C\begin{matrix} R \\ R' \end{matrix} \right]^+ X^-$$

I

General remarks. The signal of the vinylidene carbon atom, C-1, appears in the ^{13}C NMR spectrum at low field (about 370 ppm), indicating the "metal-stabilized vinylcarbonium ion" character of the molecule.

Table 13. Cationic Vinylidene Complexes of the Type $[Cp(CO)(^2D)Fe{=}C{=}CRR']^+$. Further information on compounds with numbers preceded by an asterisk is given at the end of the table

No.	Fe=C=CRR' ^{2}D	anion, method of preparation (yield) properties and remarks
*1	$Fe{=}C{=}CH_2$ $P(C_6H_{11})_3$	BF_4^- salt, I (88%), lime green solid, m.p. 170 °C (dec.) ^1H NMR ($CDCl_3$): 5.36 (s, CH_2) $CF_3SO_3^-$ salt, I
2	$Fe{=}C{=}CH_2$ $P(CH_3)_2C_6H_5$	BF_4^-, salt, I (70%), lime green solid, m.p. 130 to 134 °C (dec.) ^1H NMR ($CDCl_3$): 5.41 (s, CH_2) CF_3SO_3 salt, I ^1H NMR ($CDCl_3$): 5.41 (s, CH_2)
*3	$Fe{=}C{=}CH_2$ $P(C_6H_5)_3$	BF_4^- salt, I (90%), golden yellow solid, m.p. 154 to 155 °C (dec.) ^1H NMR ($CDCl_3$): 5.30 (s, CH_2) ^{13}C NMR ($CDCl_3$): 107.14 (CH_2), 209.40 (d, CO; J(P,C) = 26.9), 372.38 (d, C-1; J(P,C) = 29.3) $CF_3SO_3^-$ salt, I ^1H NMR ($CDCl_3$): 5.30 (s, CH_2) ^{13}C NMR ($CDCl_3$): 372.38 (d, C-1, J(P,C) = 29.3)
*4	$Fe{=}C{=}CH_2$ $P(OCH_3)_3$	$CF_3SO_3^-$ salt, I, yellow BF_4^- salt, I ^1H NMR ($CDCl_3$): 5.37 (d, CH_2; J = 3)
*5	$Fe{=}C{=}C(H)CH_3$ $P(OCH_3)_3$	BF_4^- salt, I

No.	Compound	Properties
*6	Fe=C=C(H)C_4H_9-*t* $P(CH_3)_3$	BF_4^- salt, isomer mixture (A:B = 84 to 16) ^{1}H NMR (CD_3CN, A/B): 1.16/1.42 (s, C_4H_9), 5.97/4.11 (d, CH; J(P,H) = – /14.6) ^{13}C NMR (CD_3CN, A/B): 135.7/132.7 (s, CH), 217.1 (d, CO; J(P,C) = 30), 373.4 (d, C-1; J(P,C) = 31)
*7	Fe=C=C(H)C_6H_5 $P(C_6H_5)_3$	BF_4^- salt; II, pink solid
*8	Fe=C=$(CH_3)_2$ $P(C_6H_5)_3$	BF_4^- salt, I (90%), peach colored solid, m.p. 135 to 136 °C (dec.) ^{1}H NMR ($CDCl_3$): 1.66 (s, CH_3) $CF_3SO_3^-$ salt, I, oily solid ^{1}H NMR ($CDCl_3$): 1.66 (s, CH_3)
*9	Fe=C=C$(CH_3)_2$ $P(OCH_3)_3$	$CF_3SO_3^-$ salt, I (58%), peach colored BF_4^- salt, I (96%), pink powder ^{1}H NMR ($CDCl_3$): 1.87 (s, CH_3)
*10	Fe=C=C$(CH_3)C_6H_5$ $P(C_6H_5)_3$	$CF_3SO_3^-$ salt, II (65%), dark blue solid ^{1}H NMR ($CDCl_3$): 1.80 (s, CH_3)

The complexes in Table 13 can be prepared according to the following methods:

Method I From the acyl complex $Cp(CO)(^2D)FeCOCHR^1R^2$ and $HBF_4 \cdot O(C_2H_5)_2$ (formation of the carbene cation $[Cp(CO)(^2D)Fe{=}C(OH)CR^1R^2H]^+$ as an oil) followed by addition of $(CF_3SO_2)_2O$ (BF_4^- salt) or only addition of $(CF_3SO_2)_2O$ ($CF_3SO_3^-$ salt).

Method II Alkylation or protonation of the acetylene $Cp(CO)(^2D)FeC{\equiv}CR$

Explanation for Table 13. The Cp ligand shows a doublet in the ^1H NMR spectrum between 5.10 and 5.45 ppm with J(P,H) values from 1 to 2 Hz. In the ^{13}C NMR spectrum singlets at about 90 ppm are recorded. The IR spectra of the compounds show one ν (CO) and one ν (C=C) vibration at about 2030 and 1640 cm^{-1}, respectively.

Further Information

$[Cp(CO)\{P(C_6H_{11})_3\}Fe{=}C{=}CH_2]BF_4$ (Table **13**, No. **1**) forms the (*R,R*), (*S,S*) pair of enantiomers of the cationic complex II ($^2D = P(C_6H_{11}\text{-}c)_3$) in 82%; for the mechanism proposed, see No. 3.

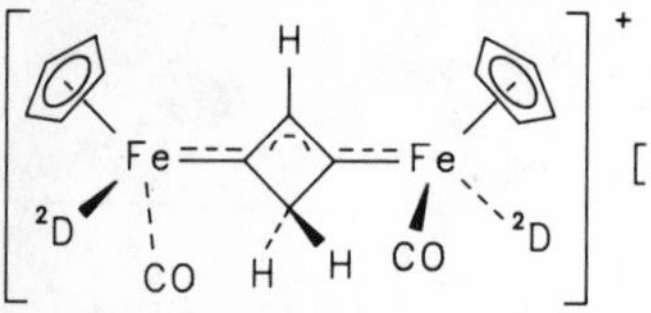

II

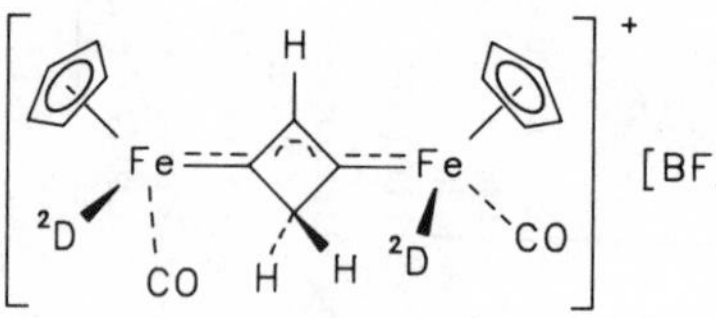

III

$[Cp(CO)\{P(C_6H_5)_3\}Fe{=}C{=}CH_2]X$ (Table **13**, No. **3**, X = BF_4, CF_3SO_3). Partial decomposition of the BF_4^- salt occurs with formation of the adduct $[Cp(CO)\{P(C_6H_5)_3\}FeC\{P(C_6H_5)_3\}{=}CH_2]BF_4$ on gentle warming a CH_2Cl_2 solution to 40 °C; see Sect. 2.4. Upon stirring the complex in THF solution for 4.5 h conversion occurs into the two diastereomeric pairs of enantiomeric μ-1,3-cyclobutylidene complexes *R,R*- and *S,S*-II, and *meso*-III ($^2D = P(C_6H_5)_3$; 1:3 ratio of II:III) with 89% yield. The II:III ratio does not alter upon repeated recrystallization or by chromatography. For the mechanism it was supposed that partial deprotonation by the base THF takes place with formation of the alkyne complex $Cp(CO)\{P(C_6H_5)_3\}FeC{\equiv}CH$, which is then trapped by the title complex in a cycloaddition reaction. The BF_4^- salt can be reduced with BF_4^- to give the vinyl complex $Cp(CO)\{P(C_6H_5)_3\}FeCH{=}CH_2$ in 40% yield. The title complex is readily deprotonated by treatment with $K[OC_4H_9\text{-}t]$ in $HOC_4H_9\text{-}t$ or $N(CH_3)_3$ (−116 °C) to give the alkynyl complex $Cp(CO)\{P(C_6H_5)_3\}FeC{\equiv}CH$ in 82% yield. The BF_4^- salt reacts with a variety of alcohols or thiols HER (E = O, S), or HCl (ER = Cl) resulting in cationic carbene complexes of the general type $[Cp(CO)\{P(C_6H_5)_3\}Fe{=}C(CH_3)ER]^+$.

Benzylamine produces the analogous cation, whereas the amines H_2NCH_3 and $HN(CH_3)_2$ form mixtures of the corresponding carbene cations with about 10 to 15% of the deprotonated product $Cp(CO)\{P(C_6H_5)_3\}FeC{\equiv}CH$; see also Sect. 1.1.2.2, preparation Method I.

For the addition of bases at C-1, see Sect. 2.4. Addition of the ether $C_2H_5OCH{=}CH_2$ generates $HC{\equiv}CH$ and the carbene cation $[Cp(CO)\{P(C_6H_5)_3\}Fe{=}C(CH_3)OC_2H_5]^+$, probably via an intermediate adduct of the ether at C-1. The BF_4^- salt can be converted with the imine $C_6H_5CH{=}NCH_3$ at $-15\,°C$ (3 d) into the carbene complex IV ($^2D = P(C_6H_5)_3$, $X = BF_4$) in 46% yield. It reacts with 1,1-dimethylhydrazine, benzophenonehydrazone, or toluene-4-sulfonylhydrazine to produce the acetonitrile complex $[Cp(CO)\{P(C_6H_5)_3\}Fe(CH_3CN)]\,BF_4$, in 62, 63, or 74% yield, respectively. The intermediate formation of a carbene cation followed by a Beckmann rearrangement was suggested.

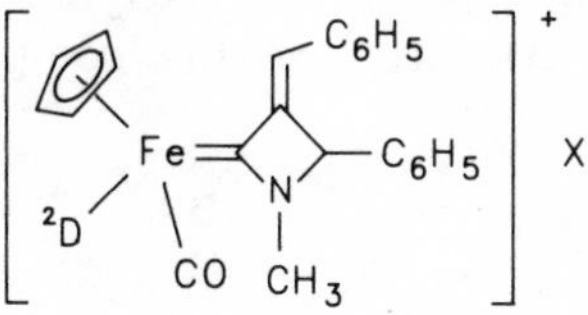

IV

$[Cp(CO)\{P(OCH_3)_3\}Fe{=}C{=}CH_2]X$ (Table **13**, No. **4**; $X = CF_3SO_3$, BF_4). Reaction with $C_6H_5CH{=}NCH_3$ in CH_2Cl_2 produces the carbene complex IV ($^2D = P(OCH_3)_3$, $X = CF_3SO_3$; 2:1 mixture of diastereomers) in 11% yield. With various allyl alcohols of the type $HOCH(R^1)CH{=}CHR^2$ (R^1, R^2 represent H or CH_3) the BF_4^- salt is converted into the corresponding carbene complexes $[Cp(CO)\{P(OCH_3)_3\}Fe{=}C(CH_3)OCH(R^1)CH{=}CHR^2]BF_4$; see Sect. 1.2.2.

$[Cp(CO)\{P(OCH_3)_3\}Fe{=}C{=}C(H)CH_3]BF_4$ (Table **13**, No. **5**). Various allyl alcohols of the type $HOCH(R^1)CH{=}CHR^2$ (R^1, R^2 are H or CH_3) give the carbene complexes $[Cp(CO)\{P(OCH_3)_3\}Fe{=}C(C_2H_5)OCH(R^1)CH{=}CHR^2]$ BF_4 in moderate yields; see Sect. 1.2.2.

$[Cp(CO)\{P(CH_3)_3\}Fe{=}C{=}C(H)C_4H_9\text{-}t]BF_4$ (Table **13**, No. **6**) is formed as a temperature-independent 84:16 mixture of isomers by proton transfer from various metal hydrides HM ($M = CrCp(CO)_3$, $MoCp(CO)_3$, or $Mn(CO)_5$) leading to an equilibrium reaction. The equilibrium constants were obtained by NMR experiments and increase with decreasing temperature. The $\Delta H^0/\Delta S^0$ values (in $kcal \cdot mol^{-1}/eu$) in CD_3CN solutions without an additional electrolyte are $-8.0/-25.7$; $-5.3/-20.5$; $-2.8/-15.3$, respectively, for the above HM compounds. In the presence of an electrolyte (0.5 M $[N(C_4H_9)_4]BF_4$) the corresponding values are $-7.4/-22$; $-4.4/-14.5$; $-2.5/-11.0$. The pK_a value in CH_3CN was found to be 13.6 ± 0.3. Thermal decomposition in CH_3CN at 57 °C releases $HC{\equiv}CC_4H_9\text{-}t$, resulting in $[Cp(CO)\{P(CH_3)_3\}Fe(CH_3CN)]BF_4$ (78% yield).

$[Cp(CO)\{P(C_6H_5)_3\}Fe{=}C{=}CHC_6H_5]BF_4$ (Table **13**, No. **7**) reacts with CH_3OH to give the carbene complex $[Cp(CO)\{P(C_6H_5)_3\}Fe{=}C(CH_2C_6H_5)OCH_3]BF_4$ in 73% yield. Stronger bases such as dimethylamine deprotonate No. 7 to the starting alkyne complex $Cp(CO)\{P(C_6H_5)_3\}FeC{\equiv}CC_6H_5$.

$[Cp(CO)\{P(C_6H_5)_3\}Fe{=}C{=}C(CH_3)_2]X$ (Table **13**, No. **8**; X = CF_3SO_3, BF_4). A low-temperature ^1H NMR study shows that the CH_3 groups are equivalent even at – 100 °C, thus indicating rapid rotation of the dimethylvinylidene ligand around the Fe–C bond axis.

Both salts react with the imine $C_6H_5CH{=}NCH_3$ in CH_2Cl_2 solution to give the carbene complexes V ($^2D = P(C_6H_5)_3$, R = C_6H_5) as a 3:1 mixture of diastereomers in about 50 to 60% yield. With the appropriate imines, the carbene complexes with R = $C_6H_4CH_3$-3, $C_6H_4CH_3$-4, or (*E*)-$CH{=}CHC_6H_5$ are similarly prepared as mixtures of diastereomeric BF_4^- salts in about 35% yield. Anionic nucleophiles can be added at the C-1 carbon atom of the BF_4^- salt. Thus, reaction with $Li_2[R_2CuCN]$ compounds at – 78 °C in THF solution (R = $CH{=}CH_2$, C_6H_5) gives $Cp(CO)\{P(C_6H_5)_3]FeC(R){=}C(CH_3)_2$ in about 50% yield; similarly, $NaSC_6H_5$ produces this complex with R = SC_6H_5 in 61% yield.

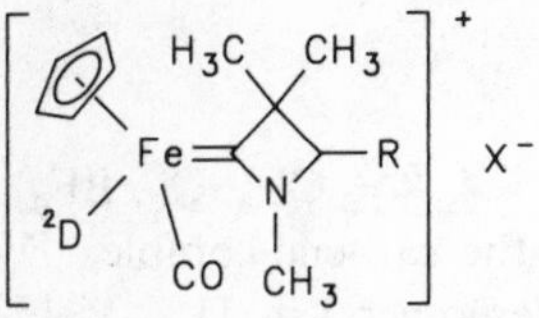

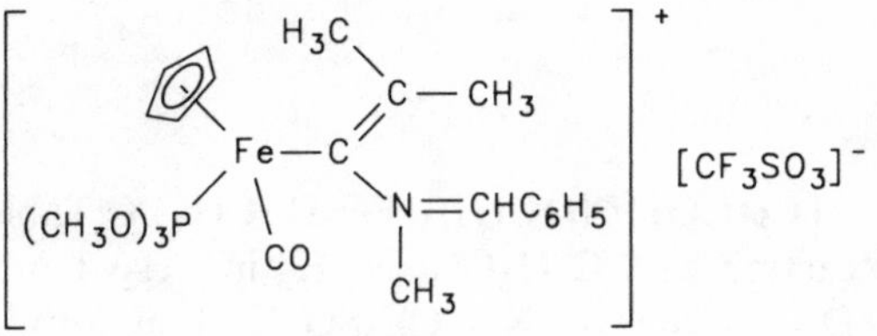

$[Cp(CO)\{P(OCH_3)_3\}Fe{=}C{=}C(CH_3)_2]X$ (Table **13**, No. **9**; X = CF_3SO_3, BF_4). The $CF_3SO_3^-$ salt produces with $C_6H_5CH{=}NCH_3$ adduct VI (30 min, 80% yield). Prolonged reaction in the dark (8 d) or 2 h reflux gives the carbene complexes V ($^2D = P(OCH_3)_3$, R = C_6H_5). A similar cycloaddition proceeds with the thiazol VII, resulting in the carbene complexes VIII (X = BF_4, CF_3SO_3; R^1 and R^2 represent various organic substituents). The BF_4^- salt is converted by $HOCH_2CH{=}CH_2$ into the carbene complex $[Cp(CO)\{P(OCH_3)_3\}Fe{=}C(C_3H_7\text{-}i)OCH_2CH{=}CH_2]BF_4$ in 21% yield and the $CF_3SO_3^-$ salt with CH_3OH into the carbene complex $[Cp(CO)\{P(OCH_3)_3\}Fe{=}C(C_3H_7\text{-}i)OCH_3]CF_3SO_3$ (96% yield). Oxidation by dry O_2 at – 78 °C generates the cationic complex $[FpP(OCH_3)_3]CF_3SO_3$ (50% yield).

$[Cp(CO)\{P(C_6H_5)_3\}Fe{=}C{=}C(CH_3)C_6H_5]SO_3CF_3$ (Table **13**, No. **10**) is unstable and is used immediately.

Reaction with $Li_2[(CH_3)_2CuCN]$ (– 78 °C in THF) produces the C-1 methylated product, $Cp(CO)\{P(C_6H_5)_3\}FeC(CH_3){=}C(CH_3)C_6H_5$, as a 93:7 mixture of the (Z) and (E) isomers in quantitative yield.

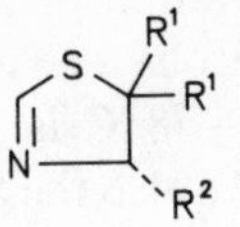

VII 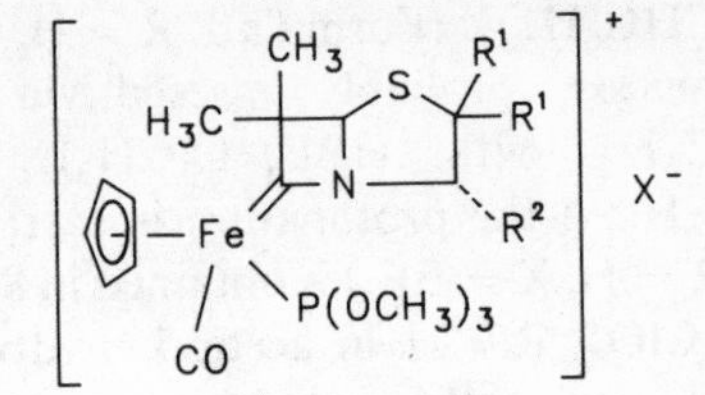 VIII

2.3 Cationic Complexes of the Type $[^5L(CO)_2Fe{=}C{=}CRR']^+$

The compounds of the general type I (^5L is always Cp) described in this section are not stable and have not yet been isolated. The compounds are generated as very reactive intermediates by the protonation or alkylation of the alkynes FpC≡CR. They are also considered to be formed via 1,2-hydrogen shift from $[Fp(\eta^2\text{-}HC{\equiv}CR)]^+$ cations; see also the proposed intermediacy of the cation $[Fp{=}C{=}CHCH_2CH_2OH]^+$, leading to cyclic carbene complexes. Powerful electrophiles, the cations readily react with various nucleophiles.

The intermediacy of the title complexes is based upon similar reactions leading to the more stable $[Cp(CO)(^2D)Fe{=}C{=}CRR']^+$ cations; see Sect. 2.2.

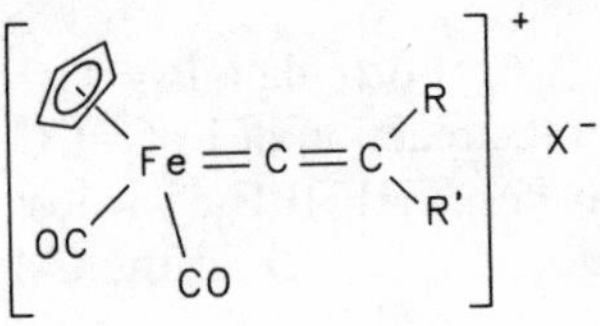

I

$[Fp{=}C{=}CH_2]CF_3SO_3$ (Formula I, R = R′ = H) is probably formed in the reaction of $FpCOCH_3$ with $(CF_3SO_2)_2O$ in ether at − 78 °C. A yellow substance precipitates and decomposes to a greenish yellow oil on warming to − 65 °C.

Addition of mercaptans HSR (R = CH_3, C_6H_5) to a freshly prepared sample at − 78 °C produces the corresponding carbene cations $[Fp{=}C(CH_3)SR]^+$; see Sect. 1.3.2.

$[Fp{=}C{=}CHCH_3]Cl$ (Formula I, R = H, R′ = CH_3) is considered to be the first stage in the reaction of $FpC{\equiv}CCH_3$ with ethanolic HCl to form $FpCOCH_2CH_3$. Further evidence of the intermediacy of the title cation is the formation of $FpC(O_2CCH_3){=}CHCH_3$ by treatment of the starting alkyne, $FpC{\equiv}CCH_3$, with anhydrous acetic acid in refluxing cyclohexane; this product is the result of the addition of $[CH_3CO_2]^-$ at $[Fp{=}C{=}CHCH_3]^+$.

[Fp=C=CHC$_6$H$_5$]X (Formula I, R = H, R′ = C_6H_5; X = BF_4, ClO_4). The following processes probably proceed via the title complex: The reaction of $FpC{\equiv}CC_6H_5$ with $HBF_4 \cdot O(CH_3)_2$ in anhydrous CH_3OH gives $FpCOCH_2C_6H_5$; if the protonation is carried out at − 78 °C in CH_2Cl_2 the complex II (R = H, X = BF_4) is obtained in 82% yield; see also the formation of II with X = ClO_4, R = H in acetic anhydride. The cation is trapped as the $P(C_6H_5)_3$ adduct when the protonation is carried out at − 30 °C in the presence of an excess of this ligand. The formation of II is explained by a [2 + 2] cycloaddition of the starting $FpC{\equiv}CC_6H_5$ with the title complex. The cation is also supposed to form via rearrangement of a terminal alkyne during the reaction of $[Fp(THF)]^+$ with $HC{\equiv}CC_6H_5$, finally giving cation II, too.

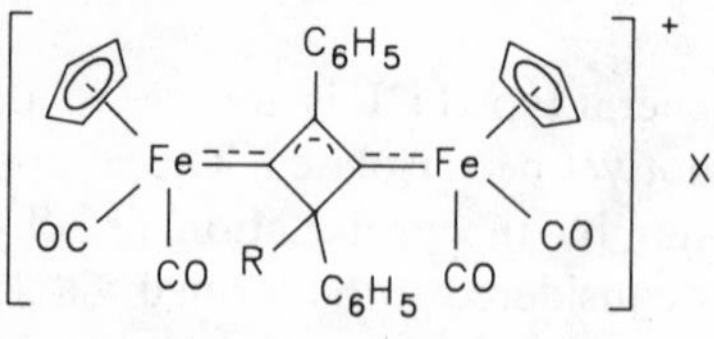

II

[Fp=C=C(CH$_3$)C$_6$H$_5$]PF$_6$ (Formula I, R = CH_3, R′ = C_6H_5) is considered to be an intermediate in the alkylation of $FpC{\equiv}CC_6H_5$ with $CF_3SO_3CH_3$ in CH_2Cl_2, followed by metathesis with $[NH_4]PF_6$ to give II (R = CH_3, X = PF_6); for the mechanism, see above.

[Fp=C=C(C$_6$H$_5$)C$_7$H$_7$]PF$_6$ (Formula I, R = C_6H_5, R′ = C_7H_7) is probably formed in the initial step of the reaction of $FpC{\equiv}CC_6H_5$ with $[C_7H_7][PF_6]$ in THF solution to give finally $[Fp(THF)]PF_6$. The loss of the vinylidene ligand is presumed to occur by nucleophilic attack at the C-1 carbon atom.

2.4 Cationic Compounds of the Type [Cp(CO)(^{2}D)FeC(^{2}D′)=CRR′]X, [Cp(CO)Fe(μ^2D-^{2}D′)C=CRR′]X, and [Cp(CO)$_2$FeC(^{2}D′)=CRR′]X

The compounds in this section are adducts of bases such as phosphines, pyridines, or imines at the electrophilic C-1 carbon atom (α-C) of some cationic vinylidene complexes described in Sects. 2.2. and 2.3, as shown in Formulas I to III. No adducts are known with the more electron-rich vinylidene cations, $[Cp(^2D\text{-}^2D)Fe{=}C{=}CR_2]^+$, described in Sect. 2.1. The compounds can also be considered as phosphonium (Nos. 3, 4, 6, and 7) or immonium (Nos. 1, 2, and 5) cations in which one of the substituents at the N- or P-bonded carbon atom is replaced by Fp or the appropriate Cp(CO)(^{2}D)Fe fragment.

For other cationic compounds of the general types $[^5L(CO)(^2D)Fe^1L]^+$ and $[^5L(CO)_2Fe^1L]^+$, see "Organoiron Compounds" B 11, 1983, Sects. 1.5.2.2.13 and B 14, 1989, Sect. 1.5.2.3.20, respectively.

I II III

General remarks. Coordination of a base at C-1 of the vinylidene cation changes the hybridization of C-1 from *sp* to sp^2. The ^{13}C NMR signal of the C-1 carbon atom shifts from about 360 ppm in the vinylidene cation to the 150 to 160 ppm region upon coordination of the base. Although in the basic vinylidene cations of the type $[Cp(CO)(PR_3)Fe{=}C{=}CH_2]^+$, J(P,C) values are recorded for CO and C-1, no coupling constants are given for the corresponding phosphine adducts No. 3, 4, and 7.

With exception of Nos. 6 and 7 the complexes in Table 14 can generally be prepared by addition of a neutral base, $^2D'$, to the vinylidene cations $[Cp(CO)(^2D)Fe{=}C{=}CRR']^+$ ($^2D'$ = N-, or P-bases).

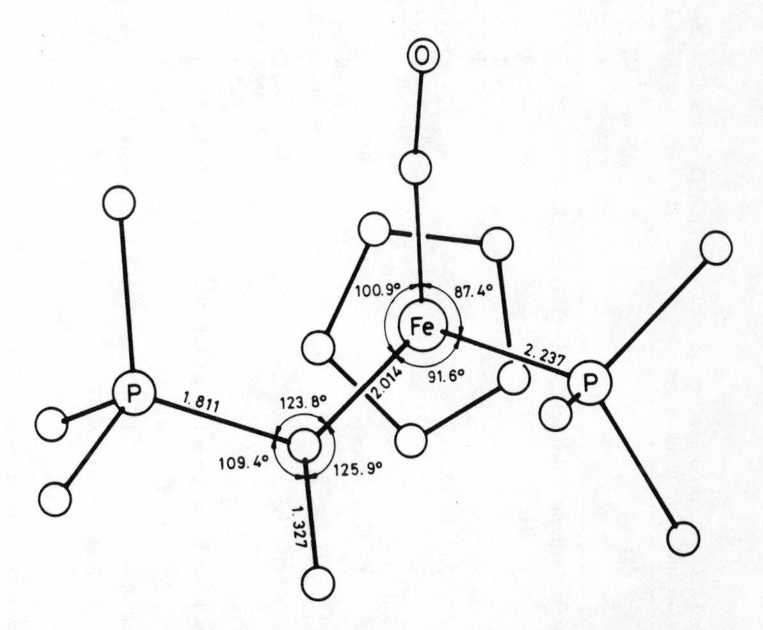

Fig. 22. Molecular structure of $[Cp(CO)\{P(C_6H_5)_3\}FeC\{P(C_6H_5)_3\}{=}CH_2]^+$; for better clarity only the P-bonded carbon atoms of the C_6H_5 rings are depicted

Table 14. Complexes of the Type $[Cp(CO)(^2D)FeC(^2D')=CRR']X$, $[Cp(CO)(\mu\text{-}^2D\text{-}^2D')FeC=CRR']X$, and $[Cp(CO)_2FeC(^2D')=CRR']X$. Further information on compounds with numbers preceded by an asterisk is given at the end of the table.

No.	$^2D'$	anion (yield) properties and remarks
the basic vinylidene cation is $[Cp(CO)\{P(C_6H_5)_3\}Fe=C=CH_2]^+$		
1	NC_5H_5	BF_4^- salt (79%), orange crystals; m.p. 165 °C (dec.) ^1H NMR ($CDCl_3$): 5.50 (m, br, CH_2) ^{13}C NMR ($CDCl_3$): 125.0 (CH_2), 162.0 (d, C-1; J(P,C) = 19.0), 210.0 (CO)
2	$NC_5H_4CH_3$-4	BF_4^- salt (82%), orange; m.p. 104 °C (dec.) ^1H NMR ($CDCl_3$): 5.43 (m, br, CH_2) ^{13}C NMR ($CDCl_3$): 131.2 (d, CH_2; J(P,C) = 4.0), 155.1 (d, C-1; J(P,C) = 19.0)
3	$P(CH_3)_2C_6H_5$	BF_4^- salt (80%), red orange; m.p. 165 °C (dec.) ^1H NMR ($CDCl_3$: 6.27 (d, H-1; J(P,H) = 23.2), 6.47 (d, H-2; J(P,H) = 53.0) ^{13}C NMR ($CDCl_3$): 121.00 (CH_2), 150.80 (C-1), 200.85 (CO)
*4	$P(C_6H_5)_3$	BF_4^- salt (40%), m.p. 174 °C (dec.) ^1H NMR ($CDCl_3$): 6.47 (d, H-1; J(P,H) = 23.0), 6.63 (d, H-2; J(P,H) = 45.8) ^{13}C NMR ($CDCl_3$): 120.18 (CH_2), 158.00 (C-1), 215.60 (CO)
the basic vinylidene cation is $[Cp(CO)\{P(OCH_3)_3\}Fe=C=C(CH_3)_2]^+$		
*5	$N(CH_3)=CHC_6H_5$	$CF_3SO_3^-$ salt (80%)
the cation is $[Cp(CO)\{(C_6H_5)_2PCH_2P(C_6H_5)_2\}Fe=C=CH_2]^+$ (Formula III)		
*6	$(C_6H_5)_2PCH_2P(C_6H_5)_2$	BF_4^- salt, orange; m.p. 106 to 108 °C ^1H NMR ($CDCl_3$): 6.37 (m, 1 H of CH_2= ; J(P,H) = 34.6, 4.0), 6.93 (m, 1 H of CH_2= ; J(P,H) = 65.4) ^{31}P NMR ($CDCl_3$): 41.0 (PC), 88.0 (PFe; J(P,P) = 65)
the basic vinylidene cation is $[Cp(CO)_2Fe=C=CHC_6H_5]^+$		
*7	$P(C_6H_5)_3$	BF_4^- salt, yellow; m.p. 199 to 200 °C (dec.) ^{13}C NMR (CH_2Cl_2): 119.3 (CH), 168.24, 168.64 (C-1), 213.41 (CO) ClO_4^- salt, yellow; m.p. 183 to 187 °C (dec.)

Explanation for Table 14. In adducts of Formula I (R = R′ = H) H-1 and H-2 are *cis* and *trans* to the $^2D'$ base, respectively. In all compounds C-1 stands for the Fe-bonded carbon atom. The cation of No. 6 is shown in Formula III (^2D-$^2D'$ = $(C_6H_5)_2PCH_2P(C_6H_5)_2$, R = R′ = H). The Cp ligands appear as doublets in the ^1H NMR spectra at about 4.50 ppm with J(P,H) = 1.3 Hz and as singlets in the ^{13}C NMR spectra centered at 86 ppm.

Further Information

[Cp(CO){P($C_6H_5)_3$}FeC{P($C_6H_5)_3$}=CH_2]BF_4 (Table **14**, No. **4**). The complex crystallizes in the triclinic space group $P\bar{1}$-C_j^1 (No. 2) with the unit cell parameters a = 9.517(4), b = 11.038(3), c = 18.965(4) Å, α = 99.81(2)°, β = 94.41(3)°, γ = 102.54(3)°; Z = 2, and d_m = 1.37, and d_c = 1.39 g/cm³. The vinyl group is exclusively σ-bonded to the iron atom without bonding contact between Fe and CH_2 carbon atom (Fe . . . C = 2.992(8) Å) as shown in Fig. 22.

[Cp(CO){P($OCH_3)_3$}FeC{N(CH_3)=CHC_6H_5}=C($CH_3)_2$]CF_3SO_3 (Table **14**, No. **5**) on standing undergoes slow cyclization to generate the carbene cation IV. Similar adducts are supposed to be formed as intermediates in the analogous reaction of the starting vinylidene cation with the 2-thiazoline derivatives V (R^1, R^2 are various organic groups) to generate carbene complexes similar to IV: see Sect. 1.1.2.2.

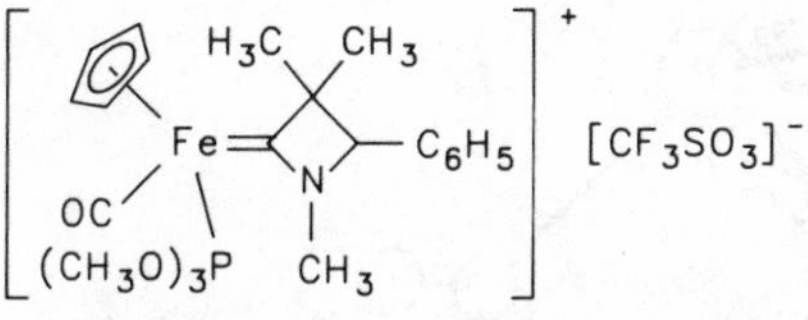

IV 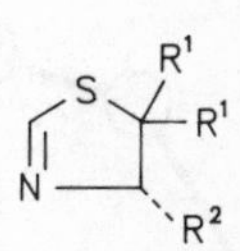 V

[Cp(CO)Fe{($C_6H_5)_2PCH_2P(C_6H_5)_2$}C=$CH_2$]$BF_4$ (Table **14**, No. **6**; Formula III; ^2D-$^2D'$ = $(C_6H_5)_2PCH_2P(C_6H_5)_2$, R = R′ = H). $HBF_4 \cdot O(C_2H_5)_2$ is added to a CH_2Cl_2 solution of the acyl complex with one noncoordinating phosphine function, Cp(CO){$(C_6H_5)_2PCH_2P(C_6H_5)_2$}$FeCOCH_3$ (quantitative yield). The formation of the title complex via the intermediate ylide complex VI (R = OH) is suggested.

Reaction with $LiCH_3$ in THF removes a methylene proton of the diphosphine to generate a neutral complex as shown in Formula VII; regeneration occurs with HBF_4. Reduction with $NaBH_4$ in THF solution produces ylide complex VI (R=H) in 52% yield.

[FpC{P($C_6H_5)_3$}=CHC_6H_5]X (Table **14**, No. 7, X = BF_4, ClO_4). The compounds are prepared by addition of a solution of $FpC{\equiv}CC_6H_5$ in acetic anhydride to $P(C_6H_5)_3$ (excess) and the corresponding HX (slight excess) in the same solvent (53 and 57% yield, respectively). The intermediate vinylidene cation is trapped by the phosphine molecule present in solution; see also Sect. 2.3.

$[Cp(OC)Fe\{C(R)CH_3\}\{(C_6H_5)_2PCH_2P(C_6H_5)_2\}]^+ [BF_4]^-$

VI

$Cp(OC)Fe\{C(=CH_2)\}\{(C_6H_5)_2PCHP(C_6H_5)_2\}$

VII

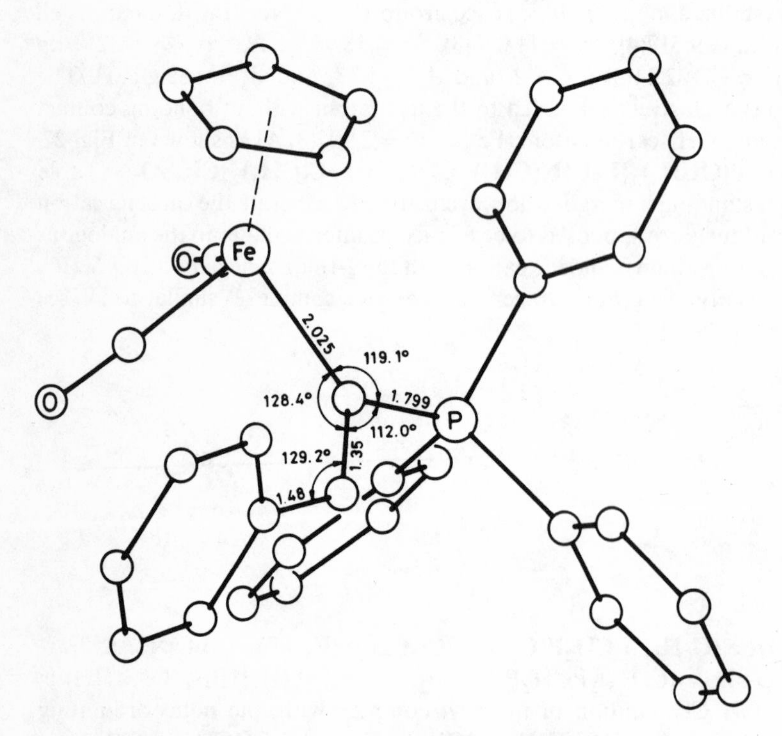

Fig. 23. Molecular structure of the cation $[Cp(CO)_2FeC\{P(C_6H_5)_3\}{=}CHC_6H_5]^+$

The BF_4^- salt crystallizes in the monoclinic space group $P2_1/n\text{-}C_{2h}^5$ (No. 14) with a = 12.050(3), b = 15.379(4), c = 16.165(5) Å, β = 95.03(2)°; Z = 4 gives d_c = 1.40, and d_m = 1.39 g/cm³. The sum of the valence angles at the vinylidene carbon atom is 359.6°, indicating sp^2 hybridization. The structure of the cation is shown in Fig. 23.